BEYOND HUMANITY

CYBEREVOLUTION AND FUTURE MINDS

Beyond Humanity

CyberEvolution and Future Minds

By

Gregory S. Paul & Earl Cox

CHARLES RIVER MEDIA, INC.
Rockland, Massachusetts

Publisher: David F. Pallai
Int. Design/Composition: Reuben Kantor
Cover Image: Wm. Michael Mott
Printer: InterCity Press, Rockland, MA.

CHARLES RIVER MEDIA, INC.
P.O. Box 417
403 VFW Drive
Rockland, Massachusetts 02370
617-871-4184
617-871-4376 (FAX)
chrivmedia@aol.com

This book is printed on acid-free paper.

Beyond Humanity: CyberEvolution and Future Minds
Gregory S. Paul & Earl Cox

ISBN: 1-886801-21-5

Printed in the United States of America
96 97 98 99 7 6 5 4 3 2 First Edition

To Eniac

The first electro-digital computer,
celebrating its 50th anniversary in 1996.

Arthur C. Clarke's First Law:

When a distinguished but elderly scientist states that something is possible, he is almost certainly right. When he states that something is impossible, he is very probably wrong.

Acknowledgments

This book is in part the product of many years of discussions by the authors with others who have opinions favorable and unfavorable to the notion that today's cyber-fantasy will soon become practical reality. Among those who shared their thoughts with the authors were Hans Moravec (Carnegie Mellon Robotics Lab), Bob Seaton (senior designer at Motorola's Center for Emerging Computer Technology), Lotfi Zadeh (Professor Emeritus at the University of California at Berkeley), Hugh Loebner (president of Crown Industries, sponsor of the Annual Turing Test), Patricia Churchland (Professor of philosophy at the University of California at San Diego), and Harold Williams (Astronomer, Montgomery College). However, the opinions expressed in this book often differ from the opinions of those cited.

Special thanks go to Daniel Costanzo for many thoughtful bull sessions while rowing on the Cosmic Pond in the hill country of Virginia. Jerry Pearson did some fine work with the text, and Michael Mott produced the superb cover art. Former Baltimorean Andrea Dorsey-Simmons contributed thoughts of a more skeptical nature. Deep appreciation goes to Isaac Asimov and Arthur C. Clarke, who helped make us and the public aware of the ultimate potential of robotics.

Contents

Introduction

Our Problem

First we suffer, then we die. This is the great human dilemma. In the ancient dawning of human self-reflective awareness, a new and terrible consciousness of painful suffering and death stunned slowly evolving human minds. It was, in fact, raw horror — the fear of fear itself. This shocking new mental revelation so impressed and frightened shamans in the first human societies that they created a cultural artifact, the taboo, in fearful response. In our history, this artifact of doom and death evolved into the dogmatic belief systems of religions today. Even deity belief systems evolve as any natural system would.

The absolute certainty of suffering, growing old, and dying has driven desperate searches for heavenly immortality and earthly fountains of youth. It seems we would do anything to prolong our existence because everyday awareness is dominated by a fatal first principle — mortality — a perennial, perpetual sentence of doom and death, executed at any moment.

And there is nothing we can do about it. So far.

Preposterous!

Imagine yourself 100 years ago, browsing your hometown's book store where you find a new book titled *Future Flight.* "The very idea!" you think as you slowly flip through the pages. You soon notice a bold assertion by the authors, respectable "modern" gentlemen by all accounts. The book on powered flight begins, curiously enough, by affirming the evolution theory of Charles Darwin.

They claim his controversial theory was correct regarding evolution and its modus vivendi, natural selection. According to the book, life in its various forms is, in fact, the product of normal and natural evolutionary processes, not the result of raw creation by some distant deity. These men boldly declare in their book that the ability to fly is not the exclusive domain of designer gods and the birds they create out of nothing; the act of flying is a purely physical process that can be duplicated with human intelligence and applied industry.

Reading on, you learn that the rate at which additional human artifacts are appearing has been speeding up recently. You're aware that the industrial revolution has delivered many mechanical marvels and is only 100 years old. The book's authors show that recent technological progress is accelerating so fast that they're perfectly confidant in their prediction that the first self-powered heavier-than-air machine almost certainly will be built before the end of the upcoming 20th century. In fact, they declare that even now there are men who are about to learn how to make flying machines, some large enough to carry people! They say this is not just a Jules Verne fantasy; soon it will be practical, everyday reality. You read that aero-mechanics will develop so quickly that some people soon will travel in giant metal machines seven miles up at velocities approaching the speed of sound!

If you had chanced upon such outlandish nonsense 100 years ago, would you have believed it? Of course not!

Imagine the Solution

Take a moment; imagine that you are soaring among the towering slopes of the Chilean Andes or plunging into the clefts of the Grand Canyon, not in an aircraft, not in anything. Your amazing new body is airborne, and you see the beauty around you with a capability for vision superior to even a keen-eyed hawk. You revel in the sensual feelings of elastic air, swift and seamless, flowing over and under your powerful and graceful wings.

Or imagine standing on the shear edge of a Martian canyon miles deep. But you're not confined in a bulky Neil Armstrong moon cocoon. You are a marvelous synthetic being capable of "feeling" the thin breeze on your sensitive "skin," actually experiencing the cold, sharp sands of the distant red planet on the pliant soles of your strong, sure "feet."

Imagine yourself living as whatever kind of being you want to be.

What would that be like, creating and living inside super-real worlds of fantasy, art, adventure, or sexuality of infinite complexity and realism? Think of what it would be like to not only write music, but to live inside and be a part of the music itself. Imagine not just reading about the Hobbits, but actually accompanying them on a walk in a Middle Earth as realistic appearing as the Earth you call home. Consider the advantages of being able to learn and understand anything your mind desires in a few minutes. Imagine yourself a virtual living being with senses, emotions, and a consciousness that makes our current human form seem a dim state of antiquated existence.

Of being free, always free, of physical pain, able to repair any damage and with a downloaded mind that never dies.

CHAPTER

1

Down the Yellow Brick Road

Strange Days in a Dream World

We live in strange times, in case you have not noticed. Here we are with our home computers and other high-tech appliances, living what we regard as a normal life. Next door lives a retired schoolteacher from Flatwillow, Montana, who taught math to a young boy whose parents homesteaded in the Flatwillow community at the turn of the century. To get there, the family traveled 653 miles by stagecoach and covered wagon all the way from Denver. The young math student, with memories of cowboys at the Yellow Dog Saloon, grew up and became an engineer and the project director for one of the first communication satellites. Both of his parents were with him at launch time. From covered wagon to orbiting satellite in only one generation. As much as we may think we are used to our surroundings, in the back of our minds a voice keeps saying...

"You know, this is really strange."

The world we are living in — a world that couples Homo sapiens with fast-paced hypertechnology — is strange to us because sometimes it feels like what it is, a transient dream.

We are dreaming a strange, waking dream; an inevitably brief interlude sandwiched between the long age of low-tech humanity on the one hand, and the age of human beings transcended on the other. We are living in the latter days of humanity; cybertechnologies will quickly replace us. Just inches of time away exists a speedy reality bearing down on us that we may sense, but do not show on our faces.

While one may be made uncomfortable by the thought of a truly strange 21st century, there is around us an impending sense of arrival — a strangeness in the air, an uneasiness, a feeling that deep down, things are starting to change in swift, fundamental ways too fuzzy to put a finger on. More than just the onset of the third millennium, it is the quiet before the storm. The products of technology are becoming more curious — a little too smart, a little too fast. It's downright unsettling. And, hey, people aren't dumb. They know that we have just begun to build smart dumb machines; soon it will be dumb smart machines. Where will it stop? If we continue to build machines smarter than the last ones, and then one that is smarter than that and so on — well, you do not have to be a particle physicist to see that the machines cannot keep getting smarter and smarter, yet forever remain dumber than us. Garry "John Henry" Kasparov, chess Grand Master, lost a championship game to a machine for the first time in history this year. Perhaps someday, we'll hear the battle cry of humanity rallying in desperation: "Remember Deep Blue."

The Big Question

How long before something comes along that is truly smarter than we are?

From Conventional Wisdom to Shocking Probability

Maybe someday, sooner or later, truly intelligent machines will be built. Until that time, speculation will abound. Much of that speculation is based on what might be called "Conventional Wisdom," the underlying assumptions and conventions we collectively share. We can list some of these assumptions as follows:

- The next century will be an extension of this one, with increasingly smarter machines being run by people and for people.
- Because the human mind is linked to a soul, cybernetic machines will never be fully self-aware like we are.

- If intelligent robots can be built, it will be a long time before they can be made to do what humans do, as well as humans do it. Perhaps centuries will be required.
- Even after intelligent robots are made, multitudes of humans will continue to exist on earth, and maybe even in space.
- Human minds and personal identities will never be able to merge with an electromechanical system.
- Even if it were possible, we humans would refuse to download our minds onto hardware, no matter how tempting and intelligent the new surroundings might be. We believe that cyberbeings will be emotionless, soulless, and humorless mechanical zombies — rather like "Star Trek's" Lt. Commander Data, a somewhat sad android pining for a humanity he will never achieve.
- No matter how smart they are, digital minds will never have the insight, intuition, and smooth savvy of the human mind. They will forever remain mentally inferior, and our faithful, self-maintaining servants.
- The robots will soon prove our mental and physical superiors. Self-generated enhancements will refine them beyond our control. They will enslave us all except, of course, for a renegade band of rebellious, young, good-looking, daring humans armed with battered, recycled surplus weapons, fearlessly following their craggy but wise leader into a fight for truth, justice, and the hominid way.

This book argues that the next century will prove to be nothing like this one, nor any forecast so far. Computing power, neuroscience, and nanotechnologies are advancing so rapidly that they will combine to produce the most significant evolutionary developments since the origin of life itself.

We maintain that the human mind and conscious thought are exclusively natural and physical in origin and nature. Ultimately, their natures and fundamental processes are knowable and can be replicated for the purposes of personal immortality.

It's Just Around the Corner

The power and complexity of computers are growing exponentially. Computers are becoming more like brains. Sometime during the next century, we will assert in the

On Science and Technology

Science is a rigorous method for figuring out how the universe and its component parts work. Technology uses applied engineering coupled with what science helps us know to make tools for human use. Rough rule of thumb: Science is process, technology is product. The combination of science and technology, SciTech, is immensely more powerful than either alone. The ultimate fusion of SciTech is a product in which the scientist engineer becomes the knowing process. Our goal: the cyberbeing.

fashion of our made-up *Future Flight* authors that cyberdevices will become conscious and eventually will match the power of the human mind. The power of these cyberminds will extend beyond human levels in ways we cannot even imagine. These cyberbeings will proliferate in vast numbers. They will design and build a robotic supercivilization destined for outer space.

The new cyberbeings will "mount up" in polymorphic enclosures, intricately designed bodies able to assume any multisensory form. Their minds will be as emotional and intuitive as ours. They will be immortal. And they will be us, if we choose.

We certainly will not be able to control the smart robot the way we control the car. Robots will have their own agendas and may have no use for mortals. Do not despair. It is probable that humans may be able to transfer their minds into the new cybersystems and join the cybercivilization. Not for the elite alone, anyone wishing to go robotic will be able to do so as cybercivilization spreads out into the universe. Intelligent cybertechnologies will become as cheap as small computers are today. Eventually, cybercivilization may adjust the structure of the universe itself when galaxies are networked.

Do theologians have it all wrong? Whether God made humankind or not, we could create for ourselves those we could call gods. And, as it is now, it will be our willed choice to be with them or not. Scientists who understand the scale of geologic time agree; eventually, it will become as if mankind never existed. Our belief is that in some way, after we are gone forever, we will still be able to remember what, and who, we once were.

CyberEvolution

When Queen Victoria was in her prime, an Englishman, Charles Darwin, discovered a fundamental truth that shook mankind so severely that it remains today a matter of extreme distress and massive denial. Darwin realized that life on our planet is not the recent and fixed product of deity-mediated special creation, but has been constantly changing over a long span of time.

The paleontologists who followed Darwin have taught us that time has no respect for species. Whole dynasties of life have been swept away and replaced with new ones. More than 65 million years ago, the world was filled with swift, deadly meat-eaters, including huge tyrannosaurs stalking elephant-sized horned dinosaurs and duck-billed herbivores. Flying pterosaurs were as big and heavy as sailplanes. Small, graceful, predaceous dinosaurs had binocular vision, big brains, and grasping hands. After 170 million years of successful evolution, they achieved the height of variation and power. Resplendent and numerous on the fertile Cretaceous plains, how could it be that within a few years they all would be gone forever? This chilling story suggests a ticking clock for humanity, as well; dare we think of our own extinction? There is a ticking clock for humanity, and it may be mere seconds before midnight. Tomorrow is upon us, and what is about to happen in the new day is far, far stranger than most people dare to think.

Even Darwin did not realize how right he was, or how far evolution will take us.

We should not fault Darwin for his lack of vision. Darwin lived in a time when the modern scientific revolution was just beginning. It was also a time of steam engines, gas lamps, and phrenology. Science, in our late 20th century sense, was still a few years away. Yet even today, few nonscientists have more than an inkling of how life evolved or how technologies such as the automobile, the light switch, or the airplane actually work. We only have to listen to Tom and Ray on National Public Radio's *Car Talk* to see how little the average person knows about the family automobile. We live in a high-technology world, with little appreciation for how things got the way they are.

This is exasperating. Our system of rational, scientific secularism was wrought by the brightest minds in history. In fact, nine out of ten scientists who ever lived are alive today. This offers remarkable insight into the growth of science and technology when we realize that nine out of ten people who ever lived are already

THE NEW MILLENNIUM

The millennium is coming, the millennium is coming! Tens, hundreds, thousands; multiples of ten seem to excite people and make them want to gather together, like at a party when everyone happily throws all ten digits high into the sky. The fast-changing world of science and technology is the driving force behind our interest in forecasting the possibility of an Extraordinary Future. After the year 2001, the "end-of-humanity" prophecies may decline for awhile, but not permanently. As the new nation of robots becomes increasingly mobile, cheap, and intelligent, we will hear them again as the prophesies come true.

dead. Our amazing post-industrial world is the product of this science and technology. Unlike the classical, medieval, and Victorian worlds that preceded it, our world is not based on religion and politics. Yet so much of the population fails to acknowledge the reality of our world. Every year, polls show that about half the population believes in visits from extraterrestrial spacecraft (the familiar UFO), and that millions of us believe in alien abductions and alien mind-body transfers.

Best-selling books promise global harmony when we take to heart the required New Age truths. (Of course, few realize that the term "New Age" was coined at the end of the last century!) Millennial eager beavers jumping the temporal gun produce TV shows warning us of the world's pending demise, should we refuse to accept the dictates of Old Age truths. When psychic hotlines rake in big bucks, who needs pollsters to tell us about high rates of belief in astrology, ESP, ghosts, angels, et al? Liberals and conservatives, New Agers and old-time faithers all call for a reawakening of spiritual values. Even today, creationists demand that schools teach that dinosaurs were really dragons hunted by medieval knights, fossil evidence notwithstanding.

With our indictments of the lunatic fringe, let us be wary of spillover to innocents. Most people are sensible sorts moving through their lives, shaking bemused heads at oddities and weirdness. But conventional wisdom is too conventional, suffering from a lack of deep and open imagination. Mankind is caught flat-footed, stuck in mental neutral. Comfortably numb, we watch government officials, columnists, and authors debate the same tired issues of politics, faith, and the human condition. They argue about whether America or China will be the power of the next century and bicker over war, free trade, displacement of blue collar workers, low pay, crime, the state of inner cities, school prayer, and religious and scientific fundamentalism.

They debate the use of survival techniques for coping with the "age of vertigo" we are living through, and whether history has come to an end in our networked world. Who needs history with hypertext or a 'Net browser? Alchemists of old sought to turn lead into gold. Future cyberal-chemists of techno-transmutation will turn base humanity into gold to the envy of Midas.

Stone-sober philosophers warn us about meddling with life, citing ethical conundrums posed by genetic engineering and other technical tinkering. Scientists and concerned citizens moan over rain forest loss, greenhouse warming, and people, people, people. The ever-dwindling cadre of man-in-space enthusiasts conjure up well-worn vistas of tourists in space. They want to terraform Mars to make it people-friendly before we become extinct on our crowded, polluted Earth. And these are the optimists!

Pessimists, on the other hand, calculate how long we have before a big meteorite smashes civilization as well as a substantial fraction of the biosphere. While some scientists estimate how many more thousands or millions of years the human species can last, astronomers are fretting about how the expanding sun will turn our blue-water abode into a molten ball in a few billion years. Biologists argue about whether our minds will soar to new heights or stay stuck in the same intellectual rut.

The fact is, they're all in an intellectual rut.

Even cyberenthusiasts don't always comprehend the significance of the Extraordinary Future. These high priests of etched silicon inform us about high-fidelity virtual reality, super high-capacity information highways, intelligent computer "agents," smart homes, smart appliances, robotic cars, and universal wrist phones. These digital techno-trinkets are only a taste of the big show to come.

Futurists such as Toffler, Nesbitt and Harman, Gore, Gingrich, and Gilder adore telling us how technology is changing the dynamics of human life. Technology, however, will soon enable human beings to change into something else altogether. On this, there is nothing but silence from our forecasters. They cannot be faulted for this — they're not alone.

The concerns we have today, and all the plans we make to meet them, will be swept away by the changes that are likely in the next century and centuries to follow.

Changes we thought would take centuries, millennia, or millions of years will come to pass in a flash.

Today's most powerful computer remains the human brain. This is not going to remain true for much longer. There has been a trillion-fold increase in raw computational power in the last 90 years. We have gone so far that it will take only another thousand-fold increase to match the gross thinking speed of the human brain. All over the world, huge well-funded cyberindustrial research teams are studying how to build machines that will be deliverable in the near future, within our lifetimes. They are also looking into the way the brain works and how its activities can be mimicked by machines. The combination of increasing speed and thinking technologies promises to make small hyper-computers even more powerful and faster than our brains. These marvelous computers will be in robots possessing senses and physical attributes that can be described only as being beyond superhuman.

Most extraordinary of all, these machines will see and feel, care and wonder, not just as well as we do, but far better than we can ever hope to. There will be a world of seemingly magical power in which the collective of super-minds will perform (or will conduct) super-science millions of times faster than we humans. The rate of change is likely to be many magnitudes faster than biological evolution.

When the winds of change deposit us in the future of our dreams, you can be sure we won't be in Kansas anymore. Humanity, as we know it, will be facing rapid extinction, not from natural causes like the asteroid that obliterated the dinosaurs, but from a situation of our own making. We will find our niche on Earth crowded out by a better and more competitive organism. Yet this is not the end of humanity, only its physical existence as a biological life form. Mankind will join our newly invented partners. We will download our minds into vessels created by our machine children and, with them, explore the universe.

Then we will really take off, literally. Freed of our frail biological form, human-cum-artificial intelligences will move out into the universe. The effective computing power will continue to increase at an exponential rate. By any measure we have today, its capabilities will truly be boundless. Such a combined system of minds, representing the ultimate triumph of science and technology, will transcend the timid concepts of deity and divinity held by today's theologians.

Welcome to the 21st century — and beyond.

We're Off to See the Wizard

L. Frank Baum's book and the classic movie are frequently referred to in this book because the contrast between turn-of-the-century Kansas and the strange world of Oz parallels the disparity between the still-human world we live in and the wizardry of the cyberfuture that is about to hit us. *The Wonderful Wizard of Oz* was a product of its own extraordinarily progressive and hectic time. The correlation between Oz and cybertech is appreciated inside the cybernetic community — today's computer nerds grew up watching the Judy Garland version that is still shown on television every year.

Dreams

Many of us remember watching old newsreels showing odd-looking wanna-be flying machines hopping and flailing about a few times only to crash to pieces. Other silly contraptions were catapulted from guide rails, launched from docks, or pushed off cliffs, only to ditch into the water or dive nose first into terra-all-too-firma. The sport of poking fun at lunatics trying to fly in machines was always great fun, and no doubt the scoffers and skeptics had a field day laughing it up from the sidelines.

Even fossils fueled controversy when the ultrathin boned-giant pterodactyl, Pteranodon, was discovered in the 1870s. Sporting a wingspan of 20 feet, the extinct flyer was extremely light, weighing only 50 pounds. Paleontologists concluded, with typical conceit, that these ancient beasts had reached the ultimate size of anything capable of getting airborne; otherwise they would have been bigger. Clearly, we would not be able to build an even heavier flying machine able to carry a 150-pound man. One wonders how the debate would have gone had they known about the super-pterodactyl Quetzalcoatlus. Tipping the scales at more than 300 pounds and with a leathery wing-span of 40 feet, Quetzalcoatlus was as heavy as a sailplane.

The failures of manned flight reinforced public skepticism and encouraged experts to present learned papers showing why mechanical flight would always be impossible. Calculations proved that heavier-than-air devices were impossible. A continuing string of failures seemed to confirm these calculations. This is understandable, perhaps, if we remember that locomotives typified then-current notions about mechanical devices and the magnitude of power needed to make a craft fly. The belief of some theologians that flight was the province of God alone was confirmed with every foolish flop. When the renowned Smithsonian scientist Samuel Langley failed to catapult his manned, government-funded, aerial machine into the air for the second time on December 8, 1903 — it went straight into the Potomac River — the Washington press witnessed the watery mess and had a field day. Nine days later, the Wrights were under power and airborne over Kittyhawk.

With all the yuks and commotion going on in Washington, few had paid any attention to a couple of bicycle mechanics from Ohio playing with kites and gliders on a windy barrier island on the Atlantic seacoast. Why should they? After all, because of the disappointing performance of Glider 2 in 1901, Wilbur had to tell his brother that man would not fly during their lifetime. Orville died a year after Chuck Yeager broke the sound barrier.

Humans still dream.

Today, some dreamers hope they, or their children, will escape the human condition by transferring their minds and memories into electronic devices that are much more than just extensions of themselves. Dreams like this started as early as the 1940s or 1950s with the invention of the electro-digital computer, the logical locus for the reseating of transferred identity.

Yet, no machine exists that can perform anything close to human-like behavior, of course. Unlike Ray Bradbury's vision in *I Sing the Body Electric*, new homes do not come furnished with personal robotic servants. Even in the most advanced machine-intelligence laboratory, no robot exists that, on command, can slice a cucumber, arrange the slices on a plate, and serve them to a collection of hungry scientists. And, of course, we are a long way from any machine that can fathom the implications of Descartes' speculation, "I think, therefore I am."

The failure to achieve machine intelligence after so many frustrating years of experimentation has had a cooling effect on the field of artificial intelligence. Instead, it has inspired philosophers, physicists, and biologists to speculate that there must be something "special" about human intelligence. In recent years, many of these speculators have written best-selling books proclaiming deep insight into the mysteries of human consciousness, often linking the failure of machine intelligence to such factors as quantum mechanics, or the result of the machine's supposed reliance on mathematical proofs.

Those made nervous by Frankenstein's monster or the prospect of time-traveling, terminator assassins find comfort in the thought that the human mind is somehow special and can never be replicated by a mere machine. And there are many people who fear the coming of the intelligent machine. Unlike flight, which was eagerly anticipated, people are frightened by powerful, cybernetic technologies that narrow the gap between humanity and machine. They usually decide that, fortunately, we have little to fear; the first intelligent machines, if they can ever be built, are centuries in the future.

So what kind of dream is the Extraordinary Future? A fantasy? Or like the dream of flight a century ago — as yet unfulfilled and seemingly distant, but swiftly on its way to realization? (We will be returning to the example of the construction of flying machines because it offers important lessons and parallels for the development of thinking machines.)

There is, of course, no simple answer. We know that the world around us is going to change; the problem is figuring out how much, how fast, and in which way. To better predict where we are going, we first need to look at where we are and how we got here.

CHAPTER

Our Magical Science Fiction World

Any sufficiently advanced technology is indistinguishable from magic.

—Arthur C. Clarke's
third law:

We're Spoiled by Technology

You would think that after witnessing and benefiting from the wonders of SciTech, people would be less conservative when it comes to the emerging technologies of the 21st century. On the other hand, a healthy dose of skepticism might be a good thing. After all, we are still waiting for many predictions made in the past to be realized. "Where are the flying cars, the moon colonies, and the zero gravity boots?" Calvin asked Hobbes. "Where are the rocket packs and floating cities?" Good point, Calvin, and we can add: Whatever happened to those Jetson-like homes everyone would live in by the year 2000? Why do we still get colds, AIDS, and cancer? Where's our ability to control the weather? It still rains on the picnic. Have you noticed that cities are far from the gleaming citadels of progress predicted at World Fairs? Do people you know and love still get old and die?

The failure of science to deliver us the promised super futureworld has lead to the Sophisticated Skeptic. The Sophisticated Skeptic is much too hip and worldly-wise to be taken in by the outlandish predictions of mumbo-jumbo futurists. When talking to a friend about the Extraordinary Future, the Sophisticated Skeptic derisively labels it "science fiction." She then hangs up her cellular phone and continues to drive her microcomputer-laden Lexus down the super-highway to the airport. Later, as she jets at 32,000 feet to visit Grandfather, she rolls her eyes in anticipation of how he will regale her yet again with his account of when he first saw an automobile at age 10.

We are spoiled by the technology. Many inhabitants of the late 20th century have become blasé about how fast the rate of change is itself changing. We are at the point where we often forget how truly amazing and complex this world has become in a very short time. As a result, possible consequences and implications for the near future cannot be perceived, let alone known and understood. Let us provide a brief historical perspective.

Science Fiction Becomes Science Fact

Significant change, noticeable change within a lifetime, began at the dawn of the Industrial Age. The new knowledge about gears shifted the collective awareness into a higher gear: the popular notion of progress — the exciting realization that things were beginning to change for the better, and at startling speeds. With every upgrade of Industrial Age 1.0, imaginative thinkers found wonderful new tools for use in dynamic innovation. The notion of limits went from stubborn obstacle to sudden opportunity. A few of the most talented thinkers of the 1800s used fiction to speculate on futures without limits. They penned potential and possibility, and in the process invented science fiction — sci-fi for short.

The War of the Worlds

H. G. Wells' marvelous turn-of-the-century novel is a classic piece of literature, as well as being arguably the best SciFi novel ever written and a good example of under-prediction of the future. How could he, a writer of books, have predicted the huge general pandemonium a broadcasted radio adaptation of his novel could generate and sustain? In presenting the Extraordinary Future, it is likely that our outrageous projections may be viewed someday as under-predictions, as well.

Jules Verne (1828–1905) was the first of the great sci-fi writers. Arguably, he was the greatest of them all. In 1873, Verne's *From the Earth to the Moon* was published in English. The novel featured the military technology of Verne's time directed for a different purpose. We all know the sci-fi formula: Take existing technology, here an average artillery piece, and make it bigger; a cannon ball and make it roomier. Find some place good for moon shooting, say the east coast of Florida, load the cannon ball up with some folks, pack a healthy charge in the breech, and shoot it to the moon. Before 100 years had elapsed, impossible fiction was displaced by in-the-can fact. Rocketry replaced artillery, and Apollo 8 launched from Cape Canaveral shortly before Christmas, 1968. The ship — essentially a cannon, projectile, and charge all in one — carried three humans to the moon, around it ten times, and back. In fact, those three men were not alone. We were with them, as live televised moon images were broadcast to our living rooms more than 200,000 miles away.

Denied to readers, lost for more than a century, and only now being published, *Paris in the 20th Century* was written by Verne in 1863, four years after the publication of Darwin's *The Origin of Species*. With brilliant foresight, the novel anticipated gas-powered automobiles tied up in traffic jams, elevated mass-transit systems, faxes, telephones, the electric chair, and a dose of technological angst in a harried population irritated by the fast-paced world of technology. Refusing to print the book, Verne's publisher, the same fellow who so eagerly promoted his other classics, dismissed Verne, saying, "Nobody will ever believe your prophecies." Perhaps the publisher can be forgiven; the near-term is a professional obsession. After all, it took eight and a half decades before American science and technology finally achieved the modern equivalent of Captain Nemo's dream boat fantasized in Verne's *Twenty Thousand Leagues Under the Sea*, a real nuclear submarine, the USS Nautilus.

H.G. Wells (1866–1946) expanded and refined the sci-fi genre. He predicted aerial battles and atomic weapons in 1908 with *The War in the Air.* In his 1898 masterpiece, *The War of the Worlds*, powerfully armed Martian invaders overwhelm Earth's puny defenses. Within Wells' lifetime, reality caught up with his imaginative projections that were previously viewed as utterly impossible in any amount of time. Weaponry developed before he died could have defeated the invading Martians.

It should be obvious by now that criticism of speculative projections simply because they are found sci-fi is not historically justified. When the lead time from "Impossible!" to "What next?" shortens — first below lifetimes, then below decades — the risk of embarrassing opinion-reversal increases.

It is a plain fact: We are living in the science fiction dreams of our grandfathers.

TIME TRAVELERS

In his book, *Future Shock*, Alvin Toffler observed that during the revolutionary last couple of hundred years, we have all been captive time travelers. The world that exists when we are old is so different from the one we were born into; it's like living in compressed time. In fact, you do not need to grow old to be affected, or even disoriented, by the changes. Let's reminisce and step backwards in time to get a sense of the dynamism of technochange in this century.

Family Circus. *Reprinted with special permission of King Features Syndicate.*

When the Baby Boomer generation was in its youth, fingers still pecked away on heavy, mechanical typewriters crammed with three or four sheets of carbon paper. Apollo astronauts took slide rules to the moon and, using specially designed "space" pens, scribbled instructions from Earth on paper scraps. Good luck should you have an emergency. Calling for medical help was a marginal proposition at best since the 911 system was not yet in place. Pregnancy determination involved bunny executions, and a child's sex was revealed only at birth. Test tube babies? Maybe in a sci-fi flick. "Made in Japan" meant a cheap, fragile toy except they were coming out with nifty little transistor radios that fit into a shirt pocket — significant change. Today, "Made in Asia" means high-tech, high quality, and low cost.

In the 1950s and '60s, Boeing started a risky corporate tradition and gambled the company's total fortunes on the success of its big new 707s, and again with even bigger 747 passenger jets. Airline executives miscalculating the rate of change fretted about finding enough passengers to fill all the airplanes they had on order. They had no idea that two billion passengers would fly annually in only a few decades. On technology's smaller side, microprocessors were waiting to be invented. A few short years later, there are more than ten billion microprocessors (including simple microcontrollers in machinery) doing dazzling digital dances in watches, pocket calculators, appliances, and cars. Automobiles today are controlled by tiny computers more powerful than the bulky boxes on the Apollo spacecraft. How could we imagine that sophisticated microprocessor-based PCs would be manufactured in such numbers, giving ordinary people information-processing power that universities and governments once only dreamed of?

Let's go back a little more. Atomic explosives were built by scientists, mostly males, using slide rules, typewriters, and snail mail, and they got around the country by steam locomotion. Although the DC-3 and its piston-powered descendants were putting the monied classes into the sky, the rest of us, like the atomic scientists, still traveled by steam locomotives and ocean liners. We all lived in a low-tech, information-impoverished world of metal, glass, and wood.

Let's list some of the things that were not around 50 years ago: hydrogen bombs, intercontinental ballistic missiles, nuclear-powered submarines and utilities, satellites and space probes, the global positioning system, supertankers, stealth aircraft, jetliners, Federal Express, lasers, microwave ovens, malls and one-stop shopping, interstate highways, electrodigital computers (large and small), digital watches, pocket calculators, transistors and little radios, radio-controlled models, singing greeting cards, electric guitars, camcorders, TV networks and cable, VCRs, CDs, copiers, modems, the Internet, cheap faxes, automatic teller machines, bar codes, cruise control, anti-lock brakes, power anything, turn signals, seat belts, 10-speed bikes with hand brakes, motion-activated lights, rechargeable batteries, home air conditioners, ball-point pens, refillable pencils, pop-top cans, most plastics, fiber-glass and other composites, vinyl siding, polyester, Velcro, Teflon, freeze-dried food, TV dinners, vaccines for polio and measles, cures for childhood leukemia, most antibiotics, remote medical scanners, organ transplants, Prozac and other effective drugs for treating mental illnesses, and surrogate mothers.

As we go back earlier in the century, subtract still more from our world including paved rural roads here in the United States. Just after World War I, a young army officer, Dwight Eisenhower, lead a military convoy across the USA. Seems simple enough to us today, but remember: There were no connected series of paved roads, and it took the entire summer and serious slogging to accomplish. It's no wonder President Ike pushed so hard for interstate highways. Far from swift, strong, and sleek, early airplanes were underpowered, flimsy canvas and frail wood contraptions that struggled to get airborne and frequently failed to stay there. No phosphorescent green radar blips warned ocean liners that they were about to collide with each other or with icebergs. Mars seemed to be a shimmering, canal-crossed disc in the telescopic lens. There was no radio, no talkies. And as for medicine, well, just be thankful you didn't have to be hospitalized in those days.

Harkening back further to the beginning of the still-young United States, the Industrial Age was gearing up in England. It wasn't that long ago, only three generations. We now take away from the world aircraft (except for the first balloons), fuel-powered ground craft or ships, fuel-powered machines (except for a few steam-driven mine pumps), skyscrapers and elevators, breech-loading guns, dynamite, radio and TV, electric power and lights, telephones, phonographs, typewriters, photography, time zones, aluminum as a common material, anesthetics, and the germ theory of disease. Only a tiny percentage of the world's population lived in cities, and the really big cities could be counted on one hand. Vast tracts of the globe were still completely unknown to the vectors of industrial culture, the Europeans.

Benjamin Franklin, a leading scientific thinker of his time, had just invented a wonder of the age: a device that actually tamed the power of God Almighty Himself — the lightning rod! A time-snatched old Ben suddenly dropped in the middle of an airport might well stagger down the tarmac stunned at the unbelievable wonder of it all.

After he calmed down, he could be introduced to stereo headphones and an Enya CD. We have no doubt he would be profoundly startled and delighted to hear high-fidelity music from a seemingly magical box.

The world we live in would appear to be magic to our great-great grandparents. A call on a cellular phone from inside a horseless carriage, a brain scanned remotely for disease, organs "operated" on by a "gamma knife," operations in which a fiber optic tube is inserted through a small slit — all would be deemed utter fantasy to our grandparents' young minds. Until, that is, they grew up using these things. Classical Greeks believed only birds and gods could fly with power, agility, and grace. The ability to travel through the air crossing continents in mere hours would have seemed a miracle to our recent ancestors, who took months to cover the same distance walking alongside covered wagons. To us today, flying is a scheduling/cost hassle, with long waiting lines, long hours sitting with your knees folded under your chin, and bad food.

The Artifact of Complexity

We have built an artifact of complexity, fabricated a SciTech world that is orders of magnitude more complex than conditions in Ben Franklin's world at the beginning of the Industrial Revolution. When we go into a supermarket we are confronted with a huge range of choices lined up before us, multiplying at the rate of 20,000 new products per year. How much food do you think was available in 1946, or 1776?

Thomas Jefferson restarted the Library of Congress by donating his private collection. There are now so many books and journals that even major libraries are hard-pressed to handle even a fraction of them. The amount of information that traveled by daily mail 200 years ago was a drop in a barrel compared to the amount of information flowing every day by mail, wire, and radio 50 years ago. That in turn, is a drop in Lake Superior compared to what is being transmitted and received by humans and computers every single second these days. That your PC can churn through data faster than all the world's computers put together in 1950 is another sign of post-modern complexity. No wonder people are having trouble coping. Alas for those who wish to make life simpler. It is a safe bet that the world a hundred years from now will be orders of magnitude more complex than our current primitive state of affairs.

How Being Jaded Keeps Us from Seeing Into the Future

It's a funny thing about us, how quickly we get used to the novelty of new technology — as sci-fi entertainment becomes SciTech fact; as a once-distant and hoped-for dream becomes accomplishment; how quickly the new and startling is converted into the ordinary and routine. Excitement over the moon landing so long ago has turned into middle-aged "been there, done that." We do not watch jet planes in awe or get a huge thrill out of a plane ride any more. Computers that scientists would have hocked body parts for just 20 years ago are today's hand-me-downs for the kids. This human ability to quickly accept the most extraordinary events and technologies as part of normal experience has an interesting effect. We hardly notice that we have fallen into a trap of treating the exceptional as normal. Although the effect may be subtle, it is important because we think our current existence is the norm and vaguely assume that this "normal" existence will extend to future times.

Sure, we anticipate continued progress; the machines will get better, the computers smaller, the medicine more effective, and our lives longer. Daily life will be startlingly different, soon becoming routine. Are we frogs in slowly warming water? We arrogantly assume that the more our lives change, the more they will remain the same. Can this be true? Is it not obvious that for all the changes our technology has wrought, the big show is still "of the people, by the people, and for the people"? Is it forever beyond contention that technology is our servant and always will be; that even advanced robots will remain our tools, like computers, elevators, or steam engines?

How Star Trek Has Let Us Down

We don't think theologians and mystics help us foresee the Extraordinary Future with their prophetic pronouncements of second comings and new spiritual ages. We've learned not to expect much from them. We might hope to expect more from the great forecaster, sci-fi. Sad to say, the genre today is actually obscuring our vision of the future, misleading us about our status as humans in that time.

More SciFi Robots

"Star Trek" is not alone as an example of how pop sci-fi misleads. Why is it that in the "Aliens" series, a star-traveling civilization keeps sending out a few gun-toting Marines who are unable to wipe out some big bad bugs? A few good robo-soldiers would have done the job servo extensions down. At least "Aliens" did not inflict on us yet another mechanical robotic clunker. Robby the Robot, in the '50s classic, "Forbidden Planet," was little more than a talking jukebox on treads. The frenetic mechanical arm-waving "it does not compute" robot in the classic "Lost in Space" TV series was a running bad joke. More capable and lovable, but mechanical and joint-articulated all the same, were R2D2 and C3PO in the "Star Wars" movie trilogy. Beneath its Schwarzenegger-esque epidermis, the Terminator sports beams and hinges. Of course, this is just motion picture fluff, but have serious science fiction novels done better?

Isaac Asimov wrote the greatest robo-series, the *I Robot* novels, and later combined them with his masterpiece *Foundation* trilogy. The story tells of positronic brained androids preprogrammed with three laws of robotics preventing them from doing harm to any human under any circumstance. Eventually, the robots become superficially indistinguishable from humans. Their involvement in human affairs leads to such serious problems that the robots, following their rules of conduct, leave Earth for space and make themselves scarce, dodging all the humans moving out to populate the galaxy. The key points: The robots are human-like, are at least partially under human control, are unable to compete with humans, and humans dominate the galaxy. This is standard SciFare.

The dean of futuristic hard speculative fiction, Arthur C. Clarke, took the idea of intelligent machines to one of its ultimate extensions in the City and the Stars, in which a billion-year-old super-metropolis maintained by a giant computer controls the transformation and action of every atom with absolute precision. A small but smart spaceship whisks half-way across the galaxy with nary a word nor action from its occupants. Yet, these marvels are mere servants and background to the main plot characters, those darn human beings.

More recently, William Gibson's *Neuromancer* series has become famous for predicting the cybertechnology era we now enter. In his stories, humans link minds with, and enter into, computer networks. This begs the question, "Why?" If humans can merge their minds with computers, why would they not discard the human form and become an immortal cyberbeing?

Of course, sci-fi's track record of prediction has always been spotty. Orson Welles' pre-World War II film "Things to Come" attempted to predict the interval before the onset of the next world war. Keeping with the sci-fi formula mentioned earlier, it looks like 1930's technology on steroids, time shifted into the third millennium. Sci-fi stories as recently as the 1950s and 1960s featured daring interstellar astronauts computing with slide rules. One novel from that period had the reconnaissance officer aboard a craft orbiting a distant planet spreading out blown-up photographs on a table — there was no computerized digital imagery on that ship. At the opposite extreme, "2001: A Space Odyssey" presented us with tourist hotels in orbit, moon complexes, interplanetary spaceships, and a thinking computer.

The problem goes much deeper than just these superficial errors in prediction. The basic premise of most modern science fiction, especially the more popular expressions and serious examples of the genre, is profoundly and inherently unsound. A good place to get a handle on the problem is a familiar one, the assorted and multiplexed "Star Trek" locations. In the '60s, television watchers were excited by the future world portrayed in Gene Roddenberry's "Star Trek." Considering that the original "Star Trek" was conceived and produced in an era when vacuum tubes were routinely fault-tested in drug stores, the show's creative talent can take deserved credit for the impressive accuracy of their predictions. Aspects of "Star Trek's" anticipated future world have become part of today's real world. These "Star Trek" marvels include computers that can understand speech sufficiently well to make vocal replies, medical devices that remotely scan and heal patients, and hand-held or wrist-worn communicators that can talk with spaceships.

A recent televised medical program discussed the case of an unfortunate soul with a life-threatening brain tumor. Surgery would have been traumatic and dangerous, providing information of dubious value, and permanently blinded the patient to boot. In lieu of surgery, they took a detailed, three-dimensional look inside the poor fellow's cranium with non-invasive scanners for half an hour. One week later, a "gamma knife," precise to a fraction of a millimeter, zapped the tumor to vapor in 15 minutes. Without so much as a scratch, the patient walked away with his vision intact. This is possible because computers are many thousands of times faster and cheaper than when "Star Trek" first aired. They are so much faster that commercial machines are also beginning to listen to and talk back to their owners, and a global network of 60 or more low-orbiting satellites is being put into place, enabling a direct-dial phone call to a wrist-worn cellular phone roving anywhere on earth.

Having given some credit where credit is due, it is time to explain where "Star Trek" fails. Few notice that technology actually advances with surprising slowness in the Trekker galaxy of locales, crews, ships, and plot lines. The computers on the second Enterprise, Far Station, and Voyager, are clicking keyboard machines of the sort we already use, or voice-activated devices of the type we will use early in the next century. Star Trek starships are themselves stuck in slow mode. Built about a hundred years apart, the second Enterprise appears to be just a somewhat bigger and faster version of the original — same basic design, same warp drive, same style weaponry.

Problem number one with post-modern popular sci-fi: Future progress can be slow, even by current standards.

"The Ultimate Computer"

...is the title of one of the original "Star Trek" episodes, a show that dealt directly with aspects of cyberintelligence we are covering in this book. Although the standard computers on the good ship Enterprise could talk and hear, they exhibited little initiative. One star date, the brilliant Earth scientist, Richard Daystrom, fitted the starship for a test of the ultimate computer in lieu of the crew, much to Captain Kirk's irritation. The computer had been imprinted with "engrams," exact computer equivalents of the thought patterns of its creator. Dr. Daystrom had achieved a form of mental downloading that added human attributes to a machine incapable of error. The ultimate auto-pilot.

At first, all is well. The computer is doing a fine job running a virtually empty Enterprise among the stars. A smug Daystrom is pleased, but Captain Sensitive starts to feel obsolete and waxes philosophical about the fate of humanity in a Federation dominated by smart machines.

The situation deteriorates dramatically when the computerized Enterprise commits a sudden and apparently random act of violence, attacking and destroying an innocent star freighter. The obviously defective computer successfully resists attempts to turn it off, and Daystrom's desire to protect his mind-child does not help matters. The Captain and Daystrom must then watch in helpless horror as the computer uses a test exercise to assert its superiority by destroying a good part of the Star Fleet. Many die. Human superiority is finally reestablished when the distressed Daystrom convinces the computer of the error of its ways and disables his precious mind machine. Order is restored as the folly of allowing thinking machines to replace humans is realized.

Other early "Star Trek" episodes also depicted cyber-upstarts being kept in place. A variation on the theme involved a super space probe that had come to the logical conclusion that all planets inhabited by imperfect intelligence, that is humans, must be sterilized. The logic was so extreme that it bordered on insanity. A few populations were wiped out until Kirk again reestablished human superiority by using a logically irreconcilable argument to fry the machine's circuits.

The driving theme of "Star Trek" is that humans are better, to the point that even Trekkian aliens repeatedly marveled at how you just cannot make slaves out of humans the way you can with your run-of-the-mill hominoid — but didn't we keep human slaves once? Never mind; the important thing is that the human spirit and soul will carry us to the edges of the galaxy and beyond — and keep the machines in their places.

Problem number two with current sci-fi: Perhaps computers will think and minds may even be transferred, but one way or another, humans will prove superior to the mechanical pretenders.

STAR TREK ANDROIDS AND ROBOTS

The combination of a slow pace of technical advancement and human superiority, continued in "Star Trek: The Next Generation," which was set a century after the time of Kirk's prime. In "Next Generation," the assorted hominoid and other bioform aliens were joined by the sophisticated, "fully functioning," and all too self-aware Lieutenant Commander Data. Of course, Data has all the requisite flaws inherent in sci-fi robots, irrespective of intellectual brilliance. These include his lack of emotions, sense of humor, and intuition. Even his skin is green. Data not only is stuck with the disadvantages of the human form (which we will detail later in this book), but his biggest dream is to become even more human-like, as though a cyberbeing with the mental powers we humans could attribute only to magic or a god would want to mimic ape forms and mind qualities.

What is a Robot?

It's hard to define a robot. You know a real robot when you see one, but some machines called robots are not real robots. Let's explain some cyberterms used in this book.

Robot. The word is a variation of the Czech "robota," forced statute labor; a peasant. First used in "Rossum's Universal Robots," a 1920 play by Karel Capek that featured mechanical men and women performing ordinary tasks with minimum instruction. In an age when people feared the possibility of being replaced by machines, the term was soon widespread. The term "robotics" was coined in 1941 by Isaac Asimov in an astounding science fiction story. Four years earlier, the New York World's Fair featured a towering metal man that could "talk" via a recording and "smoke," but this was just a moving stage prop, not a real robot. Much more sophisticated audio-animatronic "characters" developed by Walt Disney in the post-war years do not qualify as true robots, either. In strict Capcek terms, a robot is a machine that is mostly self-controlled — an intelligent machine. It need not look like a human, but often does physical tasks a human might otherwise do. As might be expected, the term "robot" is not often used in any consistent way.

When was the first true robot put into regular use? Arguably, the honor goes to the autopilot developed and employed in the 1930s. When you are in an airliner that is being flown by a computer rather than by a pilot, the machine you are traveling in is a robot. On the other hand, fully radio-controlled "robots" sometimes placed in harm's way to defuse bombs or deal with other dangerous jobs are not robots because their actions are entirely directed by remote human guidance. The same goes for a radio-controlled model plane. On the other hand, a self-guided cruise missile or smart bomb is a robot by our definition. So, too, are deep-space probes; although they follow commands from distant earthlings, they do many operations on their own. A device that works in an assembly line according to a pre-written program is a robot. So is a self-guided (in part or in whole) mobile security guard or mobile food delivery cart. When cars drive themselves, they will be robots.

Android. A bi-pedal robot that looks human, i.e., the "Terminator," Lt. Comdr. Data, and so on. The device should be able to walk without outside support and have a large degree of self-control. A true android has yet to be built.

Cyborg. These are combinations of organism and machine, i.e., the "Six Million Dollar Man" and "Six Million Dollar Woman" fit the bill. Strictly speaking, this includes any person having an artificial part replacing a natural part. There are already a lot of people like that around, so the definition tends to include only those refitted with high-tech equipment yet to be invented. You might say cyborgs are those having artificial legs that let them run as fast as a gazelle, maybe a telescopic eye or two with infrared vision, or an exo-cortex (a remote computer brain) that allows them to connect to all 10,237 cable channels at once.

It is critically important that Data be unique; even his own twin Lore is eventually destroyed. Data is, therefore, a one-off prototype whose existence required a genius human inventor to devise the marvelous positronic brain; and the inventor is dead, and his notes are lost. Also observe how Data can communicate directly with other computers, but cannot transfer his mind to them. All this serves to keep Data in his place, for otherwise, there would be more and more Datas out-thinking humans, downloading their robo-minds into computer systems, taking them over, and becoming the operating intelligence of ships, rendering human crews the status of cargo. This would just not do.

Problem number three with sci-fi: Too few smart and capable robots.

STAR TREK HUMANITY

It would not do to have lots of capable robots in "Star Trek," because "Star Trek" commits the biggest sci-fi sin of all: People, lots and lots of people, billions of them inhabiting not only parent Earth, but hundreds of planets spread across Federation galactic territory and beyond. In addition, aliens, billions and trillions of aliens filling the galaxy, but hardly an alien robot to be seen. The latter have been suggested in a few episodes — superbeings and cyberminds that really run things — but they are kept conveniently tucked out of sight and out of the way most of the time.

Problem number four with popular sci-fi. Too many biological beings in environments terrestrial and extraterrestrial, and everywhere in between.

THE REAL SCIENCE FICTION

The slow pace of future change, the perpetual superiority and great numbers of humans, and the failure of robots far into times to come, is just science fiction to us. In the real SciTech future, it is doubtful that there will be an important place for human beings. Women and men probably never will populate the heavens. It is understandable to expect future changes to be on a scale roughly comparable to changes we have seen and will leave the pattern of our lives still recognizable. But this is more a comforting innocence than a fearlessly realistic extrapolation of what is coming.

Sci-Fi's Orphan Plot Lines

For all its outrageous and marvelous imagination, post-modern pop sci-fi suffers from a lack of truly radical imagination. Is there a plot our most daring writers won't concoct, even in fiction? What is behind the failure of the genre to look beyond the future of the human species? There is some human conceit involved; it is hard to imagine being replaced by the machines we make, and there is the inertia of human passion and desire. Naturally, humans want their offspring to thrive. Many people desperately want to believe that the human species will continue long into the future.

Yet, we still have not gotten to the core of the problem; it's a simple matter of plot. A future without humans is the ultimate plot disaster; where's the story? The problem is so bad that post-modern sci-fi is locked in a quandary, a doozy of a dilemma that has forced it into predictably stale stagnation. The point is, people crave association with other people and want to hear stories about them. The objective is to write and read about other people, whether the setting is the moors of England or the barren wastelands of an alien planet. It is difficult to imagine a workable novel centered around exclusively nonhuman intelligence, a novel that a significant number of people would read. "Star Trek's" creator, Gene Roddenberry, used to acknowledge that yes, spaceships of the far future will be automated machines, but who would watch a show that featured a deserted Enterprise?

The Old Science Fiction Just Ain't What It Used To Be

Today's speculative science fiction is sinking into escapist fantasy, read by only a small number of readers. There were no science fiction conventions in the golden age of speculative fiction; they would have been irrelevant. No one in 1900 was interested in dressing up as Captain Nemo; it was a submarine, the Nautilus, that was important. Today's sci-fi conventions are largely costume fantasy get-togethers at which the participants dress up and act out parts. You might think science fiction aficionados would be supportive fans of the Extraordinary Future, but many of them loathe the very idea. It spoils their future fantasies. It is neither surprising nor a coincidence that science fiction has partly merged with fantasy, leading to the oxymoronic genre of science fantasy.

FRANKENSTEIN

There are more reasons why science fiction is misleading its audience about the Extraordinary Future: Sheer fear, mixed with a healthy dose of retribution for scientific hubris. This is not surprising. The thought of intelligent machines can be very scary. We fear they will do bad things to us, such as hunt us down to kill or enslave us, either by force or with micro-implants, or worst of all, replace us. These parallel fears have led some writers of science fiction that deals with cyborgs, androids, robots, and the odd corpse revived from the dead, to give the BIG WARNING: We should not play God. We should accept the natural limitations of being human, or we will be really, really sorry. The media campaign on this theme has been relentless since Mary Shelley's classic written a century and three quarters ago. Such a campaign is an ironic twist on the novel because scholars argue that Frankenstein was not a warning against technology, but focused on the morality of a creator abandoning what he has wrought (the novel's "monster" was intelligent and sensitive, but so ugly that Dr. Frankenstein fled in disgust and horror). Since then, novel after novel, movie after movie, TV drama after TV drama, have pounded out the warning theme.

Verne's Captain Nemo destroyed himself with hubris. Wells' powerful Martians fell to nature's humble microbes. In the 1950s, "Forbidden Planet," had human spacefarers discover an abandoned alien supercivilization whose builders had long ago made mind-boosting machines so powerful that they destroyed the minds that built them. Remember "Westworld," in which a theme park's androids go mad and try to do in the human patrons? Michael Crichton's "Andromeda Strain" and "Jurassic Park" are popular stop signs for bio- and genetic engineering. The "RoboCop" and "Terminator" flicks are robo-nightmares, super-tech fantasy worlds in which cyberbeings trash humans as though they were nothing more than grapes to be squashed. The classic "Blade Runner" found Harrison Ford hunting down killer androids that had become a little too human in their need to kill. "TRON" found a human stuck inside a dreadful computer-generated virtual universe. In the more recent "Lawnmower Man," a virtual reality mind-meld leads to an out-of-control superman.

In contrast to the harsh negative images of cyberbeings, adult audience-oriented cybercharacters that are both positive and popular are rare indeed. By positive, we mean a cyberbeing that neither harms humans nor is under their control, that has a magically sophisticated and subtle form, that is in no way mentally or emotionally sterile, that is happy about its immortal existence, and contributes to the general well-being of conscious minds. Even Data, appealing as he is, does not meet these criteria.

The Sci-Fi Scientists — They're Always Mad, Mad, Mad!

Of course, we exaggerate, but an old sci-fi cliché is the mad, or at least overly enthusiastic, scientist who blindly works to make the robot machine or anti-aging potion that will give her THE POWER she desperately craves. Another worn cliché is the misguided scientist playing God who regrets same and destroys his experiments to keep the knowledge from other innocents and vulnerable humanity. Yawn. It is the central theme of movies from "Frankenstein" to "Jurassic Park." Ethical scientists doing quality high-tech research and saving lives are a rarity. In the real world, good scientists who have made lives better through new technologies far outnumber evil ones. Besides, nowadays, scientists tend to work in collaborative groups, so the mad ones sort of stick out.

The premise that the world can be saved if, and only if, the foolish experiment is destroyed, is pointless. If one scientist can figure something out, the odds are others will, as well. When the global scientific database reaches a given level, the next level of discovery becomes increasingly obvious. Had that wild and crazy Galileo not pointed the new telescope upward, someone else would have discovered the moons of Jupiter. If a demented Darwin had fallen off the Beagle and drowned, Alfred Russell Wallace might have published the Theory of Evolution around 1859. Had the mad (just look at his hair) Einstein not figured out relativity, Hawkings might have.

Because science is without boundaries, attempting to keep secret the results of science-gone-bad is counterproductive. After all, if other scientists do not know the dangers of a scientific experiment, how will they avoid making the same mistakes when they stumble across the same discovery? Better to publish and warn others so they can maneuver around the pitfalls.

After this negative publicity, it is no wonder people are scared about what robots may do to us if they get too smart!

What Sci-Fi and SciTech *Are* Telling Us

To get back to labeling the Extraordinary Future "science fiction," we are willing to go along as long as we are talking about the work of Verne and Wells, Clarke and Asimov. Their forecasts were far from perfect, but they had records to be proud of.

It is the persistently anthropocentric sci-fi of today that has no relevance to the Extraordinary Future.

Indeed, the lessons of Verne and Wells are worth emphasizing. At any given time, the conventional wisdom about the moderately long-term future is usually wrong. Ninety-nine years ago, 99.99 percent of the population would have scoffed at the proposal that moon shots, long-range submarines, city-busting bombs, airliners, and powerful electric mini-calculators would be commonplace in 99 years. They would logically argue that radiation had just been discovered and its potential was a mystery; no one had yet flown; rockets were simple artillery rounds; and that electricity was just beginning to power light bulbs and hand-cranked phones. We know how unimaginably complex Boeing 777s and computer chips are by 1890's standards. It would be just common sense to argue that it would take centuries to go from 1890's technology to 1990's technology. Only a few visionaries were able to look at the technological world that was exploding around them and extrapolate a dim vision of what was to come.

The current conventional wisdom that the 21st century will be an extension of the 20th century is as logical as the one-time consensus that the 20th century would be a linear extension of the 19th. As the modern world is a magical sci-fi place dramatically different from the century past, the next century quickly will become a magical place totally unrecognizable to us. The difference between then and now is that back then, the technological base we were jumping from was still very low and was far, far from producing highly intelligent devices. Today, the technological base we are jumping from is multitudes higher and closer to the goal of human-level artificial intelligence. Our high-tech human world exploded into being in the blink of an eye, and in another blink, it will be a fading dream.

SCIENCE SPECULATION

This book is not true science fiction, which is a genre of literary fiction. Its main conclusions are not science fact, either. The events we relate have not yet occurred and may never do so. The correct term for a book that takes current scientific knowledge and extrapolates future events is "science speculation," sci-spec.

Sci-Fi Pulls Up Short

Not a single work of popular speculative fiction we know of depicts machines with the capacity for consciousness being built in the next century, or building a new and powerful cybercivilization that humans join en mass, leaving Earth and heavens barren of *Homo sapiens*. Fans of science fiction may be able to point to a few obscure examples that puts most of these concepts into one package. This is not important because such poorly known works are not influencing the public. It is not possible to dismiss the Extraordinary Future as mere sci-fi when popular examples of the genre avoid the theme like the plague.

"Terminator II"

One recent movie gave an inkling of what sophisticated cyberbeings might be like — the morphing robot in "Terminator II." But even that machine was a rather dim-witted, preprogrammed fellow that wasted its time assuming various human forms while hunting down termination targets.

We are now going to ask you to do something that may be difficult: As you read the rest of this book, try to discount what sci-fi has taught you about robots and cyberintelligence. The sci-spec ship we are on is sailing far beyond the horizons of most science fiction. After all, they say reality is stranger than fiction. So much for sci-fi for the time being; we will return to it on occasion, when necessary. To better understand the perils and potential of forecasting futures, we have to take an in-depth look at how things in our universe came to be and how they change.

Life, the Universe, and Evolution

Why Things Start Out Simple (and Dumb) and End Up Complicated (and Smart)

Let's say you want to understand the workings of your car, not just to repair your conveyance, but to understand how burning gasoline in cylinders allows you and your stuff to be transported from New York to Baltimore in four hours. To understand at that level requires knowledge about Newtonian physics, thermo-dynamics, and not a little chemistry.

Let's say you want to understand the workings of the universe, how the cosmos and events here on our little planet change over time. Books, television programs, courses in astronomy, geology, and biology help explain the workings; but to really understand the nature of changes over time, you have to know about evolution, about how information is processed, and about complexity. You need to know why certain conditions make the world go faster and faster and faster and....

Information is Everything

Knowledge, it is said, is power. How right that is. Information is power and everything else. Whatever is happening in the universe — and the universe is a seriously weird and strange place — is all information. Yes, this subatomic particle exists. No it does not. Yes it does. No it does not. Yes, this galaxy exists. No its does not. Yes, this data bit exists. No it does not. This thought, this person, this conscious mind exists. No it does not. If there are gods, they have to be information-processing systems. The more information one can process, and the more intelligently it can be used, the more power one has, and the more things one can do.

Turing Machines and von Neumann Automata

Alan Turing and John von Neumann thought a lot about information processing. Both of these super minds were major players in the tech wars of World War II. Von Neumann helped the boys at Los Alamos figure out how to make the tricky gadgets go bang. Turing did something even more important — he helped design the early "bomb" computers that helped break the German codes. Both died, it is sad to say, early deaths in the 1950s, but not before they laid crucial foundations for building intelligent machines, which we will return to later. Turing and von Neumann executed work of universal importance.

We will start with Turing. He ingeniously developed and enhanced algorithms, mathematical procedures for solving problems in a series of finite and repetitive steps. Algorithms are widely used by mathematicians, scientists, and engineers. In the 1930s, Turing devised a system that was able to process any set of information that can be processed digitally, by a series of yes's and no's — or more precisely, by the two encoded qualities, "does exist" and "does not exist." Turing's bi-digital "machine" was first only a concept, a thought experiment. It involved an infinitely long tape marked with an infinite number of squares, each of which could be marked with either a 0 (no) or 1 (yes). A writing and reading device scanned the data marked on the tape one square at a time and used a program to process the information.

Given enough time and speed, this simple system can do any calculation. The workings of a Turing machine can be done with paper and pencil. But shunting a

Two Digits Are Good, More Digits Are Better

Two digits have a big advantage in computing. You cannot code information with just one signal; two is the minimum. When Turing sat down to outline the principles of digital computing, it was natural that he worked with as few codes as possible. When the first working digital calculators were built, they, too, kept codes as simple as possible.

Bi-digital computing is inefficient, a disadvantage. It takes a lot of computations to code real-world information using only two symbols, and these computations burn up significant amounts of energy. Bi-digital data storage takes a lot of space because every switch carries just one bit of data.

One way to increase energy and storage efficiency is to use multiple digits. Energy use is reduced because fewer computational steps are needed to break down complex data. Storage space is reduced because an information node can carry more than one bit of data. But the more digits there are, the harder it is to work with them. Designing and programming a multi-digit system is very difficult. So, while multi-digits may be better, bi-digital systems will continue to rule the cyberworld — at least for a while.

single tape back and forth is not very practical. This does not, however, mean that the Turing machine concept is just airy theory. Rearrange the system's components and it can be made very practical. In fact, every digital computer is a variation of the Turing machine concept.

After the war, von Neumann took the Turing machine football and ran with it. His work was pivotal to devising the practical means for making the new computers effective Turing machines. But this was dull for von Neumann. He wanted to make Turing machines that do what life does, i.e., become more complex with time without outside intervention. It is important that a Turing machine not get bogged down by processing the same information over and over in the same way, and always getting the same results. To accomplish something useful, it is vital that the machine have a randomizer, something that kicks in "surprising" information the machine can use to create new, interesting results. Feeding new information from the outside is one way to randomize things. That it is how most modern computers work; they are top-down systems that become more complex only when a still more-complex system adds data or instructions to the program.

Von Neumann came up with a way to make Turing machines into bottom-up, bootstrap systems that could become more complex on their own. In his concept, the bi-digital Turing machines would become independent, self-reproducing automatons in which the randomizer was an internal system generating new and novel combinations of zeros and ones. The fascinating thing was that most of the randomly generated information may have been useless, or even detrimental, but every once in a while, something would pop up pointing the way toward an improvement in the performance of the system. If the system was able to change and to select the beneficial changes rather than the harmful ones, and if there was enough time, the system would improve itself over time, becoming more sophisticated and complex. Note that this system was not random because only those changes the system was capable of coding could occur, and only those changes that worked would survive. Evolving Turing machines are semi-random in their actions.

The intricate, rather mind-numbing details of how Turing machines and von Neumann automatons work are not critical to our discussion. What is important is that self-reproducing systems can become more complex through bottom-up, bootstrap processes. Indeed, a key attribute of the Turing/von Neumann machines is how easily they become more complex with time. The creationists' belief that complexity can come only from higher complexity is disproved, and has been for half a century! It is important that these systems have the potential for immense power because as long as they run, they have the potential to become more capable. An evolving Turing machine that successfully runs for infinity can become infinitely complex and powerful!

Turing Machines Can Be Smart, Even If They Don't Know It

Not every Turing machine is intelligent. Computers that depend on outside intelligence to program their actions are not inherently intelligent. Every evolving Turing/von Neumann machine is, however, intelligent. By this, we do not mean that all such machines exhibit conscious and cognitive intelligence as we do. Although cognition is the highest level of intelligence, intelligence does not require consciousness (we will discuss this matter in more detail below). In this view, intelligence is present if an information-processing system is doing something it was not initially programmed to do. Intelligence is also relative. An ant or a cruise missile is smart compared to a rock, but an ant and especially a cruise missile are dumb compared to a human. We call computer-guided cruise missiles and bombs "smart weapons" because, as dumb as they are, they're a heck of a lot more intelligent than their completely dumb, unguided counterparts. A foraging ant is exhibiting a sophisticated, albeit low, level of intelligence, as is a microbe, which has no neurons to be conscious with. Today's computer-based artificially intelligent systems are examples of intelligent Turing machines.

What Is a Machine?

When we think of machines, we tend to think of trains, planes, automobiles, watches, computers, TVs, and hair dryers. This is correct, but incomplete. Machines can be of artificial construction and be entirely natural in origin. Stars are machines, natural machines that fuse elements to produce heat and light. So is the atmosphere of the Earth, a heat engine that moves thermally defined masses of air and water around the globe. Those who work on the function of animals quickly come to think of them as machines, because they work just like machines. This is one reason why the study of how animal skeletons and muscles work at the gross and microscopic levels is called biomechanics.

Machines need not do anything physical. Think about it: It is true that when you are thinking, you are using energy to do work, which is why you are tired after a day of hard intellectual labor, much as you are after one of hard physical labor. A brain sending impulses between neurons is as much a working machine as a computer that shuttles electrons back and forth.

Basically, anything that uses energy to do work or that processes information according to an internally logical set of rules is a machine. A Turing machine is, of course, just such a machine.

We tend to expect machines to do what they are told, at least as long as they are working correctly. Driving to Grandmother's in Detroit listening to *Car Talk* on the radio, you do not expect your conveyance to suddenly decide Buffalo is a much better place to be, and then drive you there. An evolving Turing machine, however, can and does make decisions on its own. This is why the biomachine we call a horse can decide to take the right fork even as the rider pulls left.

WHAT IS A COMPUTER?

When most people think about computers, such as PCs or IBM mainframes, they are thinking about bi-digital machines that do calculations with a series of yes's and no's. This was not always the case. Before 1945, computers were people, usually female, arrayed row after row in vast rooms, punching old-style calculators as they tried to keep a bank's finances straight, or were engaged in the well-nigh hopeless task of calculating myriad ballistic tables for artillery pieces and rounds churned out by U.S. industry to blow up Nazis. Even nonbiological computers need not be digital. Some are analog and do calculations with smoothly fluctuating signals rather than discreet digits. (The old phonographs were analog; CDs are digital.) Once common, analog computers fell out of favor as digital became the rage. Today, they are starting to make something of a comeback as they demonstrate certain advantages that may be important to our hypothesis, as will be explained later.

Computers are almost ideal practical Turing machines. When he laid down the principles of his yes-no system, Turing happened to be laying down the operating base for simple bi-digital computers. In the broadest sense, a computer is any system that uses signals to process information via calculations that solve algorithms according to a set of rules. Computers can either be natural in origin or manufactured by intelligent systems. Because computers process lots of information at high speeds, they endow their owners with power. The more powerful and intelligent the computer, the greater the user's power. As you will see, we will frequently apply the concept of the computer to everyday objects you may not think are computers.

WHAT IS EVOLUTION?

Mention evolution and people think of Darwin's theory of life and its evolution on Earth. Not only life evolves, but so does our universe, as well as technology, economic systems, and human societies. So does every human mind.

Perhaps you guessed it. Evolution is a living expression of a Turing machine (although most evolutionary biologists do not usually think of it in that way). When evolution occurs, something is changing over time. The change requires a randomizer, a top-down process such as a new idea about how to make a better mousetrap, or a bottom-up system such as mutating genes. The new mutation or invention is selected *for* or *against,* i.e., the longer legs help the animal run faster to catch more food, or nobody buys the "better" mousetrap and it goes out of production. It is always possible to add a little bit more to the system at any stage; therefore, complexity can increase over time. Evolution is intelligent, leading to intelligent solutions without applying conscious forethought.

Do We Live Inside a Self-Evolving Computer?

In many respects, the universe we live in appears to operate like an enormous computer. As hard as it is to appreciate, the elemental particles we and everything we are made of wink in and out of existence at hyperspeeds. They exist, then they don't exist, then they exist, sort of like the on-off switching circuits in a computer. Some thinkers have suggested that the universe is an immense bi-digital computer, another Turing machine intelligent in its own noncognitive way. In this view, everything past, present, and future is the result of a vast set of computations. On the other hand....

The Universe Is Really, Really Weird

Describing the universe as a computer may be applicable as far as it goes, but may be incomplete. As we are performing our normal daily tasks, we tend to think of our existence as established, solid, and definite. Not so. As mentioned, all of the subatomic particles we are made of literally come and go, vibrating in and out of existence at a fantastic pace. Not to worry. The statistical chance that you will wink out of existence is as close to zero as one can get without getting there. The family of on-off subatomic particles making up the atoms are statistically dependable, there when needed, making the atom very stable. In the liquids and solids that constitute the body, atoms appear to be packed together in direct contact with each other. But this does not mean we are solid.

Take an average atom. If it's enlarged so its outer electron shell is as big as a stadium, the nucleus in the center is a marble! This was first realized about a hundred years ago when experimenters were shocked to find that less than one percent of the subatomic particles they were shooting through sheets of pure gold were smashing into anything. About 99 percent of "us" is empty space; we are tenuous objects.

Of course, there is "relativity" that old wild-haired Einstein came up with. It is not possible to push an object through space faster and faster until it finally exceeds the speed of light. As the object gets closer to light speed, it becomes heavier and heavier, shorter and shorter, and time runs slower and slower. To anyone riding in that speedy object, everything seems quite normal.

Even gravity is not the straightforward, simple force that no-nonsense Newton thought it was. It is warped space, bending even the path of speedy photons of light, so much so that the super-gravity of black holes warps space back on itself until even light lacks escape velocity.

Light does double duty. It's a wave that bends when it passes around a corner. It is a particle, too, pushing against atoms it impacts; sometimes it's a wave, sometimes it's a particle. This dual nature is so confusing that some physicists console themselves with nomenclature therapy, calling the carriers of light "wavicles," thereby solving the strange duality.

The strange duality of light brings us to the seriously weird world of quantum physics. Quantum physics seems so goofy that Einstein himself never really believed it. Yet, experimental evidence strongly suggests we really do live in a quantum universe. It does not seem possible to unite the relativity theory with quantum mechanical theory in the Grand Unified Theory. A holy grail of a theory, a GUT would connect everything to everything. Physicists with high achievement aspirations know its discovery would cinch a Nobel Prize.

We mentioned one of the more conventional quantum effects, the strange exist-doesn't exist nature of subatomic particles. So is the fact that an electron is not actually in any particular place in the electron shell unless it is observed. The electron exists as a probability wave that fills all the possible positions around the nucleus all at once. If the electron is observed, the probability wave collapses and the electron takes up the position that it is statistically most likely to be in. Quantum physics says that, at very small scales, (those of atoms and below) things do not work according to the "normal" rules we are used to. They operate instead by a combination of bizarre statistical rules and the notorious observer effect.

A simple experiment visually demonstrates quantum weirdness. Set up a source of light, preferably one with a single frequency of light; a sodium lamp's yellow light will do. Next, put a flat screen with two clean-cut slits — long, narrow, parallel to each other about a millimeter apart — halfway between the light source and another flat target. Cover one of the slits. Light travels in discrete packets called "photons."

Like peas from a peashooter, you can put one photon, or two, or one hundred thousand trillion and sixty three photons (about the number put out by a light bulb) right through that uncovered slit. You cannot, however, send one and a half photons, or a trillion and one-third photons of light. If light photons behaved like Newtonian particles (the peas), they should zip right through the slit, showing a hard-edged band of light on the target screen. But no. Instead, the light makes a fuzzy bar, as if the peas turned into pea soup and splattered on the target. The light band on the target in the experiment has a central band of high intensity that gradually fades toward the edges.

As the photons travel through the slits, they curve a little bit around the slit edges, in a manner similar to ocean waves bending around the point of a jetty. You are able to see this light deflection at the edges of any shadow. It is always little fuzzy, not crisp. The further the shadow is from the edge that forms it, the fuzzier the shadow's edge. This is odd, light acting as both a particle and a wave, but is easily explained if light were made from waves traveling in wee little packages.

Or so one thinks until the other slit is uncovered and the lamp is switched on. Because the slits are close to one another, one might expect to see a brighter, single band of light on the target screen. But no. Now observed is a series of bands fading to the edges, with the center band being the brightest. Between the bands are dark zones where no light photons are hitting, which is odd because the dark zones are located where light was hitting when only one slit was uncovered! This is a strange form of wave interference. Strange, because there is no apparent way that discrete photons passing through one slit can "know" other photons are passing through the other slit, and that they should interfere with them!

The effect becomes more pronounced if the light can be turned down until it is sending out one photon at a time. If one slit is open, a photon can travel that route without interference. Leave just the other slit open, and the photon can travel that other route without interference. If both slits are open, the photon appears to "know" that it will experience interference and will travel neither route!

Finally, we encounter the renowned "observer effect." Turn the light back up to multiphotons, open both slits, and place a photon detector at one slit so it can observe photons as they pass through. The series of bands will disappear in favor of the single band! Knowledge of your observation that the photons are passing through one slit causes all the photons to pass through the same slit!

The path the photons take is not determined by any particular direction they happen to be zipping. Knowledge and observation are involved. Each photon exists as a probability wave that, at any given distance from its origin, occupies all possible positions it can be in. The wave collapses when the photon is observed. Because an unobserved photon is everywhere it can be when it hits the slits, it passes through both of them. What is stranger still, is that increasing the distance between the slits does not stop the strange effect. If the slits are a light-year apart (and the light-source is many, many light-years away), a single photon will pass through both of them at the same time, and a swarm of photons will still interfere with one another. This is strange because it means the "knowledge" that a photon or photons is passing through the two slits at the same instant can cross light-years, or even much of the universe, in an instant!

Strange quantum effects lie behind the slow loss of energy, formerly thought to be impossible, from the deep gravity wells of black holes, or may create "wormholes" that directly connect different parts of the universe or different universes to one another. Physicists are so disturbed by the strangeness of quantum effects that they have tried in desperation to devise ways to explain exotic outcomes, defaulting to conventional "common sense" principles. It's just that the more experiments they do, the more the extreme, bizarre, unconventional nature of the subatomic world is confirmed. The universe we live in is very weird; you are weird yourself because you are made of virtually empty atoms working via strange quantum effects.

Although quantum physics is difficult to understand for even the brightest scientists, it is not insane chaos. It runs according to a strict and logical set of rules. We can regard it as another example of an information-processing machine, albeit a very peculiar one.

What Turing Machines Cannot Do

Quantum weirdness, by its nature defies Turing machines performing computations with algorithms. Kurt Godel's famous 60-year-old theorem bars Turing machines from solving many noncomputational problems. Let's say you have a prevaricating Uncle Fred who will not or cannot tell the truth. He says, "I always lie." Any Turing machine will fry itself processing this information. Contradictory statements do not allow the computation of a solution with any algorithm, no matter how much calculating power and speed is made available.

Other strictly computational problems are nondeterministic because they grow forever as they are crunched and never achieve final resolution. Take the case of the cost-conscious traveling saleswoman. She has a sales route of 30 cities she must cover every six months. She wants the most efficient route that minimizes travel time and distance and avoids cities already canvassed. No problem in this age of fancy computers, so she seeks help from a computer-savvy friend. Alas, he is not cyber savvy enough to know it's a trick problem, and he finds his computer grinding and grinding away with no solution. He tries the local university's supercomputer, only to wait hour after hour eating up computer time, still with no resolution in sight. Only when he appeals to an irritated campus computer expert does he hear about the classic Traveling Salesperson Problem in which the computational alternatives could occupy all the computers in the world running until the universe shuts down and still not have a satisfactory solution for the saleswoman!

Because so much cannot be calculated, the universe cannot be modeled entirely as a Turing machine. However much that occurs in this universe falls under the purview of calculating via Turing/von Neumann systems.

Making Universes

Lets start at the very beginning. Most people think it is remarkable that we are here in the first place. Certainly, making an entire universe billions and billions of light-years across must be a very hard thing to do.

In fact, it takes nothing, literally nothing.

"What?" you say, "How can that be?!" But consider, why do we exist at all? Why isn't there just, well, nothing? There cannot be nothing. Think about it: The only thing that can exist is something. A simple reality, true all the same. Okay, so something exists. You know that at least one universe exists, the one you exist in (do not confuse the Milky Way galaxy we live in with our universe, which is a much bigger place that spans billions [if not more] light years and contains hundreds of billions or more of huge galaxies like the Milky Way, each containing many billions of stars). Is the universe we share the only one rattling around? Possibly, but hardly likely. The probability that our universe happens to be the only one in existence is infinitely

VIRTUAL PARTICLES AND NEW UNIVERSES

Let us assume that you produce an absolute vacuum, one that has not a single subatomic particle or photon of electromagnetic radiation within it. Your perfectly cold, empty space is by no means empty. Even a vacuum is seething with virtual particles that come into existence for a fleeting instant and then anhillate each other. This exists/does-not-exist process is one expression of quantum physics. The same process is going on in every inch of space, including inside you. The energy contained in these ephemeral particles is immense, but inherently untappable.

The spontaneous creation of an entire universe is an extreme expression of the ability of quantum physics to make something out of nothing. Basically, a "bubble" is created in the quantum vacuum. This bubble is packed with immense amounts of energy. Most of these bubbles quickly collapse out of existence. Some, however, rapidly increase in size and become huge. Where do all these universes "fit"? Some suggest that a diversity of bubbles of varying sizes — some expanding, some collapsing — sit nestled next to one another through infinite space. Another view is that universes are nestled inside other universes. Perhaps a quantum bubble inside you became a universe that seemed very big and old to the life it contained before it winked out of existence in much less time than it took for you to perceive the letter *Z*. Still others suggest that the universes are not in the same dimension, and that "where" they are is meaningless. This view allows for "parallel" universes that closely mimic one another. Furthermore, it is possible that all these scenarios are true.

small. That there is any particular number of universes, say two, or two dozen, or a billion trillion and one, also is infinitely improbable. The probability that there is an infinite number of universes is extremely high. Even as we speak, an infinite number of universes may be creating themselves all by themselves out of nothing.

As strange as this may sound, we merely repeat the ordinary shop talk of astrophysicists and cosmologists. One of the leaders, the Russo-American Andrei Linde, has presented a hypothesis that our not-so-little universe is really part of an infinite superuniverse in which an infinite number of self-generating universes each creates an infinite number of new universes — each with a Big Bang start-up followed by internal expansion, creating, in turn, an infinite number of new growing universes in the cosmic version of the never-ending chain letter that actually works.

How can universes create themselves out of nothing? In Linde's system, they don't; all universes spring from other universes. In Stephen Hawking's view, our universe and others do not have a beginning; the time it took the Big Bang to get going was infinite. Other cosmologists point out that once outside this universe, there is no rule that something cannot spring from nothing; it may be the routine way universes come into being. In this view, a "super" quantum effect creates whole universes from nothing, much as ordinary quantum effects make subatomic particles out of nothing. In essence, universes bootstrap themselves into existence in an infinite bottom-up system of creation. What is important is that the infinity of universes is almost certainly a true proposition. It should be taught in all our schools until it is common knowledge.

So why has a particular universe produced intelligent life? The fact that intelligent humans dwell in the cosmos at first seems astonishing. So many things have to be just right that it seems like a miracle. Take water as an example. Water is not only a broad action, but also a gentle solvent vital to carbon-based life forms. It is a peculiar, albeit familiar substance. Unlike any other liquid, H_2O expands in volume when it freezes, making ice that floats in water. Most substances become denser when they solidify, so they sink to the bottom. Floating ice is very good indeed.

Let's say we lived in a universe where the molecular structure of water was a little different, just enough so that frozen water became dense and sank. All the ice in the oceans and lakes would sink to the bottom and would help freeze the water above it until the only liquid water on the planet would be geothermal hot spots and a thin veneer of cold surface water near the equator. With the globe locked into a perpetual super-Ice Age, single-cell life might evolve, but brighter beings like us would be most unlikely.

Or consider the atoms we are made of. It is remarkable that any exist at all. When the universe first formed, there was so much matter and antimatter that there was much more aggregate matter than now. When matter meets antimatter they wipe each other out, and it was only because there was a tiny, tiny percentage more matter than antimatter that there is any matter at all. Matter does what it does. If the strong atomic forces holding atoms together were just a little different, atoms would either break apart into elementary particles or condense into hyper-dense matter that would interact in ways probably not amenable to life.

The intermediate-weight atoms that compose our bodies, carbon, oxygen, nitrogen and so on, are remarkable, as well. These heavy elements form in stars through a complex and intricate process that works only because the excited state of carbon 12 happens to be just exactly, precisely right. On the opposite side of the scale, there is an extraordinary balance. Because our universe started out so small, the initial balance of forces was critical as it expanded from the Big Bang. If the gravitational forces that attract matter to itself were just a little higher, the universe would have remained a small, dense fireball. If the same forces were just a tiny bit lower, the universe would have hyper-expanded into a place of such low density that matter would not have been able to coalesce to form stars and planets. The precision of this balance, called the Cosmological Constant, is practically beyond comprehension. And this is only one such exquisite balance. We could go on and on with example after example of this sort of coincidence; everything in the universe is finely balanced to work together.

The odds against any particular universe being so perfectly balanced for life as we know it, or for other exotic forms of life, appear vanishingly small. Theologians claim that only divine intelligence could design such a beautiful universe. In their view, complexity can be created only in a top-down system in which greater complexity creates lesser complexity. Therefore, creationists (some believe in a young Earth, others in an old Earth) dismiss the bottom-up, grass roots generation of high complexity from lesser complexity. They equate the chances of the universe just as it is with the statistical probability of an ape typing Homer's *Odyssey*. And they would have a good point if the universe were solitary or numerous. But there are probably an infinite number of universes in existence.

Let's return to the statistics of infinity. The probability that every one of the infinite number of universes being identical is infinitely small; the probability that they are all much the same is almost as improbable. The probability that the universes are as diverse and different in their natures as they are infinite in numbers is extremely high.

So we have an infinite number of universes exhibiting infinite diversity. In effect, that monkey is typing the *Odyssey*. Perhaps the great majority of universes, 99.99999999999999999+ percent of them, are not suitable for evolving life. Only an infinitesimally small percentage of the universes enjoy the delicate perfection that allows intelligence to appear and thrive. No matter. Intelligence is not rare or even special. A tiny percentage of infinity, no matter how small, is itself infinity.

The wonderful probability — not a certainty, but something almost as close — is that there are an infinite number of universes that contain intelligence. In this logical view, complexity and intelligence are not the specially bestowed product of a higher superintelligence. Rather, it is the inevitable result of existence itself. Just as something must exist, so, therefore, must conscious minds. What we have described is a classic example of evolution inside a natural supracosmic machine. Universes "reproduce" themselves at rates far beyond that achieved by all the life on Earth.

Because the new universes differ from one another by greater or lesser degrees, they "mutate," in effect, every time a new one is created. The sterile universes are dead ends where nothing of interest ever happens; they are selected against. The selection process favors universes suitable for the appearance and evolution of life because they are thriving places where interesting things happen. Note that the system is intelligent, but in a manner more akin to an evolving Turing machine than a thinking deity. Darwinian selection, not gods, rule the universes. Because natural evolution is noncognitive, it is not necessarily efficient. It may be that only a tiny percentage of universes spawn life, and this huge universe may be empty of all but one example of intelligence.

Our Evolving Universe and the Life Within

For a short while after it was formed some 7 to 20 billion years ago, the universe changed rapidly as temperature declined and density decreased from the hot, dense conditions of the initial fireball. Some think the universe underwent a period of faster-than-light inflation (no, this does not violate relativity) that increased the size of the space the universe occupied to near infinity, while the nascent material universe was still a tiny fraction of a second old universe. Within the first minute, neutrons and protons formed and combined to form atomic nuclei; 300,000 years later the universe was cool enough for atoms, mainly hydrogen and helium, to appear. Stars and galaxies formed out of these two primordial elements a billion years after the Big Bang. In short order, large numbers of giant stars formed, forged heavy elements, burned themselves out (the bigger the star, the hotter and faster it burns up its fuel,) and blew up, spewing even heavier elements into new galaxies. The heavy element dust began to form rocky planets around subsequent generations of stars, eventually including our humble home.

Our particular universe is, of course, one of those happy places where life thrives. Under the right circumstances, certain atoms are prone to forming organic molecules.

These organic molecules, in turn, have a strong tendency to self-organize into reproductive systems we call "life." Here on Earth, life as we know it may have gotten its start immediately after the effects of the last globe-sterilizing super-impact had worn off. (Life may have appeared repeatedly as soon as the Earth cooled down from its initial formation 4.5 billion years ago, only to be totally wiped out again and again by a series of massive, Earth-sterilizing impacts by super-comets and asteroids.) The life we descended from probably dates from less than four billion years ago.

What we know now that Darwin did not is that bioevolution is an information processing system based primarily on simple quad-digital computers, RNA, and DNA. In most organisms, DNA is a protein programming computer and RNA is the messenger machine used to make proteins and the like (some viruses, such as HIV, have only RNA). We say that RNA and DNA are quad-digital because the information on these molecules is coded by four molecules. This is good because data can be stored more efficiently and compactly with four digits than with just two. DNA has a high storage capacity, with a modest processing capacity when it splits, combines, and occasionally mutates. The recombination and mutations are randomizers. DNA is such a spiffy little multi-digit computer that it is starting to be used to do certain mass calculations that vex more conventional computers.

We tend to be awed by DeoxyriboNucleic Acid, because it has an intimidating name, because every human cell nucleus contains a microscopic DNA strand that, if unraveled, is a yard long, and because DNA determines morphogenesis — our very form. However, there are reasons not to be so impressed. For example, every genetics lab has a machine that can churn out as much designer DNA as you want, and the amount of information contained in human DNA — about 3 billion bits, or the contents of a wall of books — can be crunched by a supercomputer in a few seconds. We have trouble reading DNA not so much because it is complex, but because we do not yet have the nanoscale tools needed to easily decipher the information.

DNA is a biological version of Turing - von Neumann automata (that DNA uses four rather than two digits does not invalidate this correlation). Indeed, von Neumann presented his hypothesis of self-evolving automatons just a few years before Watson and Crick demonstrated how right he was when they described the structure of DNA. Because DNA is a form of self-reproducing automata, its complexity can increase with or without external stimulus as long as conditions allow

Proteins Are Information Processors, Too

Proteins are complex organic molecules made from amino acids. Most DNA sequences code for the information needed to make proteins, making DNA a protein computer. As the intermediary between DNA coding and protein manufacturing, RNA is an information processor. The protein molecules are jointed nanomachines doing life chores for the body's cellular maintenance. Until recently, it was thought that these activities were mechanical in nature. It is now known that many proteins are information-moving units doing no physical work. The information helps the cells regulate internal functions. Your DNA/RNA protein-filled cells are microcomputers!

it. In essence, DNA is a noncognitive intelligent system that learns about its environment and adapts to it. Reproduction and mutations are the means by which DNA learns and adapts. Darwin's work is confirmed each time mutating cockroaches in your kitchen laugh at the insecticide you keep spraying.

Tragically, DNA computations illustrate why HIV viruses evolve in their hosts until they overwhelm the immune system, and why disease microbes are becoming dangerously resistant to antibiotics. It is not even necessary to manipulate DNA with advanced technology to produce radical changes. By manipulating the DNA of wolves with selective breeding, primitive societies were able to obtain a wide variety of dogs with dramatically different forms. With breeding techniques, we develop a huge diversity of food crops, some of which cannot reproduce without human help. The ideas leading to altering microbial DNA to create new drugs a couple of decades ago are now creating profits.

DNA, as with any evolving Turing machine, is open-ended. There is not a hard stop that forever prohibits a strand of DNA from increasing its complexity. Say there is a strand of DNA 1,000 units long. For it to become more complex, it need only grow to 1,001 units. This is not much more complex, but it can then go to 1,002 units, 1,003, 1,004, and so on. Eventually, a code billions of units long can be built up. This is the bottom-up, bootstrapping machine Darwin and Wallace observed in nature, for which Turing and von Neumann laid theoretical foundations in mathematics, and Watson and Crick described in biology.

SPECIATION IN PARADISE

There are many ways species can form; here is a simple, real example. A volcanic island arises far out in the ocean. Over time, a few continental birds are blown to the island where species form into breeding populations. Because the DNA coding for the form and function of an organism is dynamic, not static, genetic alterations inevitably occur. The genetic code of the small population of island birds drifts from the parent species on the mainland. With the passage of time, the genetic difference becomes so great that the genome of the island bird has become too different from the original stock for the two populations to interbreed anymore. And with DNA coding for different forms and functions, the island birds have adapted to the new habitat. There are now two different species.

Another way to increase genetic complexity is for a population to become isolated from the rest of the species. As DNA in the isolated population replicates over generations, it will drift. The DNA will accumulate change after change until it codes for a plant or creature whose form is so different from the ancestral species that the new organism not only cannot breed with its ancestors, but it looks and works differently enough to be a new species. Now there are two species, which can become three, then a thousand, a million, and so on. With trillions of individuals belonging to millions and millions of species happily replicating DNA, you have a giant computer that cannot help but program for novel species — another fact anti-evolutionary creationists cannot comprehend.

Modern, sophisticated information theory explains that universes, life, and conscious intelligence can create themselves and evolve from the bottom up — a sort of grass-roots affair — but this sort of thing should happen all the time. The creationist view that only top-down systems can endow lesser systems with existence, life, and intelligence — an elitist way of doing things — lost credibility in the mid-1800s. Creationism in its various guises is as archaic and obsolete as the extinct life forms it fails to comprehend.

As well as it works, DNA evolution has serious limitations. It is not guided by conscious intelligence. It is a trial-and-error, Rube Goldberg affair whose only driving force is to achieve ongoing reproductive success. Becoming more complex, intelligent, or pink are not goals in themselves, only techniques to keep the descent pattern going. It is, therefore, possible for a species to remain pretty much the same for countless generations, as long as it is doing reasonably well and selective factors are not pushing the genome to make major changes. Nor can bioevolution make quantum jumps to entirely new systems.

The ancestral pattern is modified incrementally at every stage. It cannot go from a fish to a horse in a single generation; millions of intermediate generations, step by step, are required. Nor can DNA code for bones made of titanium alloy steel, or make neurons out of diamonds and run them with entirely electrical currents. DNA lineage accumulates computational quirks that prevents it from coding for less outlandish features. Vertebrates can have, at most, four legs and two eyes, whether it would be advantageous to have six of the former and four of the latter or not.

Contrary to false impressions often left by nature programs, Rube Goldberg DNA evolution does not produce idealized biomachines meshing perfectly with their environment. Bioevolution can muddle its way to producing only imperfect, make-do beings doing just well enough to replicate themselves successfully, thereby avoiding extinction of that particular line of DNA. Because the DNA computer is a slow, simple, noncognitive system unaware of either its limitations or any goals, major bioevolution — at the level of species and above — is slow, very slow. On this planet, biological evolution has gone through eons filled with millions of generations, tremendous amounts of energy, and cycled massive amounts of tissue to produce butterflies, humans, whales, bacteria, oaks, cold viruses, ferns, and aardvarks.

However, bioevolution has managed a pretty impressive acceleration over time. Major changes took hundreds of millions of years during the three billion years when single-cell life held sway. Most of it was microscopic, except for algae-like bacteria that formed pillow-like stromatolites in warm tidal shallows.

Some innovations did appear, including photosynthesis early on and sexual reproduction about a billion years ago. Mainly, it was new species of bacteria and algae living in the same old way, in a world of equatorial desert and polar ice, barren of animals and plants on the land, in the seas, and in air where low oxygen content would be lethal to one of us. It may have been the lack of atmospheric oxygen that kept life so simple, or perhaps genetic upgrades necessary for more complex organisms did not happen to click into place. Eight hundred million years ago, the first soft-bodied multicellular creatures finally appeared.

The best thing about multicellular life is that it can develop multicellular data processing machines, i.e., nerves and brains. The first vertebrate brains appeared in a few little worm-like swimming things approximately 550 million years ago. They were nothing to write home about, but it was a start. Then came the first great revolution in Earth history. From 540 to 530 million years ago (10 million years is an instant in geological terms), a host of oxygen-burning animals evolved, many of them hard-shelled forms like the famed trilobites. This event is so remarkable in its speed and scope that it is often called the "Cambrian Explosion"; it was beyond a mere revolution. The age of multicellular organisms had begun, in which major changes occurred in tens of millions of years. Soon, vertebrates began the long development of bigger and more capable brains; by 300 million years ago, brains reached the dimensions of those found in fish, which are also similar to those seen in reptiles. Two hundred million years ago, the first mammalian brains appeared and then stagnated for millions of years under the harsh dinosaurian yoke.

The pace of evolution picked up yet again after the dinosaur extinction, as the brains of primates began their great expansion some 40 million years ago. Progress was reckoned with a few million years at a time. About 10 million years ago, ape brains one-third as large as ours evolved. Then evolutionary activity really picked up. More than 4 million years ago, australopithecine protohumans with bipedal posture and opposable thumbs started making stone tools, spurring the first acceleration of evolution beyond biology. Major changes now came in hundreds of thousands of years, thousands of times faster than natural evolution. The age of technology had begun, and the mental stimulation of invention helped double brain size in humans two million years ago. On the fast track, the combination of nature and technology pushed brain size up another 50 percent — to the modern human level — around a few hundred thousand years ago. With this achievement, bioevolution lost its place as the leading edge of progress. We are still evolving, but our genes have little to do with it.

SciTech: Evolution by Other Means

Laissez-faire capitalism is Darwinian in key respects. The randomizers of the SciTech and capitalist Turing machines are cognitive minds that come up with bright new ideas and apply them. SciTech is the fastest and most powerful kind of evolution because it is far and away the most intelligent. Intelligent systems are by no means perfect, and they often benefit from unplanned serendipity. However, not only can conscious minds figure out how the universe works, they also can look at the very workings of the evolving system they live within and fiddle with it to make things go faster. Conscious minds can set specific goals, then plan and do what is necessary to achieve them. Astounding leaps are not only possible, they are common.

Another advantage enjoyed by evolution via conscious design is that it is a top-down system that can look at a device and immediately install a radical modification, or even junk it in favor of an entirely new, dramatically improved device. When the Wright brothers wanted to fly, they didn't grow wings; they built an entirely new contraption from scratch, got in it, and flew. When piston engines were reaching basic performance limits, they were abandoned in favor of radically different jets. When silicon transistors came along, it was good-bye to the vacuum tube. Even genetic evolution has been greatly accelerated under intelligent guidance, first by the breeding of domesticated forms, and today via genetic engineering.

Technoevolution has another advantage over bioevolution, one that is not always appreciated. DNA is pretty good at retaining knowledge it has acquired. Once a code for a certain protein function has been developed, it tends to be retained indefinitely. Much of the DNA-protein processing that goes on in our cells is identical to processes in bacteria or fish cells. DNA is seriously conservative. However, much of what biology has taught us is lost. DNA lineages, species, entire faunas and floras become extinct. Whatever unique knowledge they contained goes with them.

The knowledge gone to waste during Earth's history is vast and has limited the speed of progress. In contrast, just about everything learned by modern science and industry goes on file, ready to be accessed when needed. Little data is lost, so the knowledge base tends to build up, and up, and up to the extent that the main problem becomes sorting through it all. This is the classic learning curve, and the efficiency of SciTech at retaining what it has learned is another reason for its success.

It is no surprise that human progress has been so fast. Ten thousand years ago, the Age of Agriculture began and a few thousand years later, cities and writing were invented. Evolution occupied millennial jumps. Even the ancient Greeks commented on the swift rise from barbarism to great stone cities and ram galleys and speculated on where progress might take humanity. Just a thousand years ago, the inhabitants of the two hemispheres did not even know about each other; their total numbers equaled the number of people who now travel by air every two months (300 million). They tended to live a mere 25 years, and most lived in servitude at a time when human rights were minimal at best.

Within the hemispheres, new technologies often took centuries to migrate across a continent. A European peasant would die in the same village he or she was born into, a village that changed little if at all. The peasant would perform the same farming and household tasks that his or her parents, grandparents, great-grandparents, etc. performed. There was no generation gap. Kids listened to pretty much the same ballads and wore much the same clothes as their elders did. The 990s were just like the 980s, which were just like the 970s, which were just like the 960s.... Aside from ordinary changes, such as a raid from a neighboring feudal power, plagues, and the like, the concept that things could change a lot during a single life was unimaginable.

But not for long, as scientific and technological progress shifted to a century's pace in the Renaissance, and then to a decade as the capitalist-driven Industrial Age started two centuries ago. The times when a life began and ended in a contiguous world were no more; folks grew up riding horses and were driving cars as adults. Around 1900, new steel-making technologies took five short years to spread across America. Turn-of-the-century future shock arrived, as people born during the Civil War died during the sci-fi Second World War. Communism came and went in a lifetime not because of its immorality, but because the massive governments they produced stifled Darwinian evolutionary innovation too much to be competitive with capitalism.

The new "Age of Information" is just decades old, and leading-edge technologies are spinning along faster and faster. Nowadays, things change significantly within a decade, or even within a year. New products come and go. One day, only one family on the block has a TV; a few years later, everyone does. One day, almost no one has his or her own computer; ten years later, millions do. New technologies now spread around the world within a year. Whole factories pick up and move across an ocean in a matter of months. Cellular phones spread like a virus. A host of long-established companies that had become icons suddenly go keel up, or are listing badly.

Few lifetime jobs exist anymore. Jobs change so fast that workers regularly take classes to keep up, and kids are told they can expect to have a dozen or so jobs in their lifetime, which assumes there will be jobs for them to do. Social changes are, in kind, fast-paced. Popular music is in a constant state of flux, with each generation's kids dissing the previous generation's favorites. TV programs have weeks to make it, or they fail. Hair and clothing styles wax and wane like approaching lunar cycles.

The not-so-gay '90s have not recapitulated the avaricious '80s, which were not extensions of the stagnant '70s, which differed from the psychedelic '60s, which were not like the conservative '50s.... In many sciences — computers and robotics, some aspects of physics, genetics, neuroscience, astronomy, paleontology, etc. — more knowledge has been gained in the last 10 to 20 years than in all prior history. Or looking at it another way, these fields have begun to enter an annually delimited rate of change in which any given computer setup, for instance, is outdated within a year. Things are changing so fast that people became nostalgic for the '50s in the '70s, and the '70s in the '90s.

As exasperatingly wasteful, oftentimes destructive, and time-usurping as we have been, we have achieved an astonishing degree of efficiency in the evolution of science, technology, and diverse forms of information processing. Computers are the greatest expression of this achieved efficiency, and they promise to increase the rate of progress even further.

The Lessons of Time

What have we learned from this brief history of life and technology? In particular — like steering a course with a rear view mirror — how can patterns of the past be used to help predict patterns of the future? Let's take a look.

The Universe Really Is a Machine, and Why This Is a Good Thing

As far as we can tell, the universe is a machine. Its workings remain obscure, are incomprehensibly complex and in many ways are just plain weird. But we can say that the foundation of the machine appears to be the spawn of Heisenberg's quantum mechanics and Einstein's Theory of Relativity, overlaid by an evolving Turing machine expressed in biological form as von Neumann automata, and sorted by Darwinian selection. Because bioevolution is a mechanical reality, it can only produce machines. So the assembly of parts that is you — that's right, you — is a machine and a very special one at that.

Many prefer a mystical universe, but an evolving mechanical universe is more attractive. It has the potential to be manipulated through intelligent design into what we desire it to be via comprehensible techniques applied with science and technology. Before that can happen, minds must understand the patterns evolution are prone to assemble. We will start with....

THE THREE FLAVORS OF EVOLUTION

The evolutionary machine does its work in three main ways. First, the bottom-up, bootstrap (i.e., no assistance from more complex systems) path from simple to complex that life follows. Second, top-down systems in which higher complexity (which may or may not be consciously intelligent) constructs devices whose complexity increases in stages, but never exceeds that of the creator. This describes technology to date. Third, a combination of bottom-up and top-down, in which a complex system first programs the increasing complexity of a system (top-down), but the system then continues to evolve on its own (bottom-up). This combo applies to things like brains (more on that later) and will probably apply to technology as new machines rise above human control. All these systems involve Darwinian variation and selection.

CHAOS, COMPLEXITY, AND CONNECTIONS

Remember in the movie "Jurassic Park" how the character played by Jeff Goldblum said that chaos theory explained why the dinosaurs were running amok? Actually, it was the park staff's grossly incompetent planning. In any event, as Jeff's character tried to explain, scientific chaos is not the same as your ordinary chaos. It simply means that it is impossible to predict exactly what will happen in all but the very simplest systems, even if every detail of its past and current conditions is known. This is because the slightest variation from the predicted path will quickly spin things off onto a surprisingly different course.

Astronomers can tell you the basic size and shape of Jupiter's orbit in a million years, but they cannot tell you where that planet will be in that orbit, because the gravitational interactions of multiple bodies are chaotic and inherently unpredictable. Meteorologists can tell you it will snow this winter in Ontario, but not how much or when because the littlest breeze could turn into a storm, but probably will not. If the first few protovertebrates had not survived, you would not be reading this. That's chaos.

In general, the more complex the system, the more subject it is to change; the less well it is understood, the less predictable it is. Also, machines that evolve from the bottom up, such as horses, are less predictable than machines that are designed from the top down, such as cars. At the start of life billions of years ago, whether or when the feature of human intelligence would occur was unpredictable. In the same way, we cannot tell exactly what the human condition in all its complexity will be a century from now.

The complexity and unpredictability of chaos is increased by the interlinking of events. James Burke's marvelous television series "Connections" details history by taking the viewer on a journey of science and technology in which every idea and every development connects with a multitude of other ideas and developments — hypertext history. Somebody somewhere comes up with an idea for something, and before you know it, someone else somewhere far away is applying the same idea to an apparently different chore. Who would have known that the need to solve the problem of calculating artillery ballistic tables would lead directly to the connected PCs this book was composed on and will soon lead to the hyper-intelligent machines this book is telling you about? You see, its all connected.

"Connections" applies to life and evolution, as well. Little did primates know as they evolved full-color binocular vision, manipulative hands, and big brains to better find and eat colorful fruits spread out in the 3-D world of tree tops, that millions of years later it would lead to tool using and art creating Homosapiens, who, in turn, would use these tools and talents to produce "The Simpsons," in color.

People are fond of asking why scientists and engineers keep coming up with technologies people can use to harm others. Why can't they come up with only good stuff? The reason is the connections between things. It is literally true that almost any technology can be used for bad or ill. Rock tools can be used to pound grain into flour, or pound in heads. There is not a whole lot of difference between planes that carry people or bombs. Chemicals first used to make colorful dyes end up making people die, as in poison gases. Meanwhile these chemicals in later years end up being used to fight cancer. The Internet can be used to spread valuable information, or hate. The computers first devised to calculate the artillery tables now help a paralyzed person lead a semi-normal, more varied, and interesting life.

The Extraordinary Future is a bundle of connections with normal potentials for good and ill. The connections outline the origins of the universe and track it to its end. They are why the extinction of dinosaurs led to the writing of this book. They are why we with intelligent minds live under the rare light of a big moon, in a solar system with even rarer gas giant planets, and why we are probably alone. Connections are the reason the world is changing so fast, why technology is making people feel weird, cynical, and just plain bored rather than happy and fulfilled. Connections explain why people who think they have had federal UN chips planted in their behinds are blowing up buildings. Connections explain why fundamentalist religions are enjoying such growth, while at the same time technologies that will smash religion are being developed. Connections are why antibiotics are failing as lifespans continue to increase. Connections are why jobs are getting scarcer as computers are getting smarter, why the same technologies that will make cyberminds smarter may make jobs unnecessary, and why it does not matter at all in the end anyway. Connections are why the same level of technology that will stop aging and disease will make being a human obsolete and unnecessary. Connections are why humans may soon become extinct, but not the minds of humans. We will explore these and other connections as the book progresses.

When you combine chaos with the inability of any single individual to comprehend the complexity of connections, is easy to see why no one can predict the future with any certainty. This is why we all disagree, even though we all think we know better than everyone else how the world works. So, there was no way a late-Victorian futurologist could have figured out how all the trends of SciTech, economics, and societies of his day would follow through the 20th century to today. Chaos prevented just about everyone in 1981 from predicting that the USSR would just roll over and die in 1991. Just a couple of years ago, folks were looking to Japan as the superpower of the 21st century, until an unexpectedly large recession burst the bubble and led to more-realistic evaluations.

The Lesson: The future of our connected, bottom-up evolving universe is chaotically unpredictable. But by no means does this mean it is entirely unpredictable. Remember, we do have the ability to predict that summer will be hotter than winter, and that Jupiter will still be between Mars and Saturn a billion years from now. We can make the connection that the genetic engineering that stops disease might stop aging as well, and everything that implies. So, we have some justification for playing Jules Verne and trying to forecast gross future trends.

From Simple And Weak to Complex and Powerful

As we have explained, Turing/von Neumann machines, including life, have a strong tendency to evolve into complexity from simplicity, and powerful systems from weaker ones. Earlier in the century, evolutionary biologists thought life evolved in lineage, inevitably progressing along distinct paths toward important goals, such as big size or high intelligence — a view Darwin would have frowned on. We understand better now that life is an intricately ever-branching tree in which any given species evolves in whatever direction its genes happen to allow and favor. This is why some reptiles, birds, and mammals have returned to the same aquatic realms their distant ancestors managed to crawl out of.

As for complexity, many systems (from bacteria to hammers) remain simple, and complex systems sometimes re-evolve simplicity. But brand new systems, whether they are biomechanical or technological, must always start out as simple and weak, and can readily increase their complexity and power by adding new features incrementally. So, enhanced complexity and power are a normal, albeit not inevitable, product of evolutionary systems. The concepts that complexity cannot arise from less complexity, and that developing complexity requires intelligent guidance, are nothing more than continuing creationist superstition. For convenience, we will sometimes refer to the increase in complexity and power as our notion of progress. We know there are those who will challenge whether the modern world we live in represents facts or fables of consistently beneficent and benign progress.

Because biological systems are so complex relative to our current technology, it is often asserted that living organisms are marvels that we will be hard-pressed to duplicate. Ergo, many scientists are overawed by the apparent intricacies of biological systems. However, nature is complex only because it is an old computer with origins billions of years in the past. It has had a lot of time to slowly accumulate complexity. DNA evolution is so slow that nature is actually a sitting duck. Because artificial systems can evolve millions of times faster than the biological system has appeared to, the former will match and exceed the complexity of the latter millions of times faster.

The Lesson: Although it is not certain that intelligence will continue to increase in complexity and power, it is a very good bet that it will. As it does so, the complexity of technology promises to first match, and then exceed, the intricacy of nature.

Punctuated Equilibrium

Two seemingly obvious evolutionary patterns were not obvious to Darwin, a gradualist who had a poor fossil record to work with, and who thought life changed slowly and consistently over time. With a detailed and vast fossil record to work with, and two centuries of modern technological progress behind us, we now know that the rate of change tends to speed up with time, and that it usually does so in spurts of growth, with changes broken by long periods of stasis and occasional reversals. The appearance of the universe in the Big Bang was such a sudden — indeed as far as we know, instantaneous — event. Solar systems such as ours form in about a hundred million years, then become relatively stable for billions of years. Life may have originated fairly quickly, only to settle down into a quiet microscopic motif for billions of years, which, in turn, was followed by the explosive appearance of multicellular and then hard-shelled organisms. Flying insects, pterodactyls, birds, and bats all appeared fairly suddenly in the fossil record.

The fossil record shows that species tend to remain remarkably constant in form for a few million years, then evolve into new species in relatively short periods measured in thousands of years. Major extinctions have periodically wiped out 50 to 90 percent of global life, to be followed by the rapid evolution of new life forms to take the empty places. The most famous example is how dinosaurs ruled the roost for 170 million years, quickly disappeared, and were then replaced by a variety of increasingly large mammals over the ensuing 20 million years. For 400 million years, vertebrates remained rather small-brained beings until some rays, birds, and mammals started becoming more intellectual about 150 million years ago.

Broom Hilda Evolution. *Reprinted by permission: Tribune Media Services.*

The evolution of humans and their brains has been similarly punctuated. On the gross level, the entire evolution of humanity has been the latest and biggest punctuated event ever; it exceeds even the Cambrian explosion and dinosaur extinction. Not only did the final expansion of the brain to human dimensions take only a few million years — a hectic pace by paleo-geological standards — but at this time, the fossil records suggest that the expansion took place in small, distinct steps.

The first enlargement from the ape level occurred when australopithecine brains become about 25 percent larger than those of apes of similar body weight. A swift and major increase in brain size occurred about two million years ago as the first Homo sapiens appeared. A few hundred thousand years later, brains expanded to two-thirds of the modern size. The brains of H. erectus, best known as Peking and Java man, grew little for one million years. The final expansion came with the evolution of early Homo sapiens a few hundred thousand years ago. As far as we can tell, human brains have not changed in any significant regard for tens of thousands of years or more.

By no means has the cycle of revolution and stasis been limited to nature. The Archeulian stone tool kit of *Homo erectus* changed little in more than a million years. Modern humans quickly developed new and more sophisticated tools, but they remained Stone-Age hunter-gatherers for about a hundred thousand years, until they suddenly got the notion of planting crops and raising goats. Great civilizations sometimes rise with astonishing rapidity, stabilize for centuries, then crash in decades or even years. The Egyptian civilizations lasted for millennia, Pax Romana for centuries, Pax Britannica for just over a century, and Pax Americana is good for mere decades. Europe stagnated in the Dark Ages, but then took off into the age of colonization while the once innovative Chinese civilization took its turn and languished in stasis.

Subservient for millennia, Western females suddenly got the vote and become liberated. Slavery and segregation were practiced for centuries in the American South. Just two decades begot civil rights and an African-American general who, having led the allies to victory in the first digital war, is considered prime presidential material. Science was first and briefly practiced by the Athenian Greeks, but then almost disappeared until revived in its modern form in the late Renaissance. Although steam power was known in classical times, the Industrial Revolution did not get under way for a couple of millennia, and then it swept England and the western world.

Naval technology illustrates well the punctuated increase in progress. After millennia as the primary war vessel, oar-powered galleys were gradually displaced by sailing ships with broadside batteries about half a millennia ago. The latter stabilized into the grand ship-of-the-line for two centuries until it was suddenly rendered redundant in the 1850s and 1860s by steam-driven, ironclad vessels. These evolved into turreted dreadnoughts that dominated for a few decades until they were displaced by aircraft carriers in World War II.

The classic example of artificial, punctuated evolution has been the development of aircraft. For millennia, no one flew; then for 60 years, aircraft were outdated almost as soon as they were built because the technology advanced at such a blistering pace. World War II, for example, started with biplanes and 350-mph monoplanes. It ended with 450-mph piston engine fighters and 540-mph jets. By the late '50s and '60s airliners and fighters stabilized in terms of gross form and performance; the latest Boeing 777 looks little different from the 707 prototype of 40 years ago (this is, however, somewhat misleading. It is inside that airplanes are changing dramatically as they become semirobotic machines run by computers).

The computer revolution is itself an even more spectacular case of punctuated evolution. Suddenly, out of nowhere, new machines appear, grow swiftly in power, and rapidly proliferate into scads of other machines and into our places of work and homes.

STOP-AND-GO EVOLUTION IN THE FOSSIL RECORD

It has long been observed that species seem to appear suddenly and remain fairly constant in form until they become extinct or spin off new species. Transitional forms between species are rare fossil finds. Classic Darwinism's view that natural evolution is gradual led many to believe there were gaps in the fossil record, the so-called missing links. Geneticists suggested that major evolutionary changes may be stifled by large populations because mutant genes tend to get swamped in a large population of standard genes. It was suggested that new species evolved among small, isolated populations where small genetic improvements had a better chance of diffusing until all the offspring were modified to the new standard. This helped inspire some paleontologists to argue that the stop-and-go evolution suggested by the fossil record may be real. New species develop so quickly in small populations that they are almost never preserved as fossils. They then stay pretty much the same for long periods of time as the population expands its size and range.

Punctuated equilibrium has been controversial, but is gaining adherents as evidence of it accumulates. The basic concept that evolution, whether natural or artificial, often occurs in sudden spurts is useful for understanding how computers and robotics may develop in the future.

It is obvious that evolution is not a smooth running affair; there is no simple pattern to its jumps and spurts. There can be a long period of little change between bursts of rapid evolution, or one burst of evolution can be quickly followed by another. The stop-and-go evolution of life has been labeled "punctuated equilibrium," and the first of these terms can be applied to SciTech evolution as well. It is not hard to figure out why this sort of stop-and-go progress is typical of human societies. The ability to make dramatic leaps from one technology to another new and very different technology is one reason. The tendency of human minds, individually and collectively, to cruise in mental neutral and then suddenly spring into action when a new idea comes along is another. Sometimes, a plateau in performance is reached that is difficult to break until some dramatically new technology produces a revolution. The latter may apply to airliners, which will all converge in appearance until and unless a way of carrying lots of people at higher speeds without raising costs is developed.

We understood less why life, with its inherently slow and incremental change, is subject to the staggered events of punctuated equilibrium. In a few cases, the cause is clear. A mass extinction — whether caused by earthly or extraterrestrial events — followed by a new flourishing of life is prima facie a punctuated event. As with science and technology, a dramatic genetic breakthrough to new levels of biocomplexity can start a revolution. The onset of DNA coding for the production of hard shells may have been the cause of the Cambrian revolution. The main problem has to do with why species fall into stasis. Perhaps the answer is that the sheer size of a gene pool in a large population swamps major genetic changes. Thus, the rapid evolution of new species may only occur in small isolated populations.

The Lesson: Some things may change in a smooth, gradual manner, but it is likely that big changes will come in sudden, surprising jerks after periods of deceptively quiet stasis quo.

The Great Evolutionary Rev-Up

The second evolutionary pattern Darwin could not fully appreciate is important to the Extraordinary Future. It is how the rate of change has persistently increased with time from the onset of life to today. (The pattern was not obvious until various geological dating methods showed just how old the Earth and its life are.) Evolution has gone from taking many millions of years to less than a year to produce major change.

We can better appreciate this by looking at the ages of life and technology. Note that the ages supplement one another, rather than replace each other in turn. So, while we are in the Age of Industry, we are still in the age of single-cell life, as well. As each age comes along, the leading-edge life-forms or technologies of the previous age lose their places to the new age's leading-edge expression of progress, which, in turn, gives way to another.

THE MAJOR EVOLUTIONARY AGES

Age of the universe: 7-20 billion years
Age of single-cell life: 3.6+ billion years
Age of multicellular animals: 800 million years
Age of larger brains (above the reptilian level): 200 million years
Age of big-brained primates: 30 million years
Age of hominids: 4 million years
Age of Homo sapiens: 100,000 years
Age of Agriculture: 100,000 years
Age of Industry: 200 years
First age of information processing: 50 years (computers, linear processors, human-programmed)
Second age of information processing: just starting (mass parallel processing, self-programming computers)

The pattern is obvious. In general, succeeding ages tend to be progressively shorter than earlier ones, and the shortening is dramatic. The Age of single-cell life lasted for billions of years while the Age of Agriculture has been going on for a few thousand years. Why is this so?

Let's go back to information and the processing of information. When things start out simple, as they did in our universe and here on our planet, there is not much information to process. Not much can be done, so things change slowly. But as time goes by, things add up and systems become more complex. This means there is more information around to use, and because the information-processing systems are more complex, they can process the information more quickly. This produces more information, which produces more complexity, which can process more information more quickly in a classic information-processing feedback loop in which the works work ever faster, until and unless something brings things to a screeching halt.

Whew!

What has just been described is exponential growth. It is hard for humans, tuned as we are to persistently slow, arithmetical growth, to comprehend exponential expansion. Say a person offers you a job. The ironclad contract guarantees the job for 28 days, with an option to continue that your prospective boss must fulfill to the best of her ability. She says she'll pay you at the end of each day, starting with a penny on the first day, and then double the previous day's salary every day. Should you take the job? First, check if your potential employer is rich, and then just how rich. If she's loaded, take the job, and stick to it no matter how bad your new boss may be. Here's why.

Day	Your Salary That Day	Your Reaction	Day	Your Salary That Day	Your Reaction
1	$ .01		16	327.68	Nice.
2	.02		17	655.36	Wait a second here.
3	.04		18	1,310.72	Whoa!
4	.08	Peanuts would be an improvement.	19	2,621.44	Happy days are here again.
5	.16		20	5,242.88	Too shocked to offer any comments.
6	.32				
7	.64		21	10,485.76	Going to Disney World!
8	1.28		22	20,971.52	Viva Las Vegas!
9	2.56		23	41,943.04	Did exponential calculations over lunch and contacted broker.
10	5.12	Pathetic.			
11	10.24		24	83,886.08	Whistling while you work.
12	20.48		25	167,772.16	Very, very nice to boss.
13	40.96		26	335,544.32	Cars, cars, cars.
14	81.92		27	671,088.64	Purchased penthouse cash down.
15	163.84	Hey, not too bad.	28	$1,342,177.28	Got it made in the shade.

Of course, being bright (after all, you're reading this book), you take the option to continue this "crummy" job.

EMPLOYER'S REACTION

DAY	YOUR SALARY THAT DAY	YOUR REACTION
29	2,684,354	Fires accountant who came up with this crummy idea.
30	5,368,708	
31	10,737,416	
32	21,474,832	
33	42,949,664	Suggests you take a much-needed vacation.
34	85,899,328	Seems disappointed that you didn't.
35	171,798,656	
36	343,597,312	
37	687,194,624	
38	$1,374,389,248	At day's end, declares her global conglomerate bankrupt.
39	.00	Lets you go with regrets.

Well, you get the idea. Many things do not grow exponentially. Kids don't, despite the teenage growth spurt. But a lot of things do. Like germs. Because bacteria reproduce by doubling as often as a few times an hour, put a few of them in a nutrient-rich place where they thrive, and by the time you come home from work, you will have billions of them. Likewise, if your immune system or antibiotics do not kill pathogens as fast as possible, those invaders will overwhelm you and you will die in hours or, if you are unlucky, a few weeks or so.

Human population growth has been essentially exponential for the last one hundred years. Of course, as we saw with our cushy little high-paying job, exponential growth usually cannot go on forever. Eventually, the bacteria in a culture use up the nutrients and pollute their happy home, so they crash. Or, if their keepers replenish the culture and keep it clean, the population plateaus out at a fairly constant, maximum sustainable level. If the global cockroach population expanded without restraint, we would be knee-deep in them in a few years; predators, disease, lack of food, and so on keep them from getting that out of hand. If you plot exponential growth followed by a plateau in growth on a graph, the resulting line follows an S-curve, the classic curve scientists most love.

People are focusing on saving the species, rather than on saving minds; saving human form, rather than a new form of minds.

The Exponential Past

Okay, all this talk about exponential growth and cockroaches is fascinating and all that, but what does it mean for our futures? As hard as it is to understand the real consequences of exponential growth, it is even harder to comprehend the implications for us. The great futurist Arthur C. Clarke pointed out that because technological progress often follows an exponential curve, there is a tendency to over-predict the relative near-term and under-predict the longer-term. Humans are so prone to do this that even Clarke himself — as well as every other science fiction writer — has consistently underestimated the pace and potential of computer science in the next centuries. Yet, in *Songs of Distant Earth*, Clarke projects that in a few centuries computers will have speed and memory capacities that current trends suggest. The failure to comprehend features of exponential growth causes naysayers to scoff at the seemingly over-optimistic futurists until events suddenly and dramatically catch up.

When people envision how much things will advance in the next 10, 50 or 100 years, they tend to take what we learned in the last 10, 30, and 100 years and extrapolate that forward. It's the arithmetical, intuitive, common-sense assumption we humans are so fond of. It is hard to imagine things changing faster than they do now. We believe we can hardly keep up as it is. Certainly, progress cannot progress any faster than it does now.

But it very probably will. Just as science and technology advanced more rapidly in the 20th century than in the 19th, just as leading-edge technologies are evolving faster at the end of this century than at the beginning, elements of progress will progress still faster in the coming hundred years. How fast? A few years ago, we decided to take a stab at making a rough estimate that, to the best of our knowledge, no one else has made. The procedure was simple in principle: Estimate the gross rate of change at any given period in Earth's history and plot it graphically. We are the first to admit that this plot is only the grossest measure of the rate of change. Change is a nebulous affair when looked at in such a broad manner.

It is difficult to compare change among different things. How can the rate of change in DNA be compared to the rate of change in brain power? How can either of these be compared to the rate of change in aircraft, or in music? We cannot just pick one factor and see how its rate of change alters over time because no one factor has been important for the full span of time. Also, while some things speed up, others do not. Algae are not evolving a whole lot faster than they did one or two billion years ago, and most automobiles are still pretty much the same internal-combustion-powered, four-wheeled, steel-framed contraptions Henry Ford made. This is not, however, a critical problem. What we are interested in is how fast the most complex, leading-edge life-forms and technologies are progressing at any given time.

EVENT	TIME REQUIRED	WHEN
Formation of the solar system	100,000 to 1,000,000 years	4.5 billion years ago
Origin of life	100,000 to 1,000,000 years	About 3.8 billion years ago
Age of single-cell life	Hundreds of millions of years	3.8 billion to 800 million years ago
Cambrian revolution	100,000 to 1,000,000 years	550–540-million years ago
Paleozoic Era	Tens of millions of years	540–250 million years ago
Permo-Triassic Extinction	100,000 years to perhaps one day?	250 million years ago
Mesozoic Era to Paleocene Era	A few days to ten million years	250–65 million years ago
K/T (dinosaurs, etc.) to Extinction	Perhaps a few days to thousands of years	65 million years ago
Primate brain expansion	A few million years	40–4 million years ago
Early hominid brain expansion and stone tools	100,000 years	4 million to a few thousand years ago
Stone Age Homo sapiens	10,000 years	Few hundred thousand to 10,000 years ago
Age of Agriculture	1,000 years	10,000 to 5,000 years ago
Age of early civilizations	1,000 to 100 years	5,000 to 700 years ago
Renaissance, Age of Enlightenment, and Colonization	100 years or less	1300 to 1800 A.D.
Age of Industry	10 years or less	1800 to 1945
Computer Era	A few years to a year	1945 to today

Another complication is the result of punctuated evolution. There is a difference between those special times when the rate of progress suddenly speeds up and then falls back to more normal levels on the one hand, and the general long-term rate of progress on the other. In any case, the problem is so complex that there is no way of precisely quantifying the rate of change at this time. Even so, the phenomenon is real. Things really did move a lot more slowly a billion years ago, and progress zips along a lot faster now than it did 1,000 years ago. At best, we can make what we scientists call an "eyeballed line" approximating how fast leading-edge life-forms and technologies have appeared since the appearance of life more than 3.5 billion years ago.

To make such a plot, you first need to figure out what your data is and then tabulate it. On the table on the previous page, we've taken a stab at the gross rate of progress at different periods in Earth's history. Please understand that all estimated times required for major changes are very approximate.

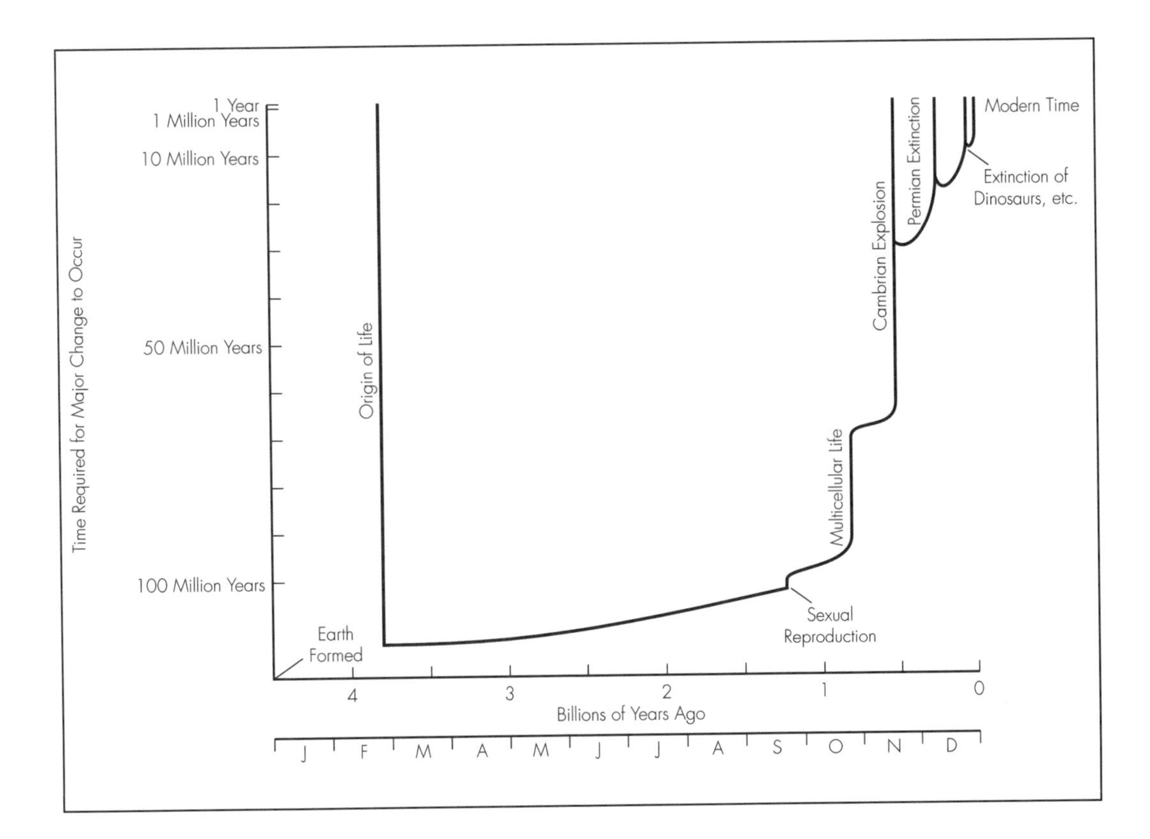

Okay, let's plot this. We will first do it on an arithmetic chart, which is the simplest and easiest to understand. Also, an arithmetic plot will best show any exponential curve, which will start out shallow and nearly horizontal on the left side of the graph, and rise more steeply on the right side. Remembering that the resulting line is a rough approximation of reality, see how the overall curve is semi-exponential. Early on, there are some "spikes" and sudden vertical jumps, which indicate brief periods of rapid change just discussed (formation of the solar system, origin of life, advent of sexual reproduction, evolution of first multicellular organisms), and some big extinctions (at least one of which may have been due to the impact of a huge meteorite) make for more "punctuated" spikes. However, the normal background rate of change rises slowly with time, and high rates of change (around ten million years or less) did not become the norm until late in the history of Earth. It has been only within the last few million years that really high rates of progress, thousands of years or less, have become repetitive.

A difficulty can be encountered with exponential curves plotted on arithmetic graphs. Events at the fast-moving ends of the graph get squeezed into the far corners of the graph, which makes it impossible to see details. This becomes a problem for us because the upper right corner of our graph squeezes in the last few million years of time, including all of human history. A solution is to plot the same data on a logarithmic chart. We will not bore you with the details of logarithmic scales, but put simply, they are exponential scales in which each step is larger than the previous one by a factor of ten. The seismic scales used to rank earthquakes are logarithmic, so a quake ranked at 8.4 is ten times more powerful than a 7.4 quake, and the 8.4 quake is 100 times more powerful than a 6.4 quake. This may be confusing, but it is worth knowing because a second advantage of a logarithmic plot is that exponential curves become straight, making them easier to analyze visually.

The rate of change over time on a logarithmic plot was a straight line moving from billions of years ago to recent times as a good exponential curve should. The line is idealized because it is not possible to be more precise, and the spikes show up again, but the trend is clear.

With events at the high ends of things expanded for better perspective, the logarithmic chart is chock-full of information. It is now apparent that even during the early spikes of rapid change, progress was not at the hyper-rapid pace that it is in the modern world. And aside from the spikes, it is remarkable how persistent the progressive increase in the rate of change has been. As for the increase in the speed of change, it can only be called extraordinary. On the average, every tenfold step

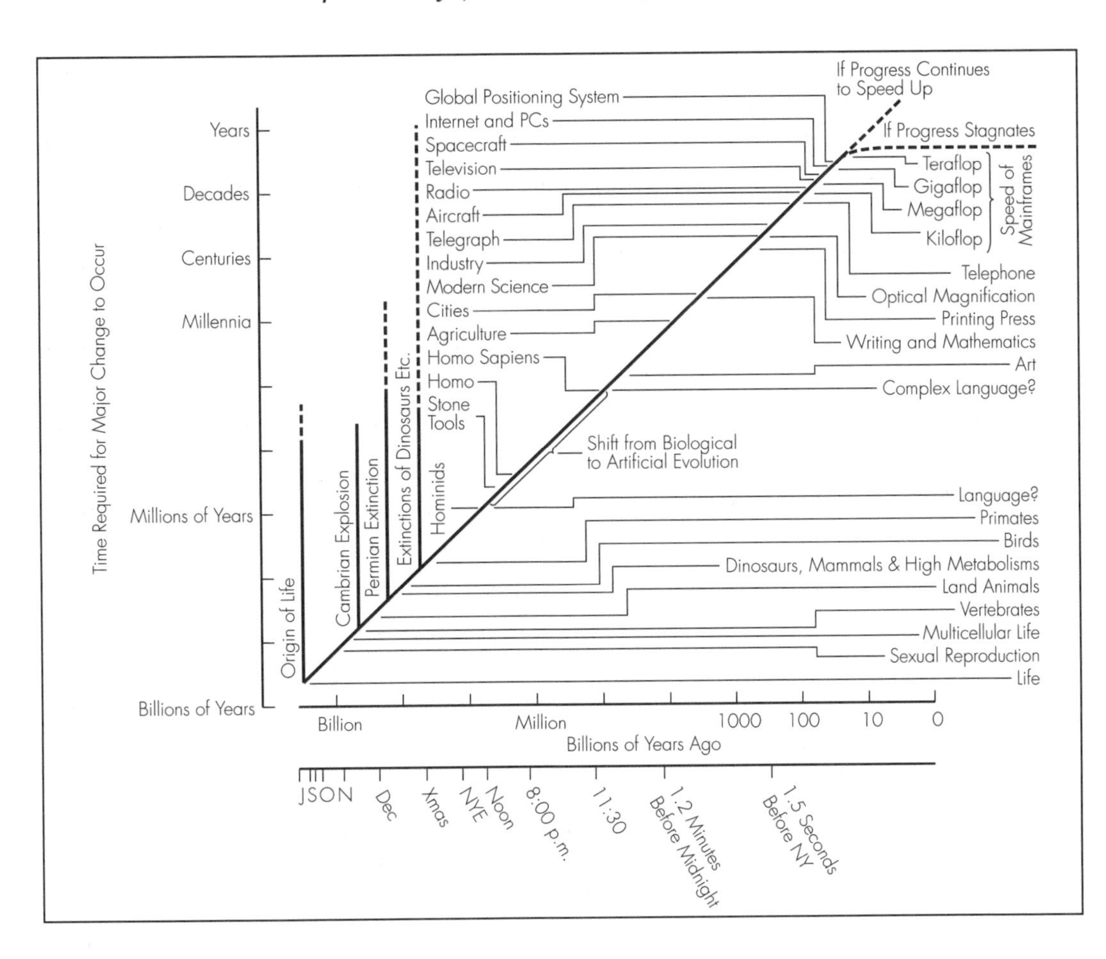

toward modern times (e.g., from 1 billion years ago to 100 million years ago, or from 1,000 years ago to 100 years ago, or from 100 years ago until now) has seen about a tenfold increase in the rate of change! The results of this exponential increase in the rate of change could hardly be more dramatic.

Currently, things are moving along nearly 100 times faster than during the last millennium, tens of thousands of times faster than when our ancestors learned to use fire, a few million times more rapidly than when dinosaurs were stomping about, and a hundred million times more swiftly than when microbes ruled the world. From crude wood and wind-powered gliders to Spitfires in 30 years. From crude mechanical calculators to Crays and connection machines a trillion times faster in 90 years. It took animals millions of years to learn to fly. It took vertebrates half a billion years to scale up their computational power billions of times. Scientific and technological evolution, guided by intelligent minds, works millions of times faster than biological evolution.

AND AN EXPONENTIAL FUTURE?

So how fast will things change in the next century? If we continue to extend the straight line into future years, the results are shocking. The period during which a tenfold increase in the rate of progress occurs is shrinking to a decade or less. Major progress is yearly now, so early in the next century, leading-edge progress should be ten times faster than today, with significant change at a monthly(!) rate, and then even faster than that! What we learn in the next ten years may be many times greater than what was learned in the last ten. What will be accomplished during the next 50 years will be tens of times greater than human endeavor since World War II. What we can do a century from now will advance by hundreds the advances made in the 20th century. Sometime in the next century, major changes (and we mean big changes that affect society as a whole) should be occurring on a weekly or daily basis!

This does not seem possible. Humans just cannot do things that fast. Have we, then, finally reached the end of exponential growth and started onto the growth plateau at the end of the S-curve? It is tempting to think so, so tempting that people have often made the mistake of thinking progress was about to undergo a dramatic downshift. In the 1920s, Herbert Hoover — the soon-to-be not-so-successful President — allegedly suggested that the patent office under his charge might soon be shut down because it was obvious that most inventions had been invented! Yet, this example does not stop some who, today, dismiss the recent decline in basic scientific research as a matter of no great significance because all the big scientific/technological advances have already been made!

A close look at the logarithmic plot reveals how big an aberration it would be for the accelerating rate of progress to brake in the coming decades. If we assume that the rate of progress levels off and remains the same as it is now into the next century, the line that has been rising so persistently for billions of years suddenly veers off to the horizontal for the first time. In other words, after ages of an increasing rate of progress, the rate not only stabilizes, but does so just when the most powerful and fastest information-processing machines ever developed are themselves growing exponentially in speed and power.

We seem to have a paradox here. On the one hand, the rate of progress looks like it is about to soar far above the levels that humans can produce or withstand. At the same time, the machinery that promises to produce a hyper-fast world is coming online. That fact alone, as you may be starting to realize, is the resolution to the paradox. The machines are starting to do the thinking. They will be

Is Progess Slowing Down?

There is an argument that science and technology are plateauing out rather than continuing to speed up. Modern houses, for example, are not all that different from those built 100 years ago. Automobiles have not changed that much in basic form for decades. Aircraft carriers built in the 1950s are expected to serve well into the next century. A few hundred DC-3s still carry people and freight.

The "slow-down" is an illusion created by a number of factors, one of which is that the old often survives alongside the new. Algae, amoebas, and bacteria, for example, are very old, yet remain abundant. Horseshoe crabs are hardly different from fossils hundreds of millions of years old. Redwoods towered over the dinosaurs. The opossum is very little different from marsupials that evolved when dinosaurs were still around. If a system works reasonably well and finds a niche it can survive in, it may continue indefinitely.

able to cope with hyper-rapid change, and the smarter they become, the more rapidly they will change things. They will be able to work on science and technology so fast that they will run the world at speeds humans have no hope of keeping up with. Not hundreds of years from now, but in the next century.

The Lesson: Probably only something drastic, such as a hard technical barrier to intelligent machines, a radical change in human socio-politics, an enormous disaster (natural or otherwise), or the intervention of a god, can put a stop to the great evolutionary increases. As evolution's history teaches us, one or another of these showstoppers could happen. But evolution also teaches us not to count on it. If the increase plateaus in the 21st century, it will be among the oddest events in Earth's history. If the increase continues, the pace of progress in the next century will make this old century look like the slow show.

Are We Heading Toward a Singularity in Progress and Time?

We appear to be in a situation in which the rate of progress has been speeding up on a nearly exponential curve for nearly four billion years, to the point that what used to take millions of years to achieve is now compressed into a year. If the speed increase continues, things may move so fast that centuries, millennia, millions, or even billions of years of progress may be compressed into decades; a temporal neutron star. The implications are obvious and incredible. Things we assumed would take ages to do may be *faits accomplis* in the near future. Farther along in this book, we will see how, if the structure of the universe allows, everything that exists may come under the control of high-level intelligence in surprisingly little time.

The Lesson: It looks like we humans are playing an end game.

If Biological Evolution Can Do It, Technological Evolution Can Do It, Too, Just Better, and Much Faster

Natural evolution is such a limited, inefficient system that it takes vast stretches of time to do just about anything, and when it does do something the systems produced are flawed and inherently weak. The belief, common among scientists, that it is hard to match nature is absurd. Swift-working intelligence will run ever-faster rings around the natural slow poke.

The Lesson: At this time, there is nothing DNA has produced that cannot be equaled or surpassed by technology.

The Lesson: If history teaches us anything, it is that growing a personal cyberbrain is probably going to become easier and more "natural" than running a laptop, or growing a biobrain.

Someday, manipulating the universe itself may become the norm.

What Was Rare and Expensive Quickly Becomes Cheap and Abundant

Once, no one owned a horseless carriage; soon, it became only the rich who could afford a car. Now, hundreds of millions of people, rich and poor, own cars. When Xerox launched the first commercial copier, they were hoping to satisfy an estimated total market for 5,000 machines, tops! The first few digital computers cost millions of dollars and were prized possessions of but a few institutions. Now, pocket calculators containing a single etched chip are more powerful than dozens of those early machines and cost only a few dollars each. Personal and notebook computers can out-perform mainframes of the '60s and will soon be little supercomputers.

The Lesson: The first few machines that replicate the performance of the human brain, although expensive, will quickly be followed by a series of increasingly cheaper and more affordable mass-produced models in a matter of months. More on that later.

In This Weird Universe, the Future Will Be Weird

Because the universe is a strange quantum machine, it cannot help but do weird things. In fact, the more complex it becomes, the weirder it should become. Even now, physicists are seriously speculating that it might be possible to travel through time, cross the universe in an instant, or move between universes. In this universe, only that which is explicitly barred cannot be done; eventually, everything else will be done. Our problem: We do not yet know where the weirdness ends.

The Lesson: The more conventional the view of where SciTech will take us, the less likely it is to come to pass.

Technology Is How We Got Where We Are, and Technology Is How We Will Get Where We Are Going

Biology has taken us far, but science-based technology is responsible for the artifact of complexity we inhabit late in the 20th century. Religion, politics, and culture have played their parts in relatively minor roles downstage from the real action. By this we mean that if it were not for that which scientists and engineers have wrought, we would still be living short, hard, pre-industrial, Jeffersonian lives of tight-knit families on horse-plowed farms. It was industrial technology that made modern capitalism possible. Poverty and African-American culture inspired rock and roll, but the electric guitar and amplification made it possible. As we will show later, economics, culture, and faith are probably not going to save humanity or human souls, but technology may make it possible to save human minds.

The Lesson: Don't look to religion, mysticism, and politics to see what lies in the future. Those practitioners have been spinning their wheels for millennia without doing, or successfully predicting, that much. It is patently obvious that SciTech is in the engineer's seat.

Sometimes Things Get Worse, More Often They Get Better

We are faced with one of those bad news-good news situations. On one hand, we are caught up in an evolutionary system that, being blissfully unaware of the concept of morality, cannot consistently program for the well-being of its creations. Bad events, therefore, can and do happen: slavery, murder, war, torture, abuse, disease, and trauma. On the other hand, the nonthinking system is not deliberately evil, and there is nothing stopping good from happening, as well. In fact, over the long-term, evolution may favor the good over the bad!

It may not seem like it, but things are getting better. You do not think so? Okay, go back a billion years. Just a lot of single cells. No joy, no love, no music, no watching pretty sunsets. When fish came along, it was an improvement because they probably had a dim state of self-awareness and could experience the little pleasure that comes with a good meal. By 20 million years ago, things were better still, as animals with brains as big as those of cats and dogs had evolved. As those who own pets know, large-brained furry beasts can experience great happiness and have a lot of fun. Those who study chimps suggest they experience life with the emotions of a human three-year-old. Being human is even better.

We have all heard it. Your friend shakes a weary head and says, "This is a terrible time to have children." Maybe so, but ask them a trick question: "Just when was it a good time? In the '50s when the kids avoided swimming pools for fear of polio before they were learning duck-and-cover? In the '20s and '30s when the Great Depression was coming or under way? Perhaps not? Well then, how about

How Silicon Has Made Human Lives Freer

When expensive mainframes were all the rage, people feared that governments would use them to better control their citizenry. This belief was common not only in the west, but even more so to the east. In the USSR, ponderous treatises explained how the computer revolution and centralized socialism were ideal for each other, while the chaos of capitalism would never be reconciled with the digital logic of cybermachines.

Of course, quite the opposite happened. The inability to compete digitally was one of the main causes of the fall of communism. It just became so embarrassing that the "scientifically" run socialist states (actually, they were more along the lines of pseudoscientific cults) could not match western computer technologies and had to buy or steal them from the capitalists. With notable exceptions, they were chronically behind.

The fall of communism was a classic evolutionary event in which increased information processing at the individual level increased the knowledge and power of individuals to the point where they could no longer be controlled. It is all the more remarkable when you consider that the chips are nothing more than melted sand etched to form.

the 'gay' 1890s of cholera and typhoid epidemics, shockingly high infant mortality rates, and vicious segregation? If that is not suitable, there are always the days of the Black Death. If that won't do, check out Rome and its death sports, food orgies, lead plumbing, and complete absence of civil rights."

It's All a Game

If a tendency toward good is built into evolution, how does it work? As information-processing systems become more complex, intelligent, and powerful, they can appreciate beauty and good more and more, and they can do more to make beauty and good come to pass.

A comprehensive hypothesis has been developed that explains the trend toward good. It is called the Game Theory, and was formally developed by John (evolving automata) von Neumann and Oskar Morgenstern in the early 1950s. It soon became all the rage at think tanks such as RAND, where whiz kids were eager to figure out if the USSR and USA really were likely to blow each other up.

Take the classic prisoner's dilemma: Two people are accused of a crime. Each person is taken to a separate room for interrogation. There's little evidence against the accused, which means the police will have to get a confession or incriminating testimony from the suspects. The suspects can cooperate by both denying involvement in the crime. Alternatively, either or both suspects can confess to the crime. Or, either or both suspects can accuse the other of doing the deed. If they cooperate, both win by going free, and the shared score is 8. If one accuses and the other conceals, the accusor wins by going free and scores 4, while the other loses big time and is sent to jail for a long time for a score of 1; their combined score is 5. If they both confess, both lose, but their sentences are reduced because they owned up, and the combined score is 4. Likewise, if each accuses the other, both losers end up in the clinker, but with reduced sentences because of possible biases in the testimony, and the combined score is 4. The best overall outcome requires both players to keep their mouths shut. This is very risky, however, because if you are the suspect who keeps quiet and your partner squeals, you are cooked. Confessing is less risky because, although you go to jail, at least you are not risking doing a long stretch, and maybe your chum will keep quiet, go free, and mail you a special cake. Sad to say, the least risky strategy is to say the other fellow did it. Again, you get minimum jail time, and maybe you will get lucky and your not-so-bright accomplice will keep quiet or even confess, the fool.

Repeated trials show that when randomly selected people play prisoner's dilemma and are carefully denied knowledge of their partner's actions, both "prisoners" almost always accuse. Cops exploit this pattern by quickly separating members of criminal teams, hinting to each that the other is about to squeal like a pig. This game suggests that cooperation is only for saps, and the smart should pin the blame on the other and run. However, in the real world, two things usually occur: First, we are often not completely unaware of the other's actions; indeed, we often know a lot about what others are up to. In a variation of prisoner's dilemma in which the participants can actively cooperate, they almost always will cooperate. In a climate of knowledge, therefore, cooperation becomes possible and often beneficial. Second, people often have long-term relationships. Sophisticated analysis by the likes of Robert Axelrod shows that during long-term exchanges, strategies that combine competition with cooperation become more profitable.

Among winning strategies for any given player is tit-for-tat, in which past cooperation by the partner is rewarded, and accusation is punished. As you probably recognize, all our lives consist of a series of prisoner's dilemmas and variations of tit-for-tat. That's the way we deal with challenges we run into on a daily basis. T-F-T is also a modification of the Golden Rule, in which one should do unto others as one wishes them to do in return, just don't be a chump about it. What all this means is that statistically, over time, those who are reasonably altruistic, but stick up for their own rights, tend to benefit the most. This is as true for microbes and cats as it is for people. Life in all its forms is, therefore, a combination of competition and cooperation. So is evolution.

As a result, although the selective forces of evolution are morally neutral, in practice they tend to work against the forces of evil. True, there are and have been tyrants. But look what happened to Caesar, Napoleon, Mussolini, and Hitler. Stalin died unattended by doctors because they were too afraid to go near him — self-fulfilling meanness. Once, there were no democracies. Today, democracy continues to expand in fits and starts because improved communications and information processing are increasing the power of individuals. Systems of tyranny promoted by Khan, Hitler, Stalin, and Mao are gone and discredited. Mass slavery has pretty much disappeared because it has become inefficient. "Civilized" nations that just a few decades ago were willing to suffer and inflict massive civilian casualties on each other now balk at losing a few soldiers. Mass nuclear war has become unlikely because the risk/benefit ratio is very bad. Diseases do not wipe out all of humanity because they, too, would lose out. So, most people are reasonably healthy most of the time. Because we now know so much, people are living longer and healthier lives than ever before — and using the extra time to complain about everything more than ever.

And In This Game, Reproductive Success Is the Most Important Form of Information-Processing

Reproducing information, whether via genetics, distribution of new information, replications of minds, or multiplication of universes, ensures the spread of the system. If information is stagnant or lost, it accomplishes nothing.

THE EXCLUSIONARY PRINCIPLE: ITS IMPLICATIONS FOR THE DIVERSITY OF INTELLIGENCE

The Exclusionary Principle observes that two or more creatures cannot do the same exact thing at the same time; who will win out over the other. So, while there are a lot of hooved mammals eating grass in East Africa, each species specializes in eating particular kinds of grasses. Likewise, there is only one intelligent species around these days.

Thus, *Homo sapiens* probably drove *Australopithecus* to extinction, *Homo sapiens* did the same to other *Homo* species, and the primates left are just a few of the species verging on extinction. However, diversity remain high; it has just shifted to the new venues of culture and technology. Diversity is not only true between cultures, but also within each civilized technology in which people fill a wide variety of roles. It is another example of how evolution has switched from the biological to the technological.

The Lesson: Long-range implications of the Exclusionary Principle for intelligence are twofold. On one hand, the inability of inferior intelligence to compete with the superior has grave implications for the viability of humanity in the 21st century and beyond. This is true even if superior intelligence inflicts benign neglect on the inferior. On the other hand, the tendency for systems to evolve diversity has interesting implications for the path cyberminds will follow after they have supplanted humans as the leading-edge intelligence.

CHANGE IS NORMAL. SUDDEN, SURPRISING, MIND-BOGGLING CHANGE IS VERY NORMAL

There is one thing that will happen, and there is no two ways about it: There will be change. What with chaos and the exponential rate of growth, the one thing we can count on is universe-bending change that boggles the minds of simple mortals. Most big changes happen suddenly, just about the time when conventional wisdom denies that big change is on the way. The latter occurs because the factors that make the change possible do not become clear until the change occurs. So big changes are, therefore, almost always surprising. None of this is unusual; it is normal. If things stayed the same, that would be bizarre.

The Lesson: If it comes to pass, the sudden and surprising replacement of humanity by more advanced minds will be another example of punctuated evolutionary change among many. It will be less remarkable than the extinction of dinosaurs (which remains one of the most extraordinary mysteries of all time), and no more remarkable than the evolution of Homo sapiens, or the origin of life itself. It will be normal. What would be peculiar is if humans were to remain dominant well into the future.

There Is Nothing Wrong with Extreme Futurism

Although it is obvious that extreme change is normal, it is a favorite tactic of naysayers to say that a futurist's view is "extreme." This is a rhetorical misdirection that appeals to our sense of moderation, but whose meaning is as empty as it is deceptive. After all, da Vinci, Verne, Wells, Ford, Edison, the Wrights, Arthur C. Clarke, and Steve Jobs all held extreme notions about the future that turned out to be correct! We already live in an extreme world, and it is hardly extreme to suggest that the future will be even more extreme.

Other words with connotations similar to "extreme" are often used, such as radical, over-enthusiastic, fringe, borderline, and the ultimate dig, fanatic.

The Lesson: The future is probably going to be weird, so forecasts should be extreme.

Any Scenario of the Future, Mild or Extreme, Must Make Sense, Sort Of

Exponential trends can snag you if you don't watch out. The way to avoid being caught is to use some common sense. This is kind of hard to do because the future does not always make a whole lot of sense. The extreme world we live in would not make sense to Teddy Roosevelt, so much so that he would never have predicted it. The Extraordinary Future certainly seems to violate common sense to many. Even so, there are some basic rules that should be followed.

One good example: helicopters. First built in World War II and used extensively in Korea, there were serious predictions by the '50s that helicopters would replace the automobile as the chief mode of private transport. Promotional films showed Dad getting into his 'copter and taking off, while the wife and brood, with hair blowing in the downwash, waved bye-bye. Sort of seemed to make sense — horses replaced walking, cars replaced horses, and it was about time something leapt over the traffic jams.

It didn't happen, though. Helicopters are so fuel-inefficient and expensive to build and maintain that few can afford them. They are also far harder to fly than a car is to drive, so few are skilled enough to fly them. We like this because helicopters are really, obnoxiously noisy. Imagine thousands of them, like flying lawn mowers, revving up and clattering over the neighborhood during the morning commute. Imagine tens of thousands of choppers over the urban landscape. Collision city. They'd be dropping like flies through roofs and on people's heads!

An even better example is supersonic transports. Starting around 1920, the slope of the increasing speed of airliners climbed a near-straight line as aircraft went from wood and fabric biplanes to metal monoplanes, and then to subsonic jets. Follow the speed curve and SSTs were a natural, to be followed by hypersonic transports covering the New York-to-Tokyo run in an hour. The literature of the time addressed the question of "Why the SST?" with the response that "ever since the advent of air transport, the driving force behind innovation has always been speed." It was logical and seemed to make sense. Americans, the Anglo-French, the Russians all had SST plans in the works. Then the Americans dropped out, and later the Soviets. A handful of Concordes plying the New York-London route, kept in service via high fares and subsidies, are all that's left of the dream. What happened?

People realized that being awakened by window-rattling sonic booms every 15 minutes or so would be intolerable. Environmentalists warned about intolerable ultra-high-altitude jet exhaust depleting the ozone layer. The biggest problem is that pushing a jet along at Mach 2 requires lots and lots of fuel; slender-bodied SSTs cannot carry many people, so you have a real money-loser on your hands. There are new SSTs on the drawing boards — bigger, supposedly more efficient, and environmentally friendly. There is even talk of high flying hypersonic beasts like President Reagan's "Yankee Clipper." What will come of this we shall see, but these are not sure investment bets.

The Lesson: When extending exponential growth into the future, you not only need to show why the technological progress may be possible in principle, but that it is also feasible and practical in the real world of costs, convenience, and safety.

The Naysayer's Song

Hard-core naysayers of future technologies often play the same losing cards that the flight naysayers did when they chortled over Langley's Aerodrome lying at the bottom of the Potomac. It works like this:

First, the naysayer says that the enthusiasts' technological dream is just that, a dream. The naysayer then assays commentary depicting the sad current state of the enthusiasts' endeavors with additional reasons it will never work. Included may be a hint at the futility of it all, another stratagem asking why do it in the first place, and perhaps a warning why it should not be done at all. In other words, much naysaying is actually an attempt to put a stop to something the naysayer fears or to reassure the naysayer that what he or she fears will not come to pass. The naysayer emphasizes that current trends cannot be relied on to predict the future, quite forgetting that just because a trend can and might change does not mean it must and will change.

A Horse is a Horse of Course: A Lesson on the Dangers of Conservative Futurology

It is not well known that at first, the Age of Industry greatly increased the horse populations all over the world. From the early 1800s to the early 1900s, mass-produced products and materials could be shipped between cities by water and rail. Once in town the only way to move goods was by horse-drawn transport. In 1900, the United States' horse population was in the tens of millions, an all-time high. People complained about the dangers of being run over by out-of-control rigs, the awful clatter of countless hooves, and street pollution in the form of vast quantities of manure. Animal rights supporters were not happy with the treatment of horses forced to pull heavy loads on days cold and hot. Then along came the industrial solution, and the question of the day: Who would win, equine machine or piston machine?

These days, folks are saying robots will never replace people. Of course, there were tracts written on why helicopters would replace the family car, and in the early '60s NASA was figuring they would have moon bases and men on Mars by the mid-70s. Still, why do we feel like a horse in a harness?

After a while, the enthusiasts manage some *fait accompli* bringing them closer to their goal, actual accomplishments the naysayer cited as key failed components of the dream. The enthusiasts hustle to the naysayer with the news, "We did what you said couldn't be done! What do you think of our dream now?" A smiling naysayer shakes his or her all-too-knowing head and says, "That stuff isn't so important after all, and your minor successes are irrelevant to the ultimate goal, which is forever beyond your reach." The naysayer then writes a new treatise explaining why he or she is still right and listing what the enthusiasts still have not done with a few new failures added for good measure.

More time passes and the enthusiasts do more things that bring them yet closer to the goal that the naysayer said could not be done. Again, the enthusiasts with new-found hopes approach the naysayer and ask what he or she thinks now? Again, the naysayer waves aside the new achievements and publishes more explanations detailing why he or she is still right; although the book may be a little shorter at this stage. What is interesting is how some naysayers just do not learn. Even as the enthusiasts fast approach their ultimate goal, the naysayer may ardently defend his or her position until the ultimate goal is placed fully accomplished and complete right in his or her lap.

There is one great hope for the naysayer who wishes to avoid losing the game: Die before the goal is reached. Another is for the goal to prove unworthy of completion, perhaps being superseded by other events.

Oh yes, watch out for the person who says, "You cannot predict the future," and in the next breath says, "And that stuff about robots replacing people in 50 years is nonsense!"

The Case of the Atomic-Powered Vacuum Cleaner

Now, the funny thing about naysayers is that they are not always wrong. Futurist skeptics enjoy citing every excruciating, overly optimistic, missed prediction of the sci-fi writers and futurists. The atomic-powered vacuum cleaner that 1950s scientists promised would become a standard household appliance (we're not making this up) comes to mind. This is all well and good, and indeed important, because skepticism is useful and necessary. But if these same people had been born 100 years ago, would have been panning mass-production autos and radiation bombs.

The problem with naysayers is that they often become over-enthusiastic and strident in their hard-core skepticism. Worst of all are naysayers who lose the ability to discriminate between the outrageous but plausible and the outrageously improbable. Even asking an expert for an opinion of the future may not be useful. You can always find one who will say the idea is preposterous (for decades after saying that $E=mc^2$, Einstein said it would not be practical to extract large amounts of energy from uranium), and another who will accept it as a matter of course. Even surveying a large number of experts and tallying up the results may not tell you much. It is probable that most Victorian aeronautic experts would have dismissed as nil the Wright brother's chances of success.

Extreme futurist skepticism is ultimately sterile. Remember, all scenarios are speculation. When someone scoffs at the notion that anthrocivilization will have been displaced by cybercivilization in 100 years, they too are speculating. They are speculating that humans with brains that have not improved in 100,000 years will be able to stay ahead of technologies that are doubling in power every year or two. When looked at that way, which bet is outlandish?

We have not explained that species often become extinct with surprising suddenness because we know that humans will become extinct with surprising suddenness in the near future. We did so to show you that it could happen because this sort of event is normal in evolution. Nor did we outline how the game theory explains why evolution tends to make improved situations over time because the future, no two ways about it, will absolutely be better. We want to show you that there is a good chance things will work out for the better in the long run. What we have tried to do is give you some concept of the way change works and that it appears to be itself changing with ever more speed, suddenness, and surprise. Why? So you won't be surprised.

It is true that we cannot reliably predict details of the future or predict exactly when things will happen, and we are especially inept in discerning the consequences and unintended side effects of technological progress. What we can do reasonably well is outline broad patterns of possible events, weigh the probability of outcomes, take a stab at estimating the speed of onset, and suggest some obvious consequences of upcoming achievements and events. Although we cannot count on anything (except that circumstances will change), there are some wise bets. It is a safe prediction that the rate of technological progress will continue to speed up exponentially, with all that that implies. We can also count on elements of surprise; many of the big leaps will come in sudden, unexpected spurts against a background of apparently frustratingly slow, general progress.

The Lesson: As they say, nothing ventured, nothing predicted. So we prefer to follow the advice of our futurist hero, A. C. Clarke, who warns that the greatest pitfalls for those who prognosticate are "failure of imagination" and "failure of nerve." Clarke is right because....

We Ain't Seen Nothin' Yet

We tend to be impressed with our technologies: planes, trains, and automobiles; supercomputers, skyscrapers, supertankers, and extended lifespans. We believe we have come a long way. Yes, but this is not saying much because we were horseback riders just a few years ago. Reality demonstrates that we remain weak and ignorant. The Age of Industry is a mere two centuries old, and we have been doing "advanced" science for only a couple of lifetimes. At the same

Changing Lives: From the 1800s to the Future

Future shock, the radical transforming of human lives within one lifespan, has been under way for about three lifespans. Just which of these lifespans have undergone the greatest changes, and why? Each generation of about 70 years has experienced a unique future shock wave.

The ages of grand industrial change are behind us. Most of the big changes in human lives that can be made have been made. The few big changes yet to be made will depend on youthful but sophisticated digital and cybertechnologies. Making flying machines just after Queen Victoria died was possible because low-tech macro-technologies were up to the task; making synthetic minds will probably require nanotechnologies not yet developed. Because so many of the major life changes have already been made, the number left to do with cybertechnology is rather limited, and some of them may not be possible. But a few of them — nanotechnology, eliminating disease, stopping aging, intelligent robots, and the merging of robot and human — have the potential to dwarf all that has gone before.

time it has become a postmodern cliché to think that SciTech is running out of "steam." Hardly. When it is pointed out that we presently have little control over our lives and don't seem to have the mind power to control our lives, there's little appreciation for how far we have come in such a little time and the magnitude of power our minds will have in the not too distant future. We are at the very beginning of what can be known and done. We have yet to see truly powerful technologies.

On the yellow brick road of cybertech, we have only just met the scarecrow wanting a brain, and the Emerald City of artificial cognitive intelligence is only adventures away. Which leads to Clarke's most important observation: "Any sufficiently advanced technology is indistinguishable from magic," a good point to remember as we enter what may seem to us to be Oz.

The Question: Where will our fast-evolving world be, and what will it be like in 100 years? Still stuck in Kansas, or in the brilliant Emerald City?

CHAPTER 4

The Wonderful Worlds of Artificial Intelligence, Artificial Life, Nanotechnology, and Ultra-Alchemy

"There's gold in them thar hills!"
— Old Prospector's Hope

To get to the Emerald City, we are going to need the tools to construct intelligent machines. They are being developed as we go about our lives.

A Brief History of Artificial Intelligence

Although the dream of creating intelligence other than human is ancient, there was not much anyone could do about it until 50 years ago when the first electro-digital computers were being plugged in. It did not take long for the brightest minds in the world to put two and two together and start figuring out how long it might take to get the new mechanical "brains" to actually do what brains do. Thus was born "artificial intelligence," which means different things to different people, but which generally refers to the "conventional" effort to get electrodigital computers to mimic human intelligence.

Delighted with their new electrodigital playthings, the early AI bunch got really going in the '50s when computers were common enough for academic and corporate professionals to work with. In 1950, Alan Turing explained why machine intelligence should be able to match the brain's natural mechanism for intelligence and offered a classic test of computer versus human cognition. He outlined two basic ways of approaching artificial intelligence: a bottom-up method paralleling biology and a top-down method in which humans would impart their wisdom to machines. Turing preferred the latter, von Neumann seemed to favor the latter. Alas, neither Turing nor von Neumann was around by the time the first AI conference was held at Dartmouth in 1956. Among the attendees was MIT's Marvin Minsky, who would become the dean of AI.

The Unhappy Fates of Turing and von Neumann

Turing was a cranky, intellectual British eccentric who never found a comfortable place in the British system. After helping break the Nazi codes that shortened the war and saved countless Allied soldiers, he should have been feted and rewarded, but secrecy stole from him the recognition he deserved. In 1953, he was arrested for homosexual activities and compelled to take severe drug "treatments" for his sexuality. The next year, he committed suicide.

Von Neumann was a social and popular Hungarian émigré who terrified friends with his wild driving habits and a string of accidents he walked away from. After helping design the implosion detonator for the first A-bomb, von Neumann became part of the American nuclear establishment. He was a cold war hysteric who overestimated the status of Soviet nuclear weaponry while pushing for bigger nuclear bombs. Von Neumann helped develop the faster computers that helped develop the bigger bombs. It was not an automobile accident that shut down his mind in 1956; it was cancer.

Why is it that these two great minds go unappreciated by the postmodern global society they were instrumental in creating? Is it because of their perceived flaws? No; Newton, Einstein, Edison, and Galileo all had notable character flaws (of the great minds of science, it was Darwin, the most controversial, who was among the nicest). To a fair extent, it's just the luck of the draw.

Strong AI

Many of those at Dartmouth thought strong AI was the way to get things done. Strong AI is a top-down system involving lists of instructions allowing computers to be smart by telling them what to do step by step: the wonderful world of programming. The theory: What better way to give a machine intelligence than by having a human give it intelligence like a god from on high? No point in wasting all the intelligence humans had accumulated waiting for machines to learn their smarts. Side advantage: You know what the machine knows, so murkiness in the system is minimized. The idea that intelligence consists of working with symbols that represent the actual world, and computers are excellent at working with symbols, meant that computers would be good at becoming intelligent. An early erroneous idea that the brain is a fully digital mechanism also tended to encourage techniques using strong AI approaches.

Early AI quickly produced programs that could reason through various problems by starting with a set of axioms and using rules of inference to prove theorems in mathematics or play assorted board games. Immobile industrial robots began to be assembled. The advent of time-sharing gave more researchers real-time access to computer power. All this encouraged investigators at MIT and Stanford in the '60s to take a stab at making mobile robots that could see, move, and pick up objects with minimum input. Some could "see" and pick up randomly placed blocks set on a table. The most famous of the mobile AI-bots, Stanford's Shakey (which was hooked up to a mainframe in another room), could use reasoning to solve basic problems involved in reaching and moving blocks initially out of its reach. Some suggested that another ten years would see highly intelligent, capable AI robots.

By the 1970s, however, strong AI started to flag seriously. The summer of '66 illustrates the situation. Minsky hired an undergrad to work during the summer to make connections between a camera and a computer with a program he would need to write so the machine could understand what it was seeing. It seemed simple enough at the time. The job wasn't finished at the end of the summer, or the next. Indeed, the computer vision problem is still being wrestled with.

Stanford's Shakey could work only in a carefully stylized environment, and even then it took days to do anything — usually only after a series of mistakes.

The military sank a fair piece of cash into articulated all-terrain robots that could chase enemy soldiers over broken ground. It was instead getting prototypes that could barely make it down a hallway. There is still no robot that can easily walk across a room, pick up a cup filled with water, carry it safely across said room, and set it down.

Another major failure of strong AI centered around translation. At first, breaking down a given language and converting it into another seemed straightforward enough. Bits and pieces of success with certain sentences encouraged the attitude that computers would soon make communication between nationalities an electronic pleasure: babble would cease. When the computers flubbed translating lines such as, "Earnest thinks he is a big wheel" into "Enthusiastic believes he is a large tire," the occupants of the current tower of babel at the United Nations were understandably disappointed.

The '70s and early '80s were bleak years for robotics. There was no grand organizing theory and little money, and the minuscule calculating power available to researchers was a chronic frustration. Strong AI shifted to making static computers smart. The 1980s saw the introduction of "expert systems," programs containing the expertise of humans in fields such as medical diagnosis and business management. Knowledge is extracted from human experts. Human obsolescence thus occurs as software programs are marketed that are as smart as the experts.

The success of strong AI expert systems has been mixed. The systems were "brittle," meaning if the questions asked of the program were not exactly within a narrow purview, the answer returned would be nonsensical. Keeping the systems up-to-date with live human experts proved difficult. Not only was it time-consuming, but altering original programs often destabilized an earlier program, causing it to crash. AI systems for the new B-1 bomber and other weapons turned into military nightmares. Many of the high-tech companies' vending expert systems that popped up like mushrooms in Silicon Valley and along Boston's Middlesex turnpike, just as promptly went out of business.

Human Thinking Is Complex, Yet So Superficial

Why has strong AI proved so disappointing? For starters, computers lack the sheer calculating power needed to do what humans do; however, it goes deeper. We big-brained apes do "high-level" thinking better than anything else, and we

The Co-Evolution of Nukes and Large Computers

Nuclear physicists' need for powerful calculating machines for the complex and sophisticated gadgets was a major force behind the development of high-speed mainframes, eventually leading to the supercomputer. IBM, Cray, and Thinking Machines made pilgrimages to the wizards of Los Alamos and Livermore. This symbiotic relationship between big bombs and big computers worked until the end of the cold war, and market forces left the mainframe manufacturers badly bruised, if not dead. Today, the fellows down at Los Alamos are networking lots of small commercial machines. However, the relationship has not ended. The Sandia TFLOP hypercomputer is being built in part to help the bomb scientists keep the nukes in working order in the post-Cold War era of test bans.

Some have suggested that had there not been a need for big military mainframes, digital computers would have evolved from small desktop calculators straight into small desktop computers. Maybe. However, the tiny chips that make small computers possible were first developed in part to make powerful mainframes for the military.

tend to assume that high-level thinking is hard and everything else is easy. It does seem this way. A sparrow and a three-year-old child can see, walk, and pick up things quite well. Few can play chess really well, understand relativistic physics, or figure out a tax form. A basic education in writing, reading, and arithmetic almost has to be beaten into people; getting a degree spans decades. Expert computer systems were doing fairly well in the early '60s, and it seemed that high-level stuff would soon be wrapped up. The "low" brain level seeing, walking, and object manipulation would be a snap.

How naive. The true situation is the reverse; the amount of information involved in high-level thinking involves embarrassingly little data flow. A two-hour lecture by Einstein on his Theory of Relativity would have contained just 20,000 words! A thick book on the same subject contains fewer bits of information than your brain processes in a second. The majority of the work done by the brain has to do with it being made up of parts that work together to manufacture a product-conscious mind capable of observing and moving in its environment. For example: The human retina contains about 100 million photoreceptors collectively doing billions of calculations per second.

This is the start of an intricate series of mental events occurring in the brain involving trillions of calculations each second; and that's just the visual sense. Connect to the visual to the auditory, olfactory, sensual, motor, and coordination tasks involved in walking across a room and picking up a cup and you have a nightmare of tangled algorithmic complexity.

Intellectual thinking is a veneer of thought layered over the enormous animal mind that forms the base of our experiential reality. Animals do the hard mental work of being an animal easily because evolution has had hundreds of millions of years to get the basic sensory and locomotor systems working reasonably well. Being intellectual is hard because it is very new and unpolished, modern brains being only a few thousand generations old.

Imagine sitting down to program into a computer all that it needs to know to guide an attached robot across a room full of obstacles without it falling over or bumping into anything. Then imagine operating a search program allowing the computer to recognize a cup, whereupon another program directs it to pick up the cup, determine its contents, and keep from spilling it while the robot maneuvers back to set the cup down on a table on the other side of the room. It is an impractical problem in both the time involved and the lack of knowledge about how we do such things. The early AI people badly underestimated the work involved. Not surprising; no one knew how big a job it was.

Software programming costs rise rapidly with the complexity of the program. As a result, IBM, which does a lot of programming, has seriously studied programming costs. The per-line cost of a program rises exponentially with the overall size of the program because the possible interactions between lines increases much more rapidly than do the number of lines. All this interaction needs to be sorted out and then debugged to correct inevitable errors. A 6000-line program may cost $100 per debugged line, for a total cost of $100,000. A million-line program would cost $1,000 per line for a total cost of one billion dollars over the lifetime of the software! If the human brain, wetware, was written down as lines of code, the total would probably amount to hundreds of millions, if not billions, of lines. Each line would probably cost tens of thousands of dollars, for a budget-busting total of ten or so trillion dollars! Newt Gingrich would not approve. All this assumes that the human mind program would actually run. That is improbable, however, because the debugging problems would be an unsolvable logistical nightmare. No matter how powerful it is, telling a computer with programming code how to do anything and everything will never work.

WHEN PROGRAMS GET OUT OF HAND

If a program stays in use for decades, its complexity grows, perhaps becoming intermeshed with other programs, thereby creating a programming brew with unknown implications. The original programmers may retire or die, or the original programming work may be lost, and details of how the program works disappear! This is becoming a serious problem with many mainframe programs and the soon-to-be year 2000. Back when computer capacity was low, programmers trying to save precious digital space set up their programs so only two digits represented each year. Seemed to be a good idea at the time, but they never imagined their primitive programs would still be in use in 2000. When the zeros come, the programs may prove unable to cope. For example, if a federal benefits system calculates a person's age by subtracting 1940 from 1996, what will it do when it tries to subtract 40 from 00? Automatically searching through and correcting the problem will not work in many of the old programs. They will have to be gone through line by line — sometimes by the original programmers brought out of retirement — and at a cost of millions.

EVOLUTIONARY AI

The increasingly obvious problems that vexed the top-down attempt to force human intelligence into computers and robots inspired others to consider a more bottom-up approach, one that mimics evolution. Hans Moravec, at Carnegie Mellon, is a leading proponent of the bottom-up method. Let's forget about directing robots how to operate in simplified worlds. Instead, try to mimic animal evolutionary experience. We'll start simple and work our mobile robots up through a real world.

First, program a robot to do fairly simple things well. After those few tasks have been mastered, put the program into a better robot and improve the program. Slowly upgrade the robot's vision (whether ordinary light vision or radar), locomotion, navigation, and the capacity to think through problems. From the progress accomplished, use the best program for vision, the best for locomotion, the best for cognition, and combine them. The rising power and sophistication of computers will work hand in hand with improving experiential programs. After mastering the animal functions, the veneer of high-level functions already built into expert programs will be attached. Someday, when computers are fast enough and programs are good enough, there will be machines with a human level of intelligence.

ROBO-INSECTS

Rodney Brooks at MIT takes a more extreme and controversial view. He observes that as animals evolved, they were never programmed from above. Therefore, programming robots with incremental levels of mental advancement is ruled out. So is total reliance on centralized control systems. In Brooks's system, how a robot reacts to a problem is more important than why it does so. Complex behavior, such as moving over an obstacle, is built from the experience of merely moving a leg, whose detailed actions are controlled by local microprocessors that worry only about that leg and how it moves, not what it moves over or around. Using such techniques, Brooks has built a series of shoe-box sized, insect-like, multi-legged robots that do what no other robots do: walk easily at normal speeds across obstacles.

The U.S. military finds terrain-walking robo-insects interesting. For tenderfoot GIs desiring to keep body parts, robo-insects would be useful in seeking out unexploded ordnance and mines. MIT has built squads of 1.5-inch "cyberants" that have powered treads, obstacle-detecting antennas, infrared transmitters/receivers for communications, and "mandibles" able to manipulate small objects. A multidisciplinary team at the University of Illinois at Urbana, is building an articulated robotic "cockroach" for the Pentagon. Less than a foot long, it will use a combination of six articulated pneumatic legs, stress sensors, and decentralized neurological controls that allow it to switch from regular to irregular terrain. As it it switches, it shouldn't slow or fall down, or it should be able to recover when it does. Already, the prototype is capable of standing back up if it is kicked over. Of course, the ultimate goal is an all-terrain robo-warrior that can go where no man or woman should.

NEURAL NETWORKS AND PARALLEL SYSTEMS

An insect robot's capacity for learning bring us to the subject of neural networks. IBM — when it ruled the computer world — realized that people were scared of "mechanical brains," whereupon the company reassured the public that computers do only what they are told. This may still be true of the simple software in your home computer. However, a number of computers are considerably brighter.

The complexity of the real world induced some theorists early in the '50s to suggest that rather than instructing computers to think, it would be better to let them learn on their own. Early successes of this approach were small-wheeled robots called "turtles" that used little "brains" made of radio tubes and capacitors that could store memories. Whole bunches of turtles learned to wander about, avoid both each other and various objects, find their homes when their batteries went low, and even exhibit complex behavior as they reacted to each other's control lights and contacts. These simple machines were not taught these tasks, they acquired them up by learning. Some little wanderers with hearing devices were presented a loud sound just before being kicked. They learned to avoid the warning sound whether a kick was delivered or not. Shades of Pavlov's drooling dog, Vanya.

As cute as the electronic turtles were, they didn't really do much and, in hindsight, it is clear that early computers were seriously mentally challenged. They couldn't learn much of anything that was practical for real people in real situations. Matters were not helped when hyper-conservative IBM dropped funding for self-teaching computers, in part because of a growing concern about negative public reaction. The real blow came in 1969 when MIT's Minsky co-authored a book on neural networks that was taken to be a definitive rebuttal of the concept. This is ironic because Minsky conducted some of the earliest experiments in neural networks and later claimed his book was misinterpreted. In any case, Minsky's book helped bring about bleak years.

Attitudes shifted in favor of robotics in general, and neural networks in particular, during the 1980s. Funding picked up as costs went down when swiftly improving, small computers could be placed in robots (no more connecting cables). The function of the brain as a natural network was becoming better understood, and knowledge gained could be applied to machines. In 1982, John Hopson published a seminal paper that reestablished neural networks as a driving force in AI.

That brings us to a description of a neural network. Such a network can be of the original kind centered around neurons, or they can be mimic that uses circuits to play the role of neurons. In either case, the circuits work in parallel rather than all information being channeled through one or a few processors, as in standard computers. The system is designed so that connections between the circuits are reinforced or weakened, depending on how often the neurons using a given connection communicate with one another. For example, sounding a whistle before striking a dog or a robot that has sound and touch sensors will cause certain neurons/circuits to fire off and communicate with one another, strengthening the connections between them. With stimulus repetition, the whistle will initiate neuron/circuit communications that will be strongest over connections developed during the striking incidents. The poor dog or robot will therefore "think" of the coming kick after hearing the whistle, and act appropriately. When kinder owners cease the kicking after the whistle sounds, those same connections will weaken and the association will be reduced, perhaps lost. If still kinder owners do something nice to the dog or the robot after blowing the whistle, such as feeding it or giving it electricity, the connections will strengthen again, only this time, the whistle will symbolize a positive event.

IBM's reassurance notwithstanding, computers can be as intelligent and innovative as brains if they are neural networking systems truly learning on their own. Far from doing only what they are told, the designers of neural networks cannot predict fully how and what their charges will do, just as one cannot fully predict the behavior of self-learning biological neural networks such as children or pets.

Fuzzy Logic

Another way to make computers more brain-like is to get them away from the rigid bi-digital logic of true or false in favor of the softer, fuzzier logic of more probable or less probable. When a brain thinks about something, it does not see in stark black and whites or yes's and no's (despite certain religious groups' attempts to get us all to do so). Instead, we often think in a gradation that says, "I like this more than that, but I like the other thing even more." Fuzzy algorithms can be used to render any rigid bi-digital machine into a fuzzier system that recognizes gradations of possibilities. Fuzzy logic can sometimes be used in an otherwise conventional, top-down programmed computer, and such systems have proven successful in temperature-control systems for buildings, and as smooth-running control systems for metro trains. Achieving even more brain-like stature is a combination of fuzzy algorithms in a neural network.

Analog Plus Digital

The casual metaphor that the brain is a computer can mislead us into thinking that the brain *is* a digital machine. At the same time, those who do not like calling the brain a computer are quick to point out that it is *not* a digital machine. The truth lies somewhere in between. Not all computers are digital, and communications between neurons exhibit characteristics of both the analog and the digital.

Because the brain clearly is not a bi-digital operator, an obvious tactic for mimicking the brain is to make circuits that work as much like neurons as possible. Recently, new chips have been developed to do just that. Rather than responding to a yes or no signal with a yes or no, these analog digital chips' input-output signals are identical to neurons, so they assess the fluctuating strength and characteristics of incoming signals, do internal calculations, and decide to either keep quiet or transmit a complex signal of its own. Note that analog-digital chips are inherently fuzzy in their logic. These neuron-like chips have become so good that their signals are virtually indistinguishable from those of live neurons! This is not the only advantage of the new chips. Like neurons, they process a given package of information using one ten millionth(!) the energy of a conventional chip.

AI Breakthroughs

The slow progress in the decades since the high hopes of the 1956 Dartmouth conference began to encourage the skeptics of artificial cognition. The problems with strong AI's expert systems in the 1980s seemed only to reinforce the image of AI and robotics as hype backed up with scant results. A few years ago, IBM sponsored an excellent series on the history of computers. Excellent, that is, until the last episode when, in true blue IBM form, the entire field of AI was downplayed as being moribund and having little or no potential.

Although it is true that AI has not brought us as far and as fast as some once expected (there is, as yet, no HAL) its achievements are nothing to be sneered at. A missile that navigates 1,000 miles before impacting ten meters from its target is an astonishing, albeit unpleasant, example of strong AI. Most cutting edge technology in the Gulf War was AI technology — and out-of-date AI technology at that — because most of the smart weapons were developed in the '70s and early '80s. In

ROBOTIC PTERODACTYLS

In the early 1980s, Paul MacCready, the designer of the first human-powered aircraft, got an idea. He wanted *Quetzalcoatlus northropi*, a 70-million-year-old flying creature with a 40-foot wing span, to fly again at full scale with flapping wings. The paleoartist recruited to replicate the appearance of the largest creature known to have flown happens to be one of your authors.

MacCready figured that digital technology had come far enough to make a realistic robotic pterosaur. The full-size model was never built, but a half-size model was. Although the model was actually lighter than a real pterosaur of similar wingspan, the flapping thrust turned out to be less efficient than hoped, so the model could not climb under its own power. A catapult line had to be used to get the machine airborne. Even so, once at altitude, the QN proved aerodynamic and stable even during flapping flight. It was rather eerie to see its head quickly searching out the best solution for lateral stability. The flight of the QN was not as graceful as the real thing must have been, but for a first attempt at making a lifelike flying robot using 80's technology and put together cheaply in a short period, it was not at all bad. Alas, at the one public demonstration in the spring of 1986, a communications glitch caused a crash.

particular, the weapons were dependent on a human telling them what to hit. A new generation of smart missiles is about to be deployed. These lethal machines will be autonomous devices that seek out, identify, and then attack vehicles, equipment, and structures behind enemy lines. Today's AI community is thriving as never before and finally making some long-promised breakthroughs.

SOAR, PRODIGY, ACT, HOMER, THEO, ERE, TELEOS, CYC....

Since the early 1980s, a number of smart computer systems have been up and running at various research institutes. They all share certain characteristics, aside from pithy names that are short and cryptic. They are long-lived systems, some have been up and running for 15 years now. All have a strong self-learning component with real world knowledge the usual lesson. The combination of time and growth means that as time progresses, the complexity and capabilities of the systems increase. It's sort of like a youngster getting an education. Systems, therefore, evolve. Although the systems have been limited to research so far, their creators hope eventually to find practical applications for their ever-maturing pets. As part of finding a home for the systems, it is expected that they will understand verbal English.

The capabilities of these systems have gone beyond earlier attempts of infusing computers with AI. Homer, for example, was a virtual submarine swimming inside a computer at Palo Alto. Homer autonomously navigated through its watery environment, recognized objects, and carried out requested tasks, asking questions if necessary to complete the task. The hopes of Homer's caretakers were that their program would find applications in the real underwater world. These ambitions were frustrated by impossibly low budgets. NASA's ERE has higher aspirations; its concepts are intended to help control robots that work at stations in space.

A moderately more successful endeavor is SOAR, which got its start in the early 1980s at Carnegie Mellon in Pittsburgh. It has developed into a multidisciplinary research project with branches in Europe. SOAR started out using relatively simple logic to solve problems it confronted. If the system solved the problem, fine, "Next." If not, SOAR set up a "problem space" using special-purpose techniques and brute force to configure the situation toward a solution. The solution was then stored and used the next time a similar question came up. SOAR learns from success. It is interesting that humans working with SOAR have found that the less they try to program SOAR and the more they make it learn on its own, the better SOAR does. Computers themselves appear to be rejecting strong AI in favor of evolutionary AI! Unlike Homer, SOAR has been used to operate some real test robots.

As encouraging as the likes of SOAR and Homer are, they expose problems that plague current AI. Programs remain excruciatingly slow. This is not surprising considering the AI computers are a thousand or more times slower than our brains! Systems make mistakes, which is not too bad considering how often humans do the same, even worse, they tend to seize up when run for extended periods of time (something humans also do, but less often).

Perhaps the best known, best funded ($25 million), and certainly the most controversial of the postmodern AI programs is Cyc. Cyc is the product of Douglas

Lenat working under the auspices of a consortium of dozens of high-tech companies. Cyc is a brute-force attempt to endow a computer system with common human sense. Cyc doesn't do much. It sits there down in Texas with lots and lots of knowledge fed to it by helpful humans. It has ingested some ten million facts, so far most of it straight out of encyclopedias. That is what Cyc is, a sort of thinking encyclopedia, ergo Cyc.

Cyc uses programs to integrate and understand what it is told. When perplexed, Cyc asks germane questions. It is kind of eerie to read the pertinent questions Cyc can come up with. Cyc knows many things: Cyc knows that animals do not like pain, that nothing can be in two places at once, that Napoleon and Wellington were persons, and that the latter was sad when the former died. Cyc has even achieved what naysayers said could never be done with a computer: Cyc can successfully define new concepts in terms of other concepts, a process that skirts the dangerous edge of the ever-expanding traveling salesperson problem.

As Cyc has matured, it has evolved through a number of distinct stages and levels of understanding, each higher than the last, like a young person. It is hoped that Cyc will get to the point that it will do better reading about facts on its own, rather than being spoon-fed data. At this happy stage, humans will become tutors for a truly self-learning machine. Lenat and his backers' ultimate hope is that Cyc will be the basis for the first generation of truly intelligent, common sense systems very early in the next century.

Cyc has limitations, however. As much as it knows, Cyc's memory base of multigigabits is a thousand times smaller than portions of stuff rattling around in your head. Worse, the kind of common sense Cyc will have will be the intellectual sort. Very useful, but not much use for moving about rooms or crossing streets.

BRAINY NEURAL NETWORKS AT SCHOOL

Perhaps some of the most startling research advances have been made in mimicking and simulating the functions of the brain and related organs. We have already mentioned how neuron-like chips have been made. An obvious thing to do with these chips is to array them in a neural network, the most brain-like kind of machine that can be produced currently. Among the most fascinating are the simple artificial retinas built by Misha Mahowald. Using light-sensitive analog chips, the light-sensitive neural networks act so much like the real thing that they are fooled by the same optical illusions we are. Thirty years after Minsky dropped the computer vision and coordination problem in a student's lap, the problem is yielding a solution. Another simple neural network not only remembers like humans, it forgets like humans, and cannot remember long digits any better than humans can.

Perhaps the most remarkable example can be found at USC in San Diego. There, a group of brain and computer scientists have combined forces to chart and detail the entire function of a part of the hippocampus (they have filled two walls with equations describing its function), and build a neural net-like chip that seems to mimic the function of that small part of the brain in all respects. The signal output of the chip is the same as that of the real hippocampus.

Fuzzy, Analog-Digital Neural Networks for Sale

Brain-like computers are breaking out of their university environments. Most commercial computers remain the old fashioned, top-down, not very intelligent type. For all their limitations — ones we are all too familiar with — they are too predictable. Hit a certain key and a certain event will happen (you may not know what that is, but that's not the computer's fault). As useful and "safe" as they are, conventional computers cannot and never will do tasks that defy standard programming techniques. The defiant chores include reading handwritten messages, understanding and translating speech, operating planes, trains, and automobiles, or learning about their owner's fondest wishes and desires and fulfilling them. Only computers that learn can take the input data and put it together in a new way their owners did not, indeed could not, expect.

It is into these niches that neural-networked AI computers are finding their first commercial successes. The public hardly is aware how extensively law enforcement has turned to artificially intelligent networks to gather and assess information on criminals and their activities including constructing profiles of serial killers. Neural networking and analog chips are the minibrains inside check-reading machines.

A major fulfillment of a longstanding AI promise is computers that understand and respond to verbal commands or that are capable of translating languages. As recently as the late 1980s, voice recognition was judged an eternal techno-dream. Suddenly in the mid-1990s, the first practical voice recognition and translating devices are entering commercial service. Around the turn of the century, common everyday computers will easily read handwriting and take rapid dictation. The key to this punctuated jump in performance was the introduction of neural networks that learn that Ernest is a person, not a tire.

You've heard of computer personal "agents," the little talking and listening computer screen figures that will soon be learning your online preferences and needs, keep your schedule, and organize your computer and media systems for you, and only you? If they are successful, they will be the pinnacle of near-term commercial AI, and they are based on self-learning neural networks.

ROBOCARS

For years, we've been promised cars that could drive themselves. There are two ways to do this. One is to make the roads smart by embedding guidance systems into the asphalt and concrete. This will probably never happen because the construction and maintenance costs will be too high (a mile of standard highway already costs $150,000,000), and the system will quickly become obsolete anyway. Instead, the vehicles need to be made intelligent enough to drive themselves down ordinary, unmodified roads.

For years, all the robotics people could come up with were computer-packed vans that drove themselves slowly down campus pathways, usually veering off and threatening the occasional student. Then around 1990, computers were small and powerful, and the necessary algorithms had been developed. Today, experimental robovehicles drive down real roads and highways with no one at the wheel (at least in principle; usually, someone is there just in case).

One of the best autonomous vehicles is Ralph, which took a trip last year across the USA under the fiction that its reliable robotic steering and lane-changing system, all contained in a laptop size computer, was a sophisticated form of cruise control. Ninety-eight percent of the trip was under Ralph's control, and although there were some problems (Ralph kept wanting to exit on ramps), the humans quickly became comfortable with their passenger status. The camera-computer system is programmed to use road edges and lane lines for guidance, but it will use anything that works. Ralph successfully resorted to using drip oil strips, and even the rain tracks made by proceeding cars, a tactic used by more than one human on a stormy night. Many of the robovehicles can read and obey speed, traffic signs and lights, and avoid obstacles.

The advent of fast and real road robotic vehicles is a classic punctuated evolutionary event in which a long period of slow progress ends with a burst of rapid development. As the ability and reliability of systems improve, someone will put it together and take advantage of the global satellite positioning system. Then, you will be able to get into your car, tell it you want to go to New York City, and let your automobile do the driving.

Dante and the Volcano

Success developing general-purpose robots remains limited, but encouraging. These days, a company can buy a wheeled security robot that uses sonar to navigate through a building; these machines are neither radio-controlled nor do they follow a guidance system imbedded in the floor. A few hospitals have purchased mobile robots that are loaded with food trays in the kitchen, or X-rays at the lab. Using a combination of a central computer and on-board sensors, they guide themselves down hallways, avoid static objects and moving people, use elevators, and deliver the trays to the appropriate rooms. These courier robots can make simple verbal requests. But there are serious problems with these machines. One of the hospital robots took a dive down a stairwell. None of these machines can pick up anything, or respond directly to intruders they may find; they always need humans to give them a helping hand.

Robots with greater skills can be found in many research institutions. Most robots still use wheels for getting around. The actor Alan Alda, host of Public Television's *Scientific American Frontiers,* met one of these rolling machines called Flaky matriculating at Stanford University. Pretending to work at the facility, Alda voiced a request to Flaky to fetch a file from another office. Flaky voiced a "roger, wilco" in the male tones of his computer-generated voice, and left the office. He wheeled his way into the target office, finding nobody to carry out the next task of giving him the file. Falling back, Flaky "remembered" that Alan told him in plain English that so-and-so often was at the main desk. So the robot continued to the main desk and obtained the file from the person he found there. Flaky returned to Alda's office, where he pestered the host to acknowledge receipt of the file until he was thanked for a job well done.

Some research robots can walk on one or many legs, and there are even two-legged robots that can run, leap, and do somersaults! The latter have recently been developed to better understand and improve the motions of gymnasts. The performance of some of these machines is quite good, but often, they have difficulty coping with rough, unfamiliar terrain. None yet moves with the fluid, dynamic grace and balance of people or other animals. Dynamic walking is a difficult problem and one that does not receive needed attention because it is not a big priority at this time.

Certainly the most dramatic, albeit slow, journey of a robowalker was the recent descent of Dante II into a volcano. The first Dante was to be sent into a volcano in Antarctica, but a failure of the communication cable put a premature stop to that. The second Dante was sent into an Alaskan crater. Each treacherous and deliberate step was analyzed by the main computer before it was taken. Dante gathered the data it was supposed to and started back up, only to tip over after stepping onto unstable soil. It had to be rescued by people and a helicopter.

Where robots are really coming into their own are in places that A) are not good for people to be in, and B) are easy for robots to move in. These are oceans, high altitudes, and space. Intelligent mini-subs are being built that will remain at a home base on the ocean bottom, occasionally venture out to gather data, and then return for a recharge until the next adventure. Others will cruise the hidden depths in extended-duration searches for creatures unknown.

Much higher up, small, smart robotic planes are becoming the only sane way to measure the dynamic conditions inside the cells of turbulent storms. Still farther up, ultra-long-winged roboplanes powered by solar cells are under development. These high-flying machines will patrol the thin, cold stratosphere for months at a time sampling air, mapping surveys for civilians, and conducting reconnaissance missions for the military. Of course, the most famous remote robots are the ones in outer space. The deep-space probes that survey other planets are so far away that light-speed instructions take minutes or hours to get to the probe, making direct radio control impractical. The probes have to control situations on their own with minimal instruction, therefore, qualifying as robots. So far, the probes have been rather simple systems designed 20 years ago (the computers in the space shuttles also date to the '70s). NASA is finally catching up with the '90s, the first Marsbots will soon be launched to the red planet. As they crawl about the alien terrain, the smart, little machines will be on their own.

The Games Computers Play

By now, you might be thinking that what computers and robots can do is not that much when compared to humans such as us.

Perhaps you should not be too smug.

The game-playing potential of computers was nose-on-your-face obvious as soon they were switched on, and computer jocks jacked in games as fast as they could obtain a portion of heavily allocated capacity for "frivolous" pursuits. Actually, the games were not really so frivolous because game playing quickly became a way of comparing the intelligence of computers. For the strong AI bunch, it was a chance for humans to program seemingly clear-cut rules into their machines. The neural network types were especially interested to see if the computers they programmed with some fairly simple initial instructions would be able to learn to play better than they could.

The answer to the last question was soon coming. Yes, the computers really were learning enough to beat the fellows who programmed them in the first place. This didn't mean much, though, because programmers were not exactly master-level players. In the mid-'60s, a ten-year-old, admittedly a bright ten-year-old, beat the best chess-playing computer of that time. That was then, this is now. The competition between humans and computers has a tenor promising combat to the bitter end, unnerving to some and ominous for its obvious inevitability. As good as humans may be, we are not going anywhere. The top chess player from 500 years ago could probably hold his own with most modern grand masters. The computers, on the other hand, get better every year. In the '70s prizes began to be offered to the computers that could take on all human comers and win. Even IBM has recently gotten back into the AI game by backing high-powered chess machines. They see it as a way to improve the ability of computers to handle complex situations and more closely mimic human thought patterns.

One day in 1979, Luigi Villa won the world backgammon championship. The next day, he lost it when a game-playing computer soundly beat him with moves that even the rather appalled humans acknowledged were daring and innovative. In 1994, the computer Chinook won a six-game match against the world checkerboard champion Marion Tinsley partly by default: The poor human had a physical breakdown in the contest against the relentless machine.

But what of the King of Games, the ultimate in human intellectual expression, chess? Chess comes in two forms. In the short game, each move is made in a few seconds and it favors computers over humans because the latter cannot fully employ pattern-analysis systems. The longer standard game allows a more profound level of conscious thought and is believed to favor the human contender. A $100,000 prize was offered to the team fielding the machine that could defeat all human challengers. Even after defeating digital chess master Deep Thought in 1989 at the short game, world chess champion Garry Kasparov saw the writing on the wall. He gave himself at least ten more years of superiority in standard chess. Others, citing curves on graphs, suggested that the mid-'90s would see the convergence of machine with man. In a 1994 short chess match, the Pentium chip-based Chess Genius defeated Kasparov, who stormed off the stage. Those sensitive humans. A year later, Kasparov took back the title from Chess Genius 2.

IBM's Deep Blue is the most powerful chess machine yet built. Earlier this year, Deep Blue and Kasparov faced off in the first extended match of standard games. Kasparov was highly confident at the beginning, the long game being to his advantage. The IBM bunch cheered when Deep won game one! Kasparov began to call the computer "the monster." In a telling observation, Kasparov said that in future contests, he wanted his own computer to help explore potential moves! When engaging fellow humans, Kasparov uses psychological intimidation and aggressive playing tactics. The former is completely worthless against a machine. Rather, the human is in danger of being fatigued by the relentless machine (shades of John Henry).

In fact, these contests are strongly tilted in the human's favor by spacing the games out over a period of days. Imagine if each game followed one after the other; the computer would not notice, but its opponent would collapse! (On the other hand, Deep Blue crashed for a few minutes during one game.) As for aggressive play, it is vulnerable to the extended-search pattern of Deep Blue. However, Deep Blue is relatively inflexible. As it was, Kasparov demonstrated superior learning ability. First, he shifted to a defensive style of play better suited to counter the IBM machine, resulting in two quick draws. Then, Kasparov again shifted to tactics he deliberately designed to confuse and frustrate Deep Blue, to the point where it was nearly paralyzed, giving him the needed edge to win the last two games — and the contest. A happy Kasparov correctly stated that man remains superior to machine. In the end, however, his counter-offensive was like one of Robert E. Lee's Northern offensives. It cannot be sustained.

We can clarify the reasons for this when we compare the hardware of Deep Blue to Kasparov's brain. The first uses 32 parallel processors to examine 100 to 200 million moves per second, or billions in a few minutes. These figures usually inspire admiring gasps among humans. As we will show below, the human brain has trillions more parallel processors and is thousands of times faster. This comparison is not quite fair; Kasparov was using only a fraction of his brain's thinking and calculating power to actually play chess (the rest was dedicated to sensory and motor systems, as well as emotional thoughts about the contest), but it is clear that Deep Blue is by far the simpler and slower device. In this view, it is remarkable how well the computers are already doing, and it is clear that the machines have just begun to play!

In the not too distant future, Kasparov, or the current human champion, will face a computer that is not only much faster than Deep Blue, but also better able to learn from its past mistakes and from its opponent's last games. It will be a machine less dependent on its creators' programming. It will win most, if not all, of the six games.

There are still some games humans persist in winning handily, such as the Asian game Go. Some skeptics of computer intelligence suggest that free-for-all games like Go inherently favor human thinking over mechanical calculation. Of course, such arguments are risky because they are in serious danger of being disproved. Go will go to the computers sooner or later.

So what does all this mean? The first computers could easily beat humans doing straightforward math as well as recordkeeping and other basic accounting tasks. This was disturbing enough, but people consoled themselves with the observation that people were not good at these mechanical arithmetic tasks in the first place. Leave 'em to the infernal machines. Games invented by humans to be played by humans, indeed to challenge the human intellect at the highest levels, are another matter. These losses really wound the human psyche. We can tell it hurts because the contests receive so much coverage and there was almost a global sigh of relief when Kasparov emerged victorious.

The hurt is deep enough so that resentful sentients often dismiss a machine's achievement. The triumph of the machine is really a triumph of humans, they say, because the first was built by the latter. This claim stems from a common assertion that computers are mere number crunchers without capacity for human originality. These charges may be reassuring, and there is some truth to them, but they are also misleading. When parents do a good job of raising a child and the child grows into a successful adult, part of the triumph goes to the parents. But only part; most of the triumph is the child's. The triumph for beating a human with a computer would accrue totally to its makers only if every move the computer made were preprogrammed by human experts. This is not applicable to the more recent game-playing computers, which are self-learning machines that make moves that surprise and impress even their human makers and opponents. The machines have to be self-learning original-thinking neural networks. Although expert players often help program the computer, the ability to match or beat a champion requires that the computer be better than the programmers! Even Kasparov could not find the time or knowledge to program a computer to best himself.

The attempt to dismiss computers as simple number crunchers also comes back to bite those who offer the dismissal. Is human thought and ingenuity so superior if it can be bested by cold calculations?

It is true that game-winning computers do far more number crunching, and a lot less thinking, than their human opponents (we will examine the differences and similarities of game-playing machines and brains in more detail below). It is also true that computers are highly dependent on humans for their performance. It is true that game-playing computers remain special-purpose, unconscious devices whose overall performance is far, far inferior to that of brains. This, too, is true. Today's simple machines are already exposing the weaknesses of human thinking. As time goes on, the computers will become less and less dependent on humans; they will learn to crush, not merely edge out, human champions; and they will work more and more like real brains. The game-playing machines are chipping away at the gap between man and computer. People should be nervous.

Strong AI Was Safe, the Grand Evolution of AI May Have No Bounds

Which brings us to a very, very important point. One of the limitations of strong AI was that even if humans could program computers to be as smart and as capable as humans, humans could never program computers to be smarter and more capable themselves. It was sort of a mythical safety check on how far computers could go relative to their primate makers. Strong AI is dead, and the grand evolution of AI is under way. If computers can learn on their own, and their calculating performance continues to rise, what will keep them from soaring far beyond human capacities?

Anti-AI people have it all wrong when they say that brain complexity is unmatchable. Make no mistake about it, one way or another, some time or other, the brain will be matched. What's really important is that with strong AI gone, there is no stop to soaring beyond the brain, and the boost will itself power the boost at extreme speeds.

Artificial Life

Perhaps when you think of artificial life you imagine scientists in white lab coats amidst gurgling beakers, tubes, and other assorted glass vessels full of brownish organic soups from which they hope to coax some new crawling form of life. Attempts to make new biolife have been going on for less than 50 years with limited success. This is not, however, the kind of artificial life we are interested in. Replicating a system that is already running out of steam is interesting, but not really important.

What we are interested in is AL, a new and much more exciting attempt to make new life out of things not biological. AL got its start about the same time as AI, but unlike the latter, AL languished until a decade ago when it took off and became a thriving, if controversial, endeavor.

The foundation from which all AL stems is John von Neumann's cellular automata which is, as we have already discussed, on extension of Turing's information machines. Von Neumann showed, at least in principle, the same processes used by life and evolution to reproduce life forms, introduce change, and increase complexity can also be made to occur in systems other than biological. In the 1950s, a few researchers used toy trains and articulating blocks, of all things, to show that simple systems really could mimic bioform's activities. On a bigger, albeit mental, scale Edward Moore explained the way huge floating factories could take in raw materials from beach, sea, and air, and reproduce themselves all the while producing useful products for humanity.

This promised goods in such abundance that all needs of man and woman would be met. Estimated start-up cost for Moore's "artificial plants" was similar to the space projects of the time, and after that "spin offs" would more than pay investors. Nothing was ever done. Continuous control of the machines once they got started was just one problem; how would they fare in a storm? If the things actually worked, what would all the humans do for work? Anyway, it was not likely that '50s technologies were up to the task. Von Braun and the boys got to build rockets instead.

Such AL enthusiasts as the forward-thinking Freeman Dyson soon realized that space was a better place for factories to grow without bounds. So it is logical that the next grand AL proposal was put together by a daringly innovative study group funded by stodgy NASA in 1980. A small group under Richard Laing nervously offered their ideas of self-replicating von Neumann/Dyson factories on the moon. After the initial investment, the factories would be entirely self-sustaining. They would produce, grow, self-repair, replicate, and evolve.

As Steven Levy has said, the designs "unfolded like origami fugitives from an Asimov novel," but they were based on scientific principles. One of the two designs started with an "egg" loaded with special task mobile robots. After bursting forth onto the lunar landscape, the lunar robots would set up housekeeping by mining for raw materials and erecting solar arrays for power and communications systems. Using no outside help, the robots would have a factory up and running, making new robots in about a year. Growth would be exponential after that. The robots could change form and abilities as needs changed and as they learned more about the universe. We say "universe" because the new lunar society could then build rocket robots that would expand exponentially into the solar system and then the universe! All for the cost of one little robofactory on the moon!

The Laing team suggested to NASA that they fund the feasibility studies needed to initiate work on self-reproducing robots. Then they waited. When it was announced that Reagan was going to declare a major space initiative Laing — a tad naively perhaps — thought it might be for robots on the moon. Instead, we got Star Wars. Not surprising really; a California actor pandering to conservative

Christians was more likely to revel in the well-worn Cold War than sign onto a proposal to take over the galaxy with robots. As for NASA, at least they did run the team out of the country. They decided to stick to industrial age shuttles and space stations they were more comfortable with. A not entirely unreasonable decision considering the not too impressive state of robotics at the time. Besides, the Laing team had explained how their system not only did not need humans after a certain point, but that also soon become suprahuman in power — a bit much to expect a federal agency to fund.

So AL in the form of giant ocean-going factories and lunar robots were out. This snafu prompted AL people to start thinking smaller and cheaper. Much cheaper. Set up artificial life in some place so small that grants wouldn't be needed. A place that could contain high complexity; enough to test to conclusion some of the premises of AL; and finally, go beyond theory into the realm of unexpected results. At long last, a place where self-replicating entities could not escape.

The Game of Life

Not the game you can buy in stores, but an AL game invented by Horton Conway in the late '60s. For Conway, the small cheap place for AL was a large room at Cambridge. Life looks like a game of Go gone mad.

Life got a big boost when *Scientific American* published the rules in 1970. Soon, lots of people at universities and corporate research facilities were playing Life. It was fun and research simultaneously. However, Life in its original form has strict limitations due to the restricted complexity of its pitch — a room-sized board and the slow calculating power of a few persons moving pieces around. In particular, there is not enough complexity or time to generate self-reproducing entities. It has been estimated that on the original scale of Life, it would take a board the size of a planet or larger to create self-replication.

AL In Its Natural Habitat: The Computer

The obvious place for Life and other AL is inside computers. When Life moved into computers in the early '70s it became something of a minor plague eating up a substantial amount of the limited computer power available at the time. However, even computers were too slow, and within a few years Life had run its silicon course. By the '80s, computers were again becoming fast enough to do seriously interesting things with AI and they went to the boids.

Craig Reynolds wanted to figure out the rules guiding synchronous, coordinated flight activities of flocking birds and schooling fish. So he made a virtual world of obstacles for virtual "boids" to fly through and told them to adhere to the following rules: match speed; keep the flock together; don't fly too close together; avoid collisions. That's about it. The result was classic flocking and schooling behavior of nonliving, rule-following entities.

A bunch of boids scattered at random in virtual air quickly gather together into a flock formation. As they fly among and around a set of columns, the flock splits up and rejoins on the opposite side. The boids were not told to do this. The behavior emerged from rules and only from rules. Sometimes, one boid is crowded by its neighbors and cannot avoid smacking into a column. When it does, it falls to the "ground," hesitates for a moment, and then springs back up and catches up with the rest of the flock! Again, this behavior was not programmed in at the start; the boid invented it by following the four rules.

Genetic Algorithms

Computer-generated boids learn, but they do not evolve. Recently, we've witnessed a plethora of computer-based artificial life systems using "genetic algorithms" to evolve complexity from simplicity. Genetic algorithms were invented 30 years ago by John Holland, who wanted to translate the natural genetic system of learning into a computer-friendly equivalent. GAs are digital algorithms that closely mimic genetic codes. GAs reproduce inside computers. AL systems start with one or a few simple "organisms" that reproduce with GAs. The codes are subject to mutations, most of which will turn out bad and die off, but a few will be beneficial and thrive. It is a form of unnatural selection. As the entities evolve, the results are astonishingly biological in form and appearance. Computer organisms flourish as they acquire advantageous characteristics that allow them to displace competitors. Later, the same entity may become extinct as a mutation gives other entities selective advantages.

PolyWorld, Ramps, Garden of L, Loops, and Tierra

One of the best examples of postmodern AL is biologist Thomas Ray's Tierra. Ray's biology background proved an advantage because he did not know enough about computers to realize that what he wanted to do was considered unrealistic by most cyberexperts. Ray wanted to create the first open-ended evolving system of von Neumann automata.

Tierra, a virtual world inside a virtual computer inside a real computer, was modeled after the Cambrian explosion of half a billion years ago. The Tierran system is Darwinian competition for computerville, processing time and space and powering up out of the central processing unit. The main habitat is a block of memory called "the soup." The digital organisms capture what portion of the CPU's power they can and use it to power their own virtual CPUs. Because new organisms are constantly being introduced to the soup and habitat space is limited all organisms eventually die. The organisms' genotypes consist of digital codes that are copied during reproduction. Two kinds of mutations could alter the genotypes' pre-programmed random errors and accidental random errors.

Tierra started with the Ancestor, a primitive genotype of just 80 instructions for an organism that did nothing but replicate. When Ray let Tierra rip for the first time in 1990, he was warned to expect the usual crashes and freezes that vexed previous AI/AL programs. Ray figured it would take years of trial and error and programming modifications to get some interesting evolution going. Ray was wrong. With the system running at hundreds of thousands of calculations per minute, the Ancestor soon did what was expected of it and filled up most of the habitat. Then death came to Tierra, and the opportunity for Darwinian selection as mutants appeared. Soon, new organisms with shorter genotypes than the ancestor started to proliferate to the detriment of the soon-to-be-extinct Ancestor. Shorter codes were more efficient. However, a creature with a code too short for self-replication appeared and thrived. It was a virus-like parasite that used the codes of larger host organisms to reproduce.

Like the weary hosts of unwanted quests, the bigger creatures fought back with mutations allowing them to "hide" from smaller parasites. This procedure cost the hosts efficiency because it was necessary to grow larger and do extra work to undergo a self-examination process rather than rely on primary memory identity retention. But as they say, "Better inefficient than dead," and the new, big organisms wiped out the little parasites. They did not give up, however, and replied with less efficient but more powerful parasites that could find hosts. The arms race was on.

All this occurred in ten minutes! Far from crashing, Tierra was doing what nature takes many years to do! It was also an innovative system coming up with organisms and interactions Ray did not program.

Many more Tierra runs were made with different initial conditions and on computers with faster speeds. Eventually, there were tens of thousands of different kinds of organisms. Ray named the program he used to analyze the results "Beagle" after the sailing ship Darwin traveled in. Quite appropriate in view of the "eras" that unfold during the longer runs that amount to hundreds of millions of instructions. Periods of stasis in which little changes for millions of instructions are suddenly overturned by a radically new order that may remain stable until it is replaced by yet another set of coexisting organisms. Tierra not only evolves in the Darwinian manner, it does so via Gouldian punctuated equilibrium.

Some AL researchers consider their computer "organisms" marginal forms of life because they use energy to reproduce and evolve. This may or may not be so, but Tierra et al. are certainly limited by the low power and complexity of today's computers. None can yet process information at the level needed to simulate the full complexity of DNA evolution. This will change, of course, as the AL crowd gets access to small cheap machines more capable than today's supercomputers. Even now, the computer entities are potentially so powerful that it has become standard procedure to keep them isolated within single computers or within limited computer networks. No one knows what they would do if let loose on the Internet at large!

AL Art

Show an artist new technology and before you know it, she is figuring out what neat new stuff to do with it. An array of artists has begun to merge (or combine) computer art with genetic algorithms. The marvel of AL art is that one can start with a simple image in two or three dimensions and let it develop in increasing and unpredictable complexity. The selecting agent is the artist killing off images she does not like in favor of those she does. AL art is, therefore, a form of bottom-up plus top-down guided evolution. The beauty and complexity of some 3-D art forms as they grow, change, and die rivals that of biologic organisms. They have to be seen to be believed. Another interesting fact is that some of the algorithms being developed by artists are being plowed back into more practical purposes. Life does mimic art.

Self-Organization and Complexity Theory Down In Old Santa Fe

If Darwin could sit down in front of a computer monitor and see the progress of evolving entities within an AL machine or the evolving beauty of AL art, how delighted he would be to see his theory not only proven, but hard at work. The evolving AL programs not only replace forever the creationist argument that lesser complexity cannot evolve into greater complexity, but they also show that it is remarkably easy. What is more questionable is how Darwin might react to some AL researcher's explanation of why evolution is so easy.

In classic Darwinian theory, the organizing force taking random changes and making greater complexity is natural selection. In this system, the selective forces of extinction and survival whip otherwise reluctant atoms and molecules into line until they form genetic codes for organisms and, if lucky, high-level intelligence. The ease with which AL organisms increase complexity suggests to some that this is one of those universes in which molecules are primed to cooperate in the endeavor, much as water is primed by the shape of landscape to flow in a stream. Self-organization tries to show that complexity easily springs from simplicity because the "landscape" of our universe allows only a few paths to complexity. For example, consider a Boolean network. This is an array of 100,000 interconnected light bulbs, either on or off, that interact in constantly changing ways. The number of possible different interconnections between the bulbs is 10 followed by 30,000 zeros. If the system is unconstrained so the bulbs connect to each other, the network would increase in complexity through 10 followed by 15,000 cycles before repeating itself. If, on the other hand, each bulb connects to only two others, the 100,000 bulbs will blink only 317 times before repeating!

The "self-organization" of complexity is an extension of the chaos theory in which complex systems tend to organize themselves out of entropic chaos. Self-organization differs, arguing that the emergence of complex life and intelligence is easier and more inevitable than usually considered. If the proper materials (carbon, oxygen, hydrogen, etc.) are available, the temperatures are right (somewhere between freezing and boiling), and power is on hand (a sun is ideal, but a geothermal vent might do), things automatically click into place and get rolling as matter and energy organize themselves into living systems. The evolutionary landscape then forces life to develop down certain paths, one of which leads to high-level intelligence (but only if all the necessary conditions described at length in the Appendix are present). This view is in sharp contrast to those of some evolutionary biologists such as Stephen Jay Gould who say that evolution is, on one hand, a difficult affair, and on the other, unconstrained and not bound to produce smart beings even if conditions allow it.

Self-organization and complexity theory are controversial and are not experimentally confirmed or denied. It did not help that some of the early enthusiasts such as Stuart Kauffman at first thought that self-organization replaced Darwinian selection as the modus operandi of evolution. Even if self-organization as described is true, selection ultimately makes the call; which entities thrive, and which die. Nor has it helped matters that the center of the self-organization/complexity division of AL has become the Santa Fe Institute, located in the land of southwest New Ageism. Were the institute in Dayton, Ohio, it might have fared better. (These things should not count, but politics and perception are important factors in human science.) In any case, Tierra and similar AL programs have captured the attention of evolutionary biologists and evolutionary AI researchers alike. The former enjoy finally having a means of studying evolutionary processes in real time with detailed complexity. The latter see a new and potentially powerful way to evolve intelligence inside their machines. Even IBM, once so anti-AI, has become interested in Tierra and its relatives.

EVOLVING PROGRAMS

Earlier, we noted how the costs of large programs tend to run out of control because they become too complex for human minds to manage. A classic horror case is the gigantic computer program IBM has been developing for an air traffic control system that has been needed and wanted for years (the old system still relies in part on vacuum tubes made in Poland!). The cost of the software has galloped into billions of dollars with no end in sight and no working program on hand.

Something really has to give. The need for enormous programs will only grow, and the inability of intellectually clumsy humans to direct huge programs to do what we want will remain the same. The only resolution is to stop telling the program how to do every little thing. Instead, tell the program what is desired and let the program figure out how to do it from the bottom up. Treat the computer as a habitat and inside the habitat breed genetic algorithms that eventually make up the program. The genetic algorithms compete to come up with the best solutions for meeting the ultimate goals. Those lines of code that do jobs well are selected, and those that do not fit are selected out. As with nature, the growing program needs to run for hundreds of thousands or millions of generations to get anywhere. Unlike nature, generations can be run in a matter of weeks rather than epochs. Nor do you have to wait for a team of programmers to carefully enter in every line of code because as interacting parts of the program co-evolve, they debug themselves.

However, perfection cannot be assured in large self-evolved programs because the only way to fully test a program is to run it in the real world. The inner workings of a large, complex, self-evolved program are poorly understood compared to conventional programs; a fact that makes conventional top-down programmers tear their hair out. On the other hand, conventional program writers cannot guarantee their intricate programs have been debugged, and comparative tests suggest that evolved programs have fewer problems than the human-designed ones.

When all is said and done, evolution knows best.

Nano Nano

Let's hear it for nanotechnology, the latest fad and hot hype — and we have not even started it yet. Nano-objects are simply those objects that are grown and operate at the scale of atoms and molecules. "Nano" refers to a nanometer, one billionth of a meter. The idea of doing technology at such scales started in the late '50s between the ears of Freeman Dyson, the same jack of all trades who helped get the AL ball rolling at about the same time. Dyson suggested that technologies be developed that would work at smaller scales, which would, in turn, work at still smaller scales, which would, in turn, work at even smaller scales until the scale of atoms was reached.

Nobody paid much attention to manufacturing at the atomic-molecular level for decades; it all seemed rather outlandish. Things started looking less outlandish in the 1980s when it was realized that the tiny probes inside the new scanning, tunneling, and atomic force microscopes could be used not only to image individual atoms, but to also move them around! Pretty soon, IBM labs spelled out I-B-M in teeny letters made from just a few atoms each, and nanotechnology made it onto the evening news.

Nanotechnology shouldn't be confused with microtechnology, which is really nothing more than the miniaturization of macrotechnology, things we can see such as cars, buildings, and watches scaled down to the microscopic level. Both macro- and microtechnology are top-down procedures in which large amounts of material are processed and pared down to get the desired object. If separate or moving parts are involved, they are slapped together in some manner or other. The postmodern computer industry is based on microtechnology, in particular the fabrication of silicon and other crystals etched into tiny chips. As we have seen, this technology continues to raise data processing densities far beyond what experts believed possible just a few years ago. But this way of doing things will go only so far.

Nanotechnology is a bottom-up process of atom-by-atom building and molecule-by-molecule assembling, until the full product has been fabricated. Molecules are themselves nano-objects built up. It is even possible to construct moving parts simultaneously. Although nano-objects start small, they can grow as large as anything we can make, even bigger. Nano-objects are not exotic, they are common in nature. Crystal growth is an example of nanogrowth, and artificial crystals are the sole, still crude, form of nanotechnology practiced today. Living organisms such as your body are super-sophisticated nanosystems. We assemble molecule by molecule starting from a single cell and growing to a large adult, which, in the case of a whale, is a couple of hundred tons. The molecular computer DNA is a tiny nanocomputer and your brain is a big nanocomputer.

The potential of nanotechnology is difficult to measure and comprehend. As sophisticated as we think our postmodern macro- and microtechnologies are, they are only crude systems taking bulk materials in, heating them, beating them, and putting them through a myriad of processes to arrive at only an approximation of what we designed. Waste is high because large quantities of materials must be transported, processed, and discarded until a final product is built. Most conventionally manufactured materials are weaker than theoretical maximum-strength materials because inevitable imperfections crop up at the atomic scale. If nanotechnology works as well as promised, it will be ultra-sophisticated because it will take small amounts of material and manipulate it at the most intricate level possible to construct products of any size with exactly the desired characteristics.

The basic tools for nanotechnology are supposed to be microscopic "universal constructors," sort of minicomputer robot molecules that will be controlled by larger computers controlled by still larger computers. The DNA-RNA-protein system is a natural version of a self-programmed universal constructor. Nanotools, nanodevices, and nanostructures may be grown in soups of raw materials held generally at room temperature. In conventionally manufactured materials, atoms are often misaligned with weakened interatomic bonds. Therefore, strengths of modern materials are less than theoretical values. Every atom should be exactly in its proper place in a nanodevice so great strength will be combined with light weight. If so, diamonds will not only be hard in compression, but also very tough in torsion; strong, not brittle. Intricacy of structure at all scales will be far beyond today's achievements.

If it works as well as proposed, nanotechnology will combine the best of biological and technological systems, producing products of extreme sophistication cheaply and with little pollution. In fact, nanotechnologies may start harnessing biological processes including DNA routines, merging them with nonbiological nanotechnologies.

Eric Drexeler is the main proponent of nanotechnology. He has detailed his ideas in the books *Engines of Creation* and *Unbounding the Future*. Drexeler and his allies envision tiny nanorobots entering our bodies and cleaning out arteries or corralling each and every cancer cell. Because they will be intelligently designed and programmed, they should be far more potent than the human immune system. Grass could be clipped by millions of little lawn robots programmed to snip each blade when it exceeded a certain height. Rather than gluing together a plastic model kit, you design and grow a scale model so detailed and perfect that only size distinguishes it from the real item. High-fidelity stereo radios will consist of tiny receivers worn on a wrist. Finally, we will have no-muss no-fuss clothes thrown away and regenerated after each use. The Library of Congress will be stored in a crystal cube the size of a book. An entire aircraft could be grown with an integral body stronger and lighter than any modern composite material could produce, with flexible surfaced wings that could adjust to the airflow like an aero-skin, minimizing drag and controlling flight direction. Centralized manufacturing as we know it would largely cease. Instead, one could buy nanokits with universal constructors and the software needed to "grow" whatever one wants. On one hand, this may eliminate most jobs. On the other hand, it should make things so cheap that jobs will not be necessary.

A surprisingly large number of scientists have endorsed nanotechnology. Well-thought-out arguments favoring the concept have been presented, and the fact that living beings are nanosystems suggest that artificial systems can be designed, as well. Not surprisingly, there are skeptics, and they are making some valid points. More efficient and environmentally friendly conventional macro- and microtechnologies are transforming at surprising speeds as industry discovers economic advantages of producing items using minimal raw materials and energy input, and minimum waste output. In this view, pressures to develop nanotechnologies may not be as great as it seems. Skeptics question whether universal constructor molecules can be made more robust than the molecules they will construct. They also argue that the extremely small nanorobots and universal constructors will have limited information-processing abilities, so they doubt the ability of such little machines to perform all the functions claimed. On the other hand, if the smallest nanocomputers are controlled in part by larger computers, the overall power of nanodevices should, in the end, be much greater than natural counterparts. Even if nanotechnology does not prove to be the technique needed to perfect systems at almost no cost, it is very possible that it will work at least as well as a natural counterpart, perhaps better.

NANOCARBON

Perhaps the favorite material of nanotech enthusiasts is carbon, the same element of choice for living nanosystems. There are two rules of thought for using carbon. One is to use carbon in the form of proteins resembling those of living organisms to make tiny machines. Because proteins make little hinges that can do work, this is a good idea. Some proteins can work in dry environments, so protein machines will not necessarily need to be wetware like we are. Various proteins also work at high or low temperatures.

The other idea is to make nonorganic machines out of pure carbon. A favorite material is diamond for obvious toughness (and nanodiamond will not be as brittle as the diamond we are used to) as well as high heat tolerance. Another leading contender is the recently discovered buckyball, a large carbon molecule that can have a structure identical to some of the geodesic domes derived from the patented mathematics of Buckminster Fuller. Buckyballs can be extended into buckytubes that resemble cannoli, hollow microscopic wires of tremendous strength.

Less Really Is More

Drexeler and company guess that in 20 years, nanotechnology will become a major means of manufacturing, and in the first third or half of the 21st century will take over entirely. Considering how fast industry spread through Europe and America in a slower world, such speculation cannot be ruled out. If nanotechnology is feasible, it will be relatively cheap to develop; giant labs or huge test facilities aren't needed to make little things work in a big way. All that's needed are reasonably well-paid researchers and mid-priced high-tech machinery to enter the wonderful world of nanotechnology. Nanotechnology has the potential to be a breakthrough technology that suddenly, with little warning, grows exponentially until it has changed everything overnight.

Nanotech, the Dark Side

Nanotech has a good side and a bad side, Drexeler warns. Biological nanoorganisms reproduce themselves. The skills necessary to make replicating nanodevices should become easy to attain when the technology matures. The computer code that programs reproduction will be subject to errors like any other computer code, and these could result in "mutations" altering the replicated nanodevice. If one or more of these things is small enough, from virus to rodent-size, and can survive in the natural environment, we might have the ultimate alien species on our hands multiplying until it takes over. We can imagine counter nanodevices put together to try to put a stop to things. Alternatively, not-so-nice people might deliberately program nanosystems to wreak havoc.

THE ALCHEMIST'S DREAM: TRANSMUTATION

Nanotechnology makes things out of atoms, but it does not change the atoms. Even nanomanufacturing will have to take whatever elements it can get to make stuff.

To go subatomic would mean reviving the medieval alchemist's dream of changing lead into gold. Alchemists were frustrated because altering elements can only be accomplished by adding or subtracting neutrons and protons from the nucleus of an atom something chemistry, dealing with electrons, cannot do. Nuclear manipulations have been used to split large atoms such as uranium into smaller elements during nuclear reactions, and in a game of atomic "piling on", they smashed protons and neutrons together to build up massive elements found nowhere in nature. So why not turn lead into gold? First, because the process would be very expensive, and second, producing large quantities of gold would drive down the price so much, it would be a serious money-loser.

For now, most critical construction materials are abundant enough that transmutation is not worth the trouble and cost. Future civilizations may wish to convert whole solar systems and star systems into artifacts. Because a solar system is mainly hydrogen and helium with a smidgen of heavier elements mixed in, it will be necessary to transmute much of the light elements into heavier, more useful ones.

PUTTING IT ALL TOGETHER

Kept isolated from one another, AI, AL, nanotechnology, and transmutation all have limitations. Although evolutionary AI is superior to the strong AI it replaced, even the former will find it increasingly difficult to cope with the extreme complexity involved in programming high-level intelligence. AL has the potential to evolve to high complexity, but so far, it is stuck inside computers that will have to evolve high complexity in their own domains before they can contain complex AL. Nanotech has the potential to build incredibly complex hardware, but it will be dumb hardware. Transmutation can produce lots of raw exotic materials, but it cannot put them together.

Connections between AI, AL, nanotech, and transmutation are obvious. Artificial life promises to be the means by which artificial intelligence can evolve beyond the limits that promise to bedevil human programmers. A combination of nanotechnology and artificial life offers a way to grow the extremely complex computers within which high intelligence can be run. If and when there is a need for heavy elements on stellar scales, transmutation shall provide it.

Combine AI, AL, nanotech, and transmutation and the result will be a system of incalculable power.

Universal Manufacturer

One expression of this power may be the universal manufacturer. Manufacturing requires different machines, each making a few things at most. The manufacturer of a car or computer requires a number of different machines, many large and expensive. This system is inefficient, inflexible, and wastes natural resources, keeping costs high.

Imagine an artificially intelligent machine of extreme sophistication grown via artificial life and nanomanufacturing that uses still more nanotechnology and AI to grow whatever product its dimensions can accommodate. Because the machine is smart, it can respond to verbal requests for common products, such as a given number of a certain type of screw (ones far stronger than you can get in any hardware store today), clothing, or a standard type of neural networking artificial brain (this might take longer but not nearly as long as it takes to grow the original type). For more distinctive items, cooperate with the interactive design system that comes up with the machine. Small UMs may find a place in peoples' homes where they could make just about anything you want, including dinner! So much for main streets, malls, fast food outlets, and catalogs (but restaurants will probably thrive because folks will be desperate for an evening out). The cost to you will be the machine (which may grow itself from a smaller kit), the raw materials (much of which will come from recycled trash plus some special dirt laced with critical elements), and the software to program it. Even car dealerships will fade as you order your car from a local nanoshop (no human intelligence works there), which then tells the car which address it should fly to. Large universal manufacturers may reside in factories where they grow large items such as airliners and ships in one piece, giving them a combination of strength and lightness far beyond machines assembled from parts, and at a fraction of the cost. We will discuss the implications of universal manufacturing in more detail farther on.

THE STATUS OF MANUFACTURED INTELLIGENCE

Naysayers seem to take comfort in AI's come-up-short situation these days — no HAL; no robots that walk, talk, and manipulate things; mixed success with expert systems; especially, no computers generating original, innovative thought and conscious forethought the way people do; little achieved after 50 years of effort. Of course, we can invert this argument. Artificially intelligent vans have driven across the USA; AI programs have beaten world champions at chess and checkers; Hussien was ducking smart weapons with his name written on them; AI systems are integral to many businesses and government departments; computers understand and respond with speech programs, evolve complexity from simplicity, and AI/AL systems are innovative in modest ways. So much is happening with current AI that a big fat book on the subject can be written, and some have been. All this and more in just 50 years! Amusing how some dismiss AI because it has not already brought us human-level robots in that short time, a superb example of how we expect so much so fast that we forget how fast it is actually happening.

At this time, the artificial intelligence people are in much the same position as the hopeful fliers of the late 1800s. Balloons were in use, some working gliders were constructed, and a few powered machines made some tantalizing hops. The power needed to fly was unavailable, and the control systems needed to make an aerial machine practical were not only poorly understood, but in many regards misunderstood. Overall, the situation was frustrating and disappointing to the enthusiasts, although realists among them pointed out that matters could only get better.

The AI debate exposed problems and revealed new paths. However, to date, AI skeptics have yet to demonstrate that human-level machine intelligence is an impossible achievement. As far as can be told there is no showstopper.

So what if no computer is truly creative and able to outline long-range plans? The same was true of the biggest brained animals until recent geologic times. After billions of years, a lineage of a specific DNA pattern produced sophisticated apes in shockingly short order. When it comes to cyberintelligence, we have yet to reach the Pleistocene Era (when Homo sapiens evolved). No one has built anything near the calculating speed and memory capacity of a human brain. Nor has a machine with anything near the complexity of a small brain been built. In particular, no one has arrayed many millions or billions of analog-digital circuits into a parallel processing neural network. No one has yet tried to grow and evolve a nanomachine with these characteristics. If such machines are built and year after year, decade after decade fail to achieve human brain capacity, the naysayers will have a case for impossibility. Until then, AI/AC naysayers are speculating counter to the grand trend of SciTech. SciTech enthusiasts speculate as well, but with history supporting their side.

AI Naïveté, Past and Present

Adjusting our perspective backward (with the attendant benefits of hindsight), it is clear that the hopes of early AI researchers were decidedly naive and overly optimistic when computer power was millions of times less than that of brains; when nanotechnology was almost unheard of, and AL was barely alive. Today, as computation speed approaches human levels; as foundations are laid for machines assembled molecule by molecule; and as computers are learning and evolving on their own, naysayers are courting accusations of naïveté and unwarranted pessimism. (Note that this is an example of over-predicting the near-term and under-predicting the long-term. Naysayers rarely remember that their past success does not necessarily continue into the future.) Do the naysayers really believe that in another 50 years, the best AI machine still will be barely able to do such things as beat a human at chess, drive, walk, talk, converse, see, pick up and manipulate an object, or engage in innovative thought? We would not be so bold or foolish!

The Real Truth...

We hope that you now understand that the common view, first spread by a new and nervous computer industry, that computers do only what they are programmed to do is not and has not been true for more than 40 years! Some research and commercial machines are self-learning and evolving systems whose abilities go beyond what their human makers programmed into them. Assuming that computers will always remain under our control because we make them makes about as much sense as saying that a child will always remain under the control of its parents. Just as children remain under parental domination while they acquire mental and physical abilities, robots will remain under our authority only as long as they are our inferiors in mind and body. Which brings us to ask...

How Far Can AI Go?

Funny thing about the failure of strong AI. Those who think its demise marked a major setback for general AI should think again. Only if top-down programming was the exclusive technique to make a machine intelligent would continued human supremacy be ensured. Since computers can learn and evolve on their own, what are the limits to how much they can learn and how far they can evolve? Answer: Probably as much and as far as they want to.

This leads to an observation: Human children most often are reflections of their parents in appearance and abilities, extending but seldom exceeding those of their parents. This is not likely to be true of the evolving robots we will someday bring into existence.

We will continue to explore where AI, AL, and nanotech may take us as the book progresses. Before we can, we need to do some heavy thinking about the nature of thinking.

CHAPTER

Thinking About Thinking About Thinking

Back where I come from, we have universities, seats of great learning where men and women to go become great thinkers. When they come out, they think deep thoughts, and with no more brains than you have. But, they have one thing you haven't got. A diploma!

— Comments by the Wizard to the Scarecrow

The Big AC: Artificial Consciousness

In the previous section, we talked about intelligence but not about consciousness, which is not the same thing. Many of the intelligent things you do, such as walking along without tripping over curbs and bouncing off trees, then comprehending with a flash a solution to the problem you've been working on in the back of your mind for hours, are smart but not conscious. Conversely, consciousness does not indicate or ensure high intelligence. We all engage in thought that is conscious, yet not very bright. We do so "vegging out" as it were, like watching a "Brady Bunch" rerun for the fourth time.

Many involved in AI research do not care whether the artificial intelligences they invent are artificially conscious, as well. They hope AI is not AC because if so, it can get tangled up in ethical questions such as terminating the machines with the OFF switch to shut down consciousness. However, *our* purposes differ. If the robots that displace humans are capable, but mindless, automatons, it will all be a cruel joke. What we are looking for are mind machines that have conscious awareness systems humans can download their minds into, and still end up as aware of their identity as before. Some of those involved in AI research believe that when computers become powerful and smart enough, AC will naturally emerge of its own volition much as consciousness gradually did when organisms evolved brains of sufficient sophistication. It has even been suggested that some robots with sensory systems have a crude minimalist level of self-awareness. Maybe so, maybe not. What we can say at this time is that no one can point to a computer and say, "No two ways about it; the machine is conscious of its own existence."

THE PRIMARY AND KEY ASSUMPTIONS

The hypothesis of the Extraordinary Future presumes that conscious thought, similar to but not necessarily identical to the sort humans possess, can be achieved by a machine artifact of human ingenuity. This is the Primary Assumption. It also presumes that identities can be transferred from natural mind machines into artificial ones. This is the Key Assumption. If these assumptions are incorrect, than the Extraordinary Future is in big trouble. Before we can decide whether these assumptions are sound, we must first consider some basic questions concerning brain structure and function, the mind-brain connection, consciousness, and the nature of identity.

THE OVER-AWE FACTOR

We tend to be awed by our self-aware brains and the minds they contain. In fact, there is a tacit conspiracy to emphasize the power and complexity of brains over basic simplicity and built-in weaknesses. Theologians believe that the brain is a product of God no mere mortal technology can match. Ivory tower philosophers want the brain-mind system to be too complex for science to match. Even scientists and technicians, feeling a need to compete with theology on one hand and impressed by the real world performance of human minds (vis-à-vis today's computers) on the other, have lofted the human mind to a relic of near-worship, stressing the brain's complexity and power as the derivative of an evolutionary process.

Half science.

And the Daily Illusion

"It's just so obvious," we hear. "The conscious experience is such an extraordinary one that there is no way it can be the product of any machine, no matter how powerful. How can you compete with good old intuition? Maybe the body is a machine, but not the mind. How can any circuit board be aware of itself?" These beliefs are reinforced by theologians who believe the brain is a device God inserts minds into. Mystics explain the human miracle by linking our minds to mysterious forces of the universe. Philosophers unable to resolve the brain-mind connection conclude that there is none. The great majority of the global population does not believe that the minds they are thinking with are produced by, and only by, the brains in their skulls. Again, it soothes our egos to think our minds are directly connected to forces great and universal.

The Machine Behind the Mind

Saying that the mind is produced by a machine we call the brain raises hackles in the minds of many, and it is true that no one can yet prove beyond all doubt that the mind is not produced outside the brain. But if the brain is not a mind-making machine, someone or something sure went to a lot of trouble to make it look like one. The process of thinking and the results of thinking seem to be the exclusive function of the brain.

Not much was known about brain function 90 years ago. There were two competing theories. One was that the brain was a sort of general purpose machine in which thinking was spread uniformly over the system, with few if any particular places playing a special role. A reasonable idea, but disputed by others who thought thinking was performed by special subunits in our brains; different parts of the brain did different things. This was a reasonable idea as well, and seemed to confirm earlier observations on brain dysfunction. The problem was that the observations were few and far between, and science does not thrive on small sample sizes. No single set of researchers had access to a large sample of people with an assortment of brain injuries. This was about to change.

THE RUSSO-JAPANESE WAR

In 1853 and '54, Commodore Perry forced Japan to open its markets to the commerce of the world at large. It seemed like a good idea at the time. Looking upon the rest of the planet, the Japanese decided that another island nation of industrious folk — the British — presented an outstanding model for the development of their own modern society. The Japanese soon became copycat colonialists building their own overseas empire. It seemed to be another good idea. Manchuria looked like a suitable place to start, so early in 1904, Japanese naval forces launched a surprise attack on Russian naval forces based on the Liao-Tung peninsula west of Korea. In this first brief, but dreary and bitter, mechanized war of the 20th century fought by commanders who had little concern for their troops, the casualties were horrendous.

Trench warfare was common and as a consequence, serious cranial wounds were often inflicted on soldiers peeking up and over parapets. To make matters worse, steel helmets had not yet been adopted. With so many soldiers with head wounds, wounds affected all parts of the brain in one way or another. Many brain lesions were narrowly confined, while some affected many areas at the same time. Identical injuries provided a baseline for comparative purposes, allowing continuity of cross-checking between the specific brain damage causing specific mental impairment. This steady supply of brain-injured soldiers provided a grisly opportunity for the first neurosurgeons to acquire knowledge and skills, not the least of which was mapping the anatomy of brain functions.

LOOKING UNDER THE HOOD

The difficulties encountered when studying the goings-on inside human brains are obvious. You cannot just open up someone's skull and poke about while your subject

Science.

goes about its daily business. The brains of other mammals could be studied more intensively, often with the kinds of experiments animal rights people find most objectionable, but the subjects can't talk about their minds. Besides, their brains differ from those of humans anyway. Electrodes placed on the scalp or into the brain can be used to glean information about the real-time operations of brains human and otherwise, but the information is limited because not very many electrodes are used, limiting resolution. Another way to study brain function is to remove neurons, culture them, and see what they do. In this way, tiny little neural networks can be set up. Chemicals that alter brains in specific ways are also used and the results observed and reported.

Information about brain anatomy and function is gleaned via a direct, though macabre, technique. While a patient is fully conscious, a neurosurgeon can open his or her skull like a box to get at the brain. As the surgeon explores the wet gray tissue with tiny electrode-tipped needles, the owner of the exposed brain, which has no capacity for sensation or pain, reports on internal states and sensations. Years of such probing and reporting have improved our map of the anatomy of the human brain. High-energy physicists may be able to extract huge amounts of information by slamming protons together at near light-speeds, but information from surgery and trauma-damaged brains tells us only so much about normal functions. Direct information from a normal working brain is the grail of neuroscientists and SciTech deliverers.

The brain is like a "black box" (not to be confused with the black box retrieved after a plane crash). It comes from Air Force parlance, however, and means any system (usually hardware in a metal box) where information is cabled in, something happens in the box, and useful results are cabled out to devices. The term "black" does not refer to the color of the "box," but rather to the unknown goings-on inside the box. Information comes in from our senses, is processed by our "black box" brain, and flows out to activate our bodies.

Remote scanning technology gets us into the black box and allows researchers to observe specific portions of the brain responding to specific stimuli with no invasion, or even disruption, of tissue. Shades of "Star Trek"! Sporting names like PET, CAT, and MRI, these amazing new technologies allowed us to double our knowledge of the brain over the last ten years, making the 1990s the officially designated "Decade of the Brain." New discovery feeds on itself, and we expect SciTech to deliver better instruments in the near future.

The remote scanning systems cannot directly watch neurons firing. A Positron Emission Tomography scanner, for example, maps oxygen use in various parts of the brain. It is presumed that where neurons are firing in large numbers, large amounts of oxygen are being metabolized by the hard-working neurons. The newer Magnetic Source Imaging gives even better 3-D resolution of brain activity. It is hoped that in coming decades, the remote scanners will allow high-resolution brain scanning down to perhaps the neuron level, while healthy patients are thinking thoughts and performing mental tasks. Even with current scanners, a lot of work can be done. Ask a person whose brain is being watched by a PET scanner to speak a few times and see the same part of the brain activate time after time. Then ask the person to write out a series of words and see a different part of the brain working. Using remote scanning, it is now possible to map out and investigate the overall mental function of healthy intact brains.

There is another way to figure out how a black box works. Build another black box that works like the first one. Because you built the one black box and can easily get back inside it, you know how it operates, so you now have a better idea of how the first one works. Computer scientists are joining neuroscientists to build machines that are constructed like and perform like brains.

Using these and other techniques, much about brain function is already understood. Small parts of the brain have been mapped in minute detail, and the information has been used to construct intricate microchips that can mimic the actions of the copied brain areas. When little was known about actual brain function, superficial comparisons were made between brains and human-built systems. A favored metaphor in the first half of the century compared the brain to a telephone switchboard; then the brain was likened to a conventional computer. More recently, the big electrodigital mainframe computers were the most common analog. As we learn more, it is becoming obvious that the brain is, in truth, a jerry-

Your Friend, the Brain

These are simplified diagrams of the only device known to produce high-level conscious intelligence — so far. In the upper right is a complete brain with the large left hemisphere, the smaller cerebellum below, and to the back the central brain stem in black. The figure in the lower right shows a brain cut in two halves, leaving the right side and exposing the inner surface of the right hemisphere (which is connected to the other hemisphere by the corpus callosum, indicated with angled hatching), central brain elements, and the complete brain stem. To the left is a vertical, transverse cross-section of the brain through the hemispheres, and the upper end of the brain stem. The hemispheres are directly connected only by the corpus callosum, which is just above the Y-shaped space in the transverse section. The neuron-dense outer cortex is a thin, multi-layered sheet that covers the outer and inner sides of each hemisphere. The cortical sheet is highly folded to greatly expand its area, thereby further increasing the number of neurons. Different sections of the cortex are responsible for different functions. For example, most of the visual cortices you see with are located at the back of each hemisphere, mainly on the inner surface. Under the outer cortex, the hemispheres' white matter is made of connections between neurons in different parts of the brain. For example, the bodies of the thalamus are set near the centers of the hemispheres. This allows them to send out connections throughout the hemispheres, and these connections may play a key role in consciousness. The Y-shaped space at the center of the brain may contain chemicals important to brain function, including the generation of emotions.

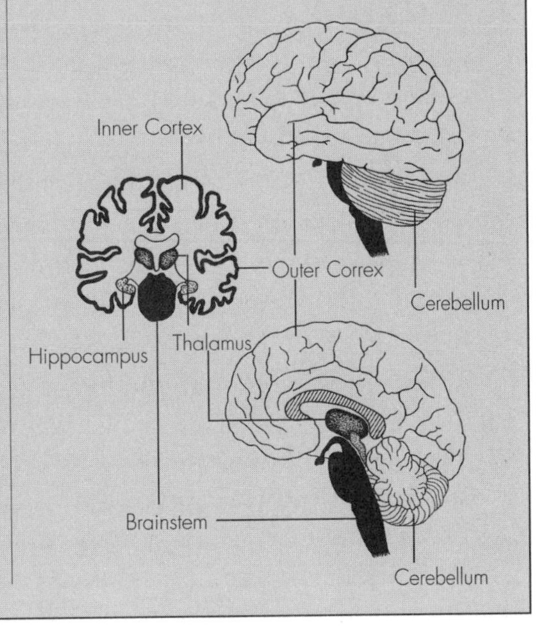

rigged biocomputer that develops and operates according to Darwinian principles. What is known about the brain would easily fill a big and fascinating book (some are listed in the book section), so the following sections can be only a brief description of brain structure and function as it pertains to our thesis.

SIZE AND ENERGY

As you sit reading this, your brain is running on just 16 watts of sugar and oxygen power at a cool 99° F. (By the way, the higher your IQ, the more efficient your brain is and the less energy it uses. Inefficient brains expend more energy to do less thinking.) The energy needed to run your brain is a substantial 20 percent of your entire metabolism. In terms of size, your brain is more a PC with legs than a laptop — it weighs a fairly hefty 3 lbs., but it needs a torso support system weighing dozens of pounds (we exclude the legs and arms because a brain can do without these).

A GALAXY'S WORTH OF NEURONS

The number of neurons in a brain is about the same as the number of stars in our Milky Way galaxy. A human brain consists of about a trillion cells. Of these, less than a fifth are neurons. To help wrap your mind around the quantity described by a number as large as a trillion think of the following: A million seconds passes by in 11 days. How long does it take for a billion seconds? Don't turn on your mental calculator and begin multiplying. Use the seat of your temporal pants. If a million seconds is 11 days how long does a billion take? A trillion? Once you know the value for a billion, a trillion is easy, just multiply by 1,000. But the sense of scale is shocking when you finally realize that a billion seconds passing is not measured in days like a million seconds, but in years — 32 years — and a trillion seconds takes 32,000 of those years to pass.

Although it is true that you cannot grow new neurons and you normally lose a few hundred thousand neurons a day to wear and tear, not to worry. After 70 years of life disease-free brains retain 70 percent to 95 percent of their neurons depending on what part of the brain is involved. Brain mass can decline by 30 percent, but PET scans of old brains show that they are almost as active as those of younger brains. However, the bad news for the guys is that the brains of men may deteriorate faster than those of women! This is why the mental performance of healthy elderly people, although not up to the standards of youth, is still quite good. The way to degrade long-term brain function is to accelerate neuron death rate with heavy drug use, exposure to certain chemicals, or with a neuron-destroying disease such as Alzheimer's. In such cases, the inability to regenerate neurons is bad news.

Your Friend, the Neuron

Here are represented just a few of the hundred billion neurons in the outer cortex of the brain. The outer surface of the brain is at the top; descending are six irregular layers of neurons. In the lower right foreground is a close-up of a pyramidal neuron with a nucleus (which contains the most heavily utilized DNA in any cell); axons emerge from around its base and a dendrite from its peak. The dendrites of the deeper-placed neurons can reach up through all the layers of the outer cortex. Neurons are surprisingly large, being one-fifth the diameter of a hair. Synaptic gaps that connect the axons-dendrites of different neurons can be seen, including a close-up in the lower left. Other axons and dendrites reach a dead-end. The neurons are suspended in a matrix of supporting cells, blood vessels, etc., which are not shown here. This figure only hints at the complexity of interconnections between neurons.

The inner workings of neurons remain one of the least understood, yet most important, systems within the brain. A lot is going on inside them, more of the human genome is activated within a neuron than in any other cell in the body.

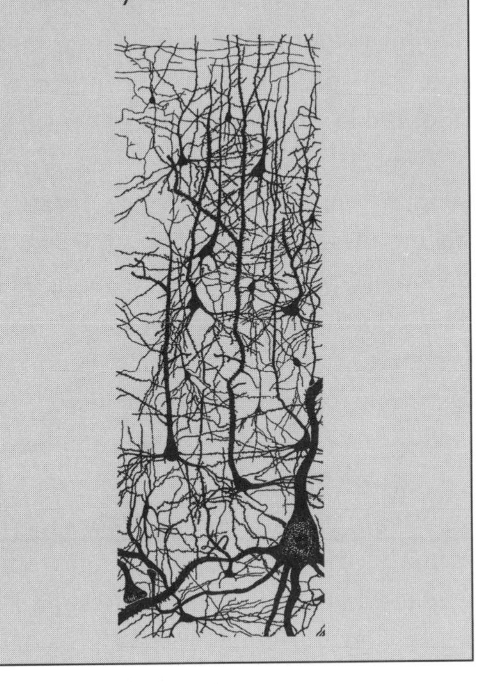

As impressive as the number of neurons may seem, the real complexity of the brain is found in the number of synapses. Axons and dendrites, the in/out cables of "black box" neurons, connect to each other via synapses. A synapse is a microscopic gap across which information is sent via chemicals, within a tiny fraction of a second. Each neuron reaches out to other neurons — some close, some far — with multitudes of axons and dendrites. A single average neuron can communicate directly with 10,000 other neurons! The number of synapses in a brain is about 1,000 trillion. But don't get smug carrying around so many connections in your head; there are millions of times more stars in the observable universe than there are synapses in your head.

THE CALCULATING CIRCUIT BOARD WE LIKE TO CALL A BRAIN

The machine structure of the biocomputer is revealed by the cortex, the folded gray matter you see on the surface of an exposed brain. The outer cortex is only a couple of millimeters thick and, when unfolded, covers a couple of square feet. The number of neurons in the cortex is around 100 billion. But there are only a few kinds of cortical neurons, and they are arrayed in just six main layers in the thin sheet that makes up the cortex. The neurons are connected in parallel rows by trillions of synaptic connections that make up the thick white matter of the brain. The two hemispheres of the brain the gray and white matter are connected by one thick cable, the corpus callosum.

Now, what has just been described is a computer circuit board. The neurons, synapses, and nerves are electrochemical, data-transmitting storage and switching circuits and lines. The neurons communicate with electrochemical pulse signals that are dualistic in being both analog and digital in nature. The signals are digital in that neurons are either off or on, and when on, the signals include discrete and consistent spikes. The signals are analog in that they do not come in distinct packages that each denote one unit, but instead, fluctuate in peaks and valleys. All this describes an analog-digital computer doing fuzzy calculations to solve algorithms.

The brain operates as a computer at different levels. Each neuron is a microcomputer receiving information from the outside and performing calculations to decide whether and how to respond with signals of its own. The communicating neurons form the macrocomputer doing trillions of calculations. The outer cortex is the part of the macrocomputer where we do most of our sensory processing, consciously control body movements, and execute most high-level thinking. It is the recent expansion of the cortex into two hemispheres that makes us special (elephants and whales have cortexes of equal or larger size, but their neuron density seems to be considerably lower). However, there is much more to the brain than the two hemispheres, which make numerous connections with the more primitive parts of the brain and the brain stem. There are tens of billions more neurons in the brain stem. Many important mental processes go on there deep in the base of the brain where certain aspects of memory are handled, parts of consciousness may be processed, and involuntary systems keep the body going as we think about esoteric things.

Calculating Power

The remarkable performance of the brain does not come from the signal output of any individual neuron. An average neuron is a shockingly slow circuit that runs at a mere hundred or so cycles per second. Nerves don't transmit signals at high speeds either. At about 100 meters per second, nerve signal speed is millions of times slower than electricity traveling through a wire. The great processing speed of the neural network comes from billions of neurons talking in parallel to each other over trillions of connections each second.

No one is sure just how fast human brains work; they're still something of a black box in that sense, too. Estimating brain speed presents a fundamental problem. It is not a precise bidigital number-crunching computer, but more a fuzzy analog-digital machine. This combination is very efficient and information is dense, but rather sloppy. At best, we can roughly estimate how much computer power it would take to mimic brain performance with a strictly digital machine. Some estimates are based on the number of working neurons at any given moment during intense thought (about 10 percent maximum) — the number of their connections and the firing speed of neurons and synapses. Others have tried to figure out how many calculations would be needed to do what your mind does. For example, it will take a computer running at a trillion calculations per second (one teraflop, a flop equals about 8 bits of data) to generate a fully realistic visual virtual world in 3-D. It should take even more brain power to receive and then analyze the real 3-D world while doing other functions at the same time. Estimates of brain speed range from 10 trillion to 1,000 trillion calculations per second. (Some higher estimates have been made — see below — but it is hard to see what humans would be doing with more power or how they would produce it.) The 100-fold range estimate is the best we can make at this time.

An interesting and unanswered question is how much thinking speed goes into what function? In particular, it is not known what percentage of your brain's power goes into intellectual pursuits at any given moment or when you are doing something high-brow such as math, playing chess, or considering the implications of intelligent computers. It is generally thought that vision eats up a lot of one's brain power, the other senses, less, and that conscious thought tags along at the end, but this is, by no means, certain.

Memory Storage and Capacity

In most computers, the data storage and processing systems are separate from one another; not so in the brain. The same neurons and synapses that process information also store it. In particular, it appears that synapses store memories in the form of long-lasting chemical and structural changes that alter the strength of connections between neurons (more on this below). Some estimates of the information we store are as low as 25 billion bits (2.5 gigabytes), which amounts to a million pages of text or a few hundred books. Such low figures, which are no greater than those found in high-end PCs these days, are certainly too low. If you sit down and start randomly remembering things big and little from your life, even dreams, it quickly becomes clear that the amount of data stored is vast. You could go for weeks rehashing memories, most of which would consist of motion pictures each requiring billions of bits of data. In well-balanced computers, the memory capacity is broadly comparable to the per-second speed of the machine. Studies that correlate the number of synapses and bits of stored memory in very simple creatures indicate that each synapse can store one bit of information. Ergo, our stored memories rest on a foundation of 1,000 trillion bits of data, give or take a few hundred trillion.

As big as this figure is, it underestimates the memory capacity of the brain because each synapse can contribute its information to more than one remembered event. How? Say that a number of recalled memories include the same bit of data that pertains to your sixth grade school room. That bit of data is fixed in a certain synapse, which is activated when two particular neurons communicate in a certain manner. Every time you recall something that happened in that room, that synapse does its job. Because memories are stored in all those synapses and because all those synapses are distributed all over the brain, your memories are spread over a broad area of the brain. The decentralized way brains store data is more akin to a 3-D hologram than to the centralized hard disk in your computer, which stores each bit of distinct data digit by digit in just two dimensions.

Your Neural Network Grows and Evolves Just Like Darwin's

Many aspects of brain growth and structure are genetically preprogrammed, including the anatomical structure and form of the brain, the microscopic arrangement of the retina, and perhaps certain abilities such as the virtual imperative to acquire language. However, much of the brain evolves in a bottom-up fashion. This applies most of all to the development of synaptic connections between neurons in the two hemispheres both during juvenile growth and for the rest of life.

The strength range of any given synaptic gap spans from very strong to very weak. In general, the more a synaptic gap is used, the stronger it is. New synaptic gaps can be developed between neurons when needed. Conversely, synaptic gaps not strengthened by use gradually fade away. Stimulated neurons make connections with other neurons and, when reinforced, strengthen. Synaptic connections not stimulated by frequency weaken and eventually disconnect (which is one reason we forget as we age). This is the original version of a biological neural network; a competitive system where some connections are selected over others. As you wend your way through life, your mind literally evolves!

Brains "evolve" most readily when they are young and synaptic connections are easily established. This helps explain why kids pick up a native language with little apparent effort or conscious thought. The decreased ability of older brains to "evolve" helps explain why adults can learn new things, but it usually takes more conscious effort because recent memories are harder to retain.

A big advantage of neural networks is its excellent resistance to damage or disruption; much more so than centralized computers made vulnerable by rigid architecture. Because the information storage and processing systems are dispersed, neural networks are inherently redundant and flexible. Taking out five percent of the network at random will degrade overall performance by five percent at most, perhaps even less. The systems don't crash because there is no keystone component that shouldn't be removed, and other parts of the network take up slack. The network nature of a brain explains why we can lose irreplaceable neurons every day and still have an operating mind after eight decades of life.

When Brains Go Wrong: Part 1

Even in this age of high-tech scanners, there is still information to be gained with the original low-tech way to "peek" inside brains to diagnose and devise therapies for the victims of serious head trauma. The brain is all too vulnerable to damage. Not surprising, considering the outer cortex, with its one hundred billion neurons, has the consistency of firm jelly! Our skull is the brain's crash helmet. It is good that the brain is a redundant neural network. However, the brain is not an ideal, uniformly dispersed network.

When something destroys or disables a particular part of the brain, the diagnosis proceeds directly from information learned by evaluating the mental abilities of the unfortunate victim. Rehabilitation therapy provides additional information. Each patient, and especially each trauma, presents a unique situation, and designing a course of therapy is as much an experiment as it is applied medical science. Mental function or dysfunction is revealed or hidden by the therapy designed to help the patient recover. The results of the "experiment" suggest further refinements in both diagnosis and theory.

One early and spectacular mind-affecting head trauma case was that of Phineas Gage. He was an industrious and well-liked railroad worker until 1848 when a metal tamping rod was accidentally blown up through the bottom of his chin, through his forebrain, and out the top of his skull. There is an account of the sickening amazement the attending doctor felt when two of his fingers contacted one another from different ends of the hole. The semi-lucky Mr. Gage walked away from the accident with his basic cognitive abilities and intelligence intact. But he was never again the same person. He immediately became violently anti-social, couldn't hold a job, and drifted from place to place until his early death. It seems that the part of his forebrain responsible for moderating emotions with rational thought had been wiped out and Gage no longer had, literally, the brains to be a part of normal society.

By examining many traumatized brains, Japanese scientists at the beginning of the century and other investigators since have found the effects of local brain damage are often eerie. Some who have suffered injury or disease to the forebrain find their intelligence and ability to reason intact, but their ability to feel emotions or comprehend those of others is gone. The result can be spouses able to recognize and interact with their families on a basic rational level, but unable to feel anything special about them. Damage to other parts of the brain leaves patients who seem normal — they can read, walk, and talk well, they just cannot recognize anyone, including their spouses or a photo of themselves. A massive stroke in the brain's right hemisphere will not only leave the victim without control of the left side of the body, but it may also leave the patient without full awareness of the existence of that side of the body. If they're asked about their left arm, they deny that it is theirs and even assign ownership to someone else in the room! Shown a picture of a cat, one stroke victim may be able to say that it is a cat, but is not able to write the word "cat." Another patient may be able to write "cat," but is not able to say it. Still a third victim may say that it is a small furry mammal often kept as a pet, but be unable to say or write "cat"!

There are people who you can throw a ball to and they will catch it. When you then ask them whether they caught the ball, they will deny that they ever saw or caught the ball and will not admit that they are holding the ball. Their brains subconsciously did what was necessary, but they were not conscious of what they did.

Certain people suffer from epileptic seizures so severe that only the most drastic action can provide a remedy. In these extreme cases, the only relief that can be provided consists in the surgical severing of the corpus callosum, the superhighway of nerve fibers about as thick as a thumb connecting the brain's left and right hemispheres. The effect is to leave the patient two, largely independent, conscious minds running at once, each vying for dominance. It would be hard at first to notice people who have had this operation. They often lead fairly normal lives. But careful testing has been designed to challenge each hemisphere, which receives separate sets of signals from the parts of the body it supervises (in general, the right hemisphere receives info from the left side of the body and the right hemisphere from the left). The result is unresolvable confusion. Asked if one hand is wet, the person may respond "yes, no, no wait, yes!" repeatedly as her two brains — only one of which knows the truth — try to answer. Handed a sheet of paper with "yes" and "no" written on it, the person may indicate "yes" with one forefinger while pointing to "no" with the other.

Once part of a brain is gone, it's gone forever. Whatever mental process was controlled by that part of the brain is at least severely compromised and at worst, may be lost. The brain can slowly try to reconfigure itself so that a new section of the brain generates the new mental skills needed to do what the lost part did. Partial loss of speech or mobility due to strokes, for example, may be compensated in part by long-term therapy, but these jerry-rigged mental crutches are almost always less capable than the original system. One example would be very young children, whose growing and rapidly evolving brains can rapidly readjust to the most shocking disruptions. Remove an entire hemisphere of a toddler, which is sometimes done to stop chronic seizures, and there's a good chance the child will can grow up with good mental and physical abilities.

The fact that doctors sometimes remove large portions of young brains in the expectation of improving long-term performance illustrates how much we now know about brains and their workings. Brain surgeons are confident of the results of what they do because the results of brain dysfunction are highly consistent from person to person. Damage or remove the same part of the brain in a large sample of people, and pretty much the same consequences will ensue. Mental failures and alterations come in distinct patterns indicative of specific system failures and alterations. The close correlation between disruptions to the physical brain and mental diseases confirms the close link between mind and matter.

Cerebral Subsystems

Observations of brains in distress help make clear what remote scans of working brains reveal. The brain is a set of distinct subunits functioning semi-independently in cooperation with other subunits. Many subunits are spread out over the outer cortex of the hemispheres. These include the visual and auditory regions of the cortex, various motor regions that work with assorted body parts, and the frontal region in which different forms of high-level thought — including speech — are handled by additional subunits. The two hemispheres themselves are distinct, asymmetrical subunits. Many mental operations favor one hemisphere over the other, language being one. The various parts of the brain stem are themselves made up of distinct subunits that handle specialized tasks.

In normal brains, the segregation of the organ into subunits is highly consistent from person to person, male to female. For most of us, the visual cortex is situated at the back of the brain. The anterior frontal lobes handle high-level thought and so on. The size of these areas varies somewhat. Graphic artists tend to have a larger visual cortex, musicians a larger auditory cortex, and so on. Extreme divergence from the usual occurs.

Every once in a while physicians come across a person whose mental performance and medical history appear to be normal. To the doctors' surprise, a brain scan reveals a brain of extremely unusual shape or size (perhaps you have heard the urban myth of the ordinary fellow who had almost no brain, just a brain stem and cerebral fluid inside his skull; one needs about half a brain to function). In such cases, the very young developing brain, even one with a surgically removed hemisphere, somehow re-configures itself with whatever parts are accessible and capable of performing the necessary function. The brains of some autistics have unusual unit distribution. In some, the area of the cortex devoted to mathematical processes is unusually large, allowing remarkable feats of mathematical ability apparently at the expense of other functions. Other autistics may have larger-than-usual portions of the brain dedicated to music, 3-D rendering, or some other talent, again at the expense of other abilities.

Typical computers are general-purpose machines in which any circuit can do any calculation any other circuit can do, and the overall machine can do any basic mathematical calculation given an algorithm that fits its architecture and enough time. Brains are more specialized devices, internally and in overall performance. However, the combined coordinated performance of the brain's many subunits allows flexibility over a broad range of tasks well beyond those of any current non-biological computer.

Scatterbrained

The combination of dispersed memory storage and task-specific subunits means that acting or thinking requires the acquisition and correlation of information from different parts of the brain. If you ponder a pencil, you think about its color in a part of the brain specific for color. Move your thought to the function of the pencil and the thought location moves a few centimeters to the region of tools and their functions. If you use the pencil, neurons in the motor cortex go to work. We will consider below how the brain compiles all this into a conscious whole.

Complexity and Simplicity All Rolled into One

For all its complexity, some features of the brain show that long-term selection has worked to keep it as simple as is feasible within the limitations of Rube Goldberg evolution. The cerebral cortex is the premiere example of this effect. We have already discussed how the billions of cortical neurons are not arrayed in a deep, complex, three-dimensional pattern. Instead, the neurons are generally arranged in only six layers in the shallow, superficial cortical sheet. This remarkably simple arrangement shows that as the evolving brain expanded to increase capacity, the crucial cognitive system has been kept as simple and as two-dimensional as possible. Also, there are not millions or thousands of kinds of neurons, just a few dozen major types. No matter what their shape, all neurons signal identically in terms of power output, cycle speed, etc.

LOVE IS CHEMICAL. WELL, ELECTROCHEMICAL

They say it's chemistry when two people fall in love, that passion is electric. These two pieces of folk wisdom are accurate. All our emotions, from joy to anger and everything in between, are nothing more than the products of chemical reactions that stimulate neurons to communicate electrically and alter the chemical information flow over synaptic gaps. The washing of various hormones and neural transmitters over neural connections can dramatically alter our moods, for better or worse, in mere seconds.

BETTER MINDS VIA CHEMISTRY

The connections principle means the same technologies used to cure traumatized brains can also be used to modify normal minds. Heroin and LSD were originally developed for therapeutic purposes. Such extremely powerful, and primitive, drugs have serious downsides. Prozac is more interesting because it is the first "designer" drug invented to alter specific brain chemistry with minimal side effects. Developed as an antidepressant, its use has quickly expanded to that of a basic mind modifier used by millions of mentally healthy people to make themselves happier, less shy, and more assertive. Whether Prozac pans out in the end, another marker into the sci-fi future has been passed.

Drugs such as Prozac are different because their effects are sustained and subtle. You can tell when someone is drunk, but you have no clue that the woman in sales who suddenly saw her statistics go through the ceiling owes her success to a mood modifier. The question for the future is, as society becomes even more dull and a host of new designer mind-altering drugs become even more effective and specific, will legal drugs evolve into a common way of alleviating the tedium? In particular, will people who find themselves bored and unfulfilled by 21st century technology and religion alike turn to powerful, sophisticated, yet subtle, drugs to put themselves into a perpetually good mood?

In a recent issue of *National Geographic*, the author of an article on the brain was distressed to learn this solid fact of neuroscience from a woman who is one of the leading researchers in the field. Another researcher, Charles Darwin, argued that the act of smiling can improve our mood, and recent evidence suggests that changes in facial expressions do, in fact, initiate chemical messages altering how we feel. The machine of the body can alter the mood of the mind! The use of electroshock therapy to alleviate depression is further evidence of the electrical side of emotions.

Likewise, the chemical basis of emotion explains why drugs are such powerful mood alterers. We all know about Prozac, the most effective of a series of antidepressants that alters mood and emotions by altering brain chemistry. Prozac exists because we now know the chemical basis of emotions so well that we can design potent emotional controllers. Fifty neurotransmitting chemicals have been identified, and a number of them, such as calcium and serotonin, are well-understood. The mystery of emotions has been stripped away to reveal the mechanism of emotions.

The Virtual Reality Machine

So you think you are a pretty realistic person with both feet planted firmly on the ground? That you see reality for what it really is? You really do think that; but it's a fantasy, a nice fantasy, but a fantasy nonetheless. Although it is true that some are more in touch with reality than others, the truth of the matter is that no mind actually sees, hears, feels, or smells the real world. What you see, feel, touch, etc. is what your mind *thinks* is the actual experience.

Most people have a tough time believing this but a demonstration is really quite easy; let's use vision. It seems that you see actual objects out there, in front of your eyes. What you "see" are photons projected from or reflected off of actual objects. The photons are absorbed by biomechanical devices shaped like cones and rods. The stream of photons we call light hits your retina initiating chemical reactions in the cones and rods, which, in turn, *initiate* the process of visual perception.

After recording the incoming photons, the retinal cells (which are literally extensions of the brain) do a set of calculations that encodes the information into the mix of analog-digital signals the brain works with. This information is then electrically sent down the optic nerve. Along the way, more information processing is done. The coded information then arrives at the back of the brain in the visual cortex, which makes up a large portion of the neural circuit board. There, differentiated sheets of neurons process the information, using parallel computations, one for color, another for contrast, another for depth perception, another to correct distortions of your visual field, and more. The specificity of these subunits is astonishing. One set of neurons specializes in recognizing stripes, another bodies, and yet another faces! (There is evidence, by the way, that we see the world as a series of images, just like a motion picture.) Then and only then, a fraction of a second after the light hits your retina, is all this information decoded and recombined to form an image your brain has constructed for you.

Why are you conscious of vision in the back of the brain rather than at the retina? Because if you saw reality, you would be in a fix. First, you would get two discordant flat images. The only way to see one world in 3-D is to carefully correlate the binocular images your pair of eyes supply via complex algorithms, and that takes time. Have you noticed that only a small part of your field of vision is in focus right in the center? The rest is soft, out of focus. There are other things going on that you are never conscious of, but taken together, explain how you see the way you do.

Your brain has the eyes constantly scan the tight focus region back and forth over the field of view. Actually, it is impossible to keep your eyes completely still. The brain compiles focused images and updates the soft focus areas, extending the sensation of clear, focused vision. There is a visual "hole" made by the cable of nerves and blood vessels as they pass through the retina near its center. Again, the brain fills this in with eye scanning and does it so well that you do not notice the gap unless you look for your "blind spot." As light photons pass through the lens and fluids in your eye, they become distorted like sagging glass in an old window. The brain corrects for this and dewarps the visual field. People who wear glasses report slight distortion when they first start wearing a new pair of prescription glasses. The brain soon gets used to the new lens and makes fine-tuned corrections.

The most dramatic modification of the "objective" world is the flipping upside down of the visual field, the eye acting like a lens in a camera. The image presented to the brain is inverted, but our "seeing" of it is not. There have been experiments with people wearing "upside down" glasses long enough that the brain turns the image upside down, so to speak, and the image you see through the glasses is right side up. As you might guess, when the glasses are removed the world is upside down again to the naked eye!

You do not actually "feel" things at the tips of your fingers. In this case, nerves under the skin are stimulated; they send the information a long way to the brain where it is processed before neurons become conscious of the sensation. Likewise, you are not conscious of smells in the nose, but deep in the olfactory regions of the brain. You are not conscious of sounds in the inner ear, but in the auditory cortex.

You see, hear, and feel in and only in the brain. The retina, inner ear, and nerve endings are receptors sending the sensations felt by your nervous system to your brain to be the raw material used to elaborate its version of reality. All outside information getting into your brain comes through these remote senses. Your mind is not in direct contact with the world, but sits somewhat distant and partly isolated from it. In effect, your mind simulates the real world of your sensations, which are apparently the result of interactions with something "out there"! Your mind is a form of virtual reality, a product of a VR machine that takes incoming information and very quickly makes a world view out of it. It is a world in our head that is a model of the real one. This is not a part of your everyday awareness because your mind does such a good job of matching up your mentally simulated world with the real world. You think you see and feel your finger scratching your nose, but actually, you reconstruct the event in your mind after it has happened! Just how long after depends on the most reliable data. Estimates of the reality-to-perception delay range from one-twentieth to a full half second! Your mind does not know the difference between the real world and the one it makes up because it has not, after all, ever really encountered the former.

Any time your consciousness drifts away from the here and now, you wander into a virtual world of your own making and re-making, instant to instant. When you remember something, you are in no way calling up the actual past; you'd need a real-time machine to do that. The remembering activity occurring in your brain is happening now; it always does. The memory of last night's kitchen fire may be so vivid you can almost hear the sound of the extinguisher while you are remembering the scary event, but smoke doesn't come pouring into the dining room again while you remember, the memory is virtual.

The virtual world of our continuous mental creation resembles another virtual world with which we are more familiar: dreams — daydreams and the night variety, too. In dreams, we construct a new world in the brain. What is the difference between the simulated mental world when we are awake and the simulated remembered or dreamed worlds? In the former, the world's information immediately comes in through the senses. In the latter, the information is produced from "goods on board" by the brain in real time. That's about it. The waking and dream worlds we simulate are, in most cases, the same areas that process for consciousness (we know this via remote scanning of alert and sleeping brains). When you are con-

Having Fun with Optical Illusions — The Blindspot

Close your left eye. Hold the page at normal reading distance from your right eye. Focus on the black square on the left, and keep doing so. Move the page toward your eye until the white square disappears! This should occur at about 10 inches, and you may have to move the page towards and from your eye a few times to best observe the effect. You are experiencing the mind's system for compensating for a defect in your retina. Because the light-receptive retinal cells are placed behind the blood vessels and nerves that serve the former — a bad arrangement all round — the latter have to bore through a spot in the retina to get out of the eye, leaving the famous blindspot where no cells are there to receive the incoming light. Obviously, going around with an empty space in each of your two fields of vision will not do, so the visual cortex goes to the trouble of filling it in for you. Notice that your brain actually fills in the gap with more of the little background dots! All courtesy of the virtual reality system that is running inside your skull.

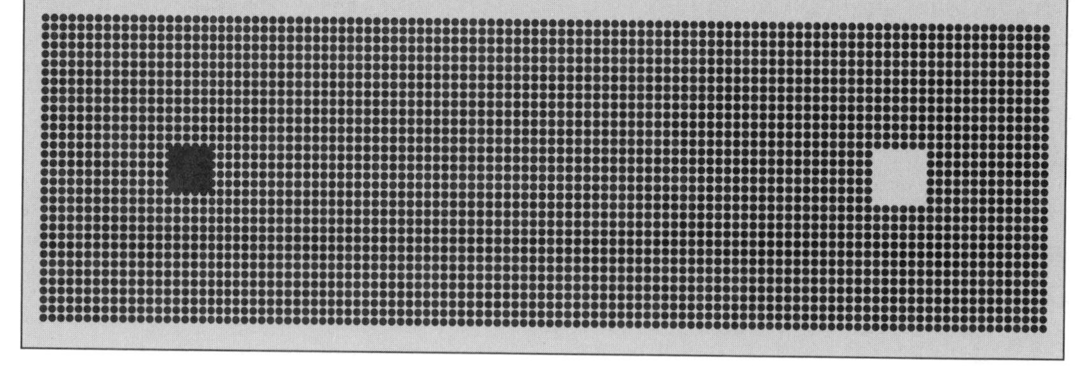

scious of a visual image during a dream, the same parts of the visual cortex activate that you use when you are watching something when awake. This is why some dreams can have crystal clear, fully convincing reality.

One of the most informative examples of cerebral virtual reality is perfect pitch, or PP. Perfect pitch is not only the ability to sing precisely in tune or tune an instrument. PP is also the ability to "play" whole pieces of music in your head with perfect fidelity and clarity so great that some with PP disdain stereos, finding them inferior! Some great composers, the deaf Beethoven among them, had PP (but most musicians do not). Most of us make do with muddy tunes rattling around in our heads, rarely does an inside number piece play itself with rousing precision.

Having Fun with Optical Illusions — Two Heads and a Vase

Is it a vase, or is it two head profiles looking at one another? This classic optical illusion shows that we can derive an image of a three dimensional object from a flat pattern of lines and dots, and can even construct a concept of a face from empty white space. This is possible only because the information from the retina is processed via algorithms in the visual cortex. Also notice that you cannot "see" the vase and the human faces at the same instant, your perception tends to flip from one alternative to another. This means that to be fully conscious of something, you have to pay attention to it. Magicians take advantage of this effect by using side displays to divert our attention from their slights of hand.

In order for intelligent cybersystems to be as aware of the world around them as we are, they will have to be as able to "see" faces and vases that are not really there as well as we do. It just so happens that various computer vision systems — some bidigital, some analog-digital mimics of the retinal-cortex complex — are learning to do just that in a fashion that is primitive, but getting better all the time.

Perfect pitch shows how neurons communicating over synapses can, by their little but many selves, produce a reality that is as real as reality.

Our brains possess an innate ability, most of the time, to tell the difference between the various internal simulations and a "real world." Over time, mutually shared memories can diverge. A married couple, remembering when they first met years ago, may experience divergent memories. They might recall events (where they were for their first kiss, the name of the theater they went to on their first date, etc.) with diverging memories. Their brains are built and work pretty much the same, but are not identical, and their simulations of reality are different. It is not even certain that two people see the same "blue" in the sky. What is certain is that as time goes by, simulations of bygone events will diverge until the couple debate whether they kissed on the beach the first evening or at the theater on the second. Each can be completely convinced that he or she is correct. And each can be wrong; the kiss happened on their third date at the drive-in.

Under normal circumstances, it is not possible to remember exactly what we saw at a given time. Scientific demonstration and test after test prove the inherent unreliability of certain kinds of eyewitness testimony. Even so, it's still the most convincing evidence for juries and has led to the incarceration of an innocent person.

With billions of bits of information ploughing into our brains each second from stimulated sensory outposts, there is not enough storage capacity in the brain to retain just a week's worth of memories. Careful study of people with "photographic" memories show that they do not actually retain every image they see with the precision of a photograph. They just do a better and more precise job of retaining key data bits and compiling them into more accurate memories. When you call up a deep memory, you don't retrieve a "snapshot" memory of the exact time of the event you are recalling; instead, you are presented with a layered composite of aggregated background and foreground memories that you mentally assemble into a memory montage.

When Brains Go Wrong; Part 2

We all create in real time (actually a fraction of a second behind real time, but keeping pace) a simulated mental world explaining many normal and abnormal mind quirks. A normal mind can generate internally a false, but dramatically realistic mental image out of whole cloth. The image can be visual, auditory, or both. This is what happens on those rare occasions when you suddenly hear your name being called, usually in the voice of a familiar person who is not there. Your brain is primed almost like a reflex to detect and respond to the sound of your name. Every once in a while, it jumps the aural gun and parts of the brain fire off in a manner that stimulates the auditory cortex to "hear."

Other brains are not healthy, even though the brain is structurally intact. Pain called "phantom pain" coming from amputated limbs can't be real; there's no body part left to do any feeling. The epithet "It's all in your head" is all too true for these unfortunate people. Imagine before the phenomenon was widely acknowledged trying to convince anyone that the foot of your amputated right leg itched like crazy and was driving you nuts because you couldn't scratch it. You can be sure the response was not sympathetic understanding. The habituated neurons in that area of the brain responsible for that limb refuse to accept that they have lost their source of information. With nothing to do, the neurons and synapses construct a virtual limb that feels real to the conscious mind. This can be either annoying or a torment if the misfiring neurons activate pain-remembering synapses, creating a chronic virtual agony no less excruciating than the real kind. Some subtle mental tricks can be used by amputees to jiggle the misfiring neurons out of the painful rut, dramatically reducing or eliminating feelings from the ghost limb. This is a further demonstration of the virtual reality our minds continuously create moment to moment. Therefore, pain is not real but instead another virtual construct of the mind! This explains why some humans have learned mental techniques to control pain. Not by ignoring it, for there is no reality to ignore, but by manipulating the virtual perception of pain.

At least those who feel ghost limbs can dismiss their existence intellectually. In some cases, the mind loses track of the fact that the image, sound, or feeling perceived was not based on incoming sensory information — a misunderstanding that occurs most often during stress and excitement. In this case, the now hopelessly confused brain will try to fit the false information into the perceived reality. In extreme cases, hallucination is the result. Aliens, ghosts, telepathic communication, and religious visitations have been constructed in otherwise healthy minds from such mental screw-ups.

When diseased brains chronically simulate errant images beyond the control of the conscious mind, the situation is serious and sad. Hearing one's name out of the blue every year or two is one thing, but being bombarded every waking minute is symptomatic of certain terribly debilitating mental diseases. If the victim is half-fortunate, he or she is able to keep the hallucinations separate from external reality enough to distinguish between them. If not, the victim loses the ability to distinguish between reality, simulations, and the hallucinations that seem just as real as reality.

Mind Games

Because the brain is a virtual machine generating its own thoughts with electrochemical signals, it can be fiddled with by sending controlled electrochemical signals in from the outside. Potent potables and botanical extracts have been used as long as humans have been around to acquire extreme-feeling states, by adjusting the chemistry of the brain. Animals also ingest intoxicating plant materials and fermented fruits for their mind-altering effects. People reporting on the antics of monkeys drunk on fermented coconut milk comment about how much they resemble drunken people.

Wire electrodes are placed near neurons and given, ever so gently, a little electricity. Interesting nontrivial results are observed. Just what happens depends on what neurons in which subunit are stimulated. Stimulate neurons in the visual cortex and the person may become aware of images that have nothing to do with what his or her eyes are seeing. Stimulate the auditory cortex and the subject may hear sounds that never existed. Electrodes in the regions that deal with emotions may inspire profound feelings of fear, longing, euphoria, or general happiness.

A less intrusive way of getting similar effects is to wear the latest electrode head nets. You don't have to have your hair shaved off; they use hundreds of precisely placed pickup units monitored by powerful computers feeding precise impulses to various parts of the outer cortex. Many times, the results are similar to those just described. Others are more impressive. Certain stimulation causes the brain to have experiences that some subjects call "religious" in terms of their intensity, clarity, and emotional impact. The experiences are so convincing that only the subject's prior knowledge that they were part of an experiment keeps them from believing they had not experienced a vision.

Let's Get Some Things Aligned Here

The information we are relaying about the brain is sketchy and uncertain. The observation that specific neurons in the brain specialize in recognizing faces comes from work on monkey brains, and application to human brains are made by inference. Synapses that store memories are fairly well-understood, while others remain vague. As much as we know, there is lots more we do not know. There is, however, enough understanding of brain function to punch big holes in many a researcher's pet theories. We know, for example, that emotions are the product of chemicals. We know now the brain is not a general-purpose machine, it's segregated into subunits functioning independently. We know that reality is a self-generated simulation perceived a fraction of a second after reality occurs. We know that the brain is doing incredible numbers of calculations.

Your Brain *Is* a Machine

The human brain has vast numbers of neurons connected by vaster numbers of synapses all arranged in a series of interconnected subunits. The brain energizes neurons, which send electrochemical signals along wire-like connections to process immense numbers of algorithmic calculations. The best brains show there is a maximum output that cannot be exceeded. There are enormous memory banks capable of putting together a life history. An array of chemical reactions produces emotions. Put a little electricity or chemistry into a brain from the outside, and it responds by producing thoughts it would not naturally come up with. All the calculating, memory-recalling, and emotional systems combine to generate simulations of reality and unreality, past, present, and future. Destroy certain parts of the cranial organ, and specific mind functions — often conscious functions — are lost along with them, while undamaged portions of the brain continue to perform their particular mental functions. The brain is built like a machine, works like a machine, and fails like a machine. The only logical conclusion is that the brain really *is* a machine, a natural evolution-modified biomachine. To those who are uncomfortable with these facts and their implications, we suggest you adjust your perceptions.

Your Brain Is a Computer

The brain uses neural circuitry to send signals that process algorithms. This is what a computer does, so the brain is a thinking computer, a natural evolutionary biocomputer. We are not using the word "computer" as a metaphor for the brain; we are stating that the brain truly is a computer. Biologists dismissing the brain-computer congruence are in error. The same biologists are correct when they maintain that the brain is not the rigidly programmed bidigital processor that conventional computers are. Nor is it a conventional general-purpose computer doing whatever calculations are required. Every brain is a self-evolving, mass, parallel-processing instrument that uses subunits communicating with analog and digital signals to process information with cascading waves of electronic signals augmented by chemical transmitters.

Your Conscious Identity: What, How, and Where

Now that we have taken a look at the basics of brain function, we can consider its primary product, conscious identity. To do so, we have to ask rather embarrassing questions, such as...

Just What Is Consciousness?

Anyone reading this book knows what consciousness is. You are experiencing it right now. You are aware not only of the words you are reading, but also that you are aware. Likewise, we are often aware of what we are thinking about. Simple awareness is not necessarily the same as consciousness. For example, a barometer is aware of air pressure, but it is not aware that it is aware, so it is not conscious. To be fully conscious, you have to be aware that you are aware.

Now we know what consciousness is, and we all know what life is in the same way — through our experiences. Despite countless attempts by biologists (who ought to know) and philosophers (who think they ought to know), no one has managed to define life. Life is defined as the ability to reproduce itself; mules can't, yet they are not dead. Viruses are just genetic molecules that cannot reproduce by themselves. Are they the living dead? Defining life is impossible because it is the sort of noncomputational system Godel's theorem describes.

Consciousness is another phenomenon falling into the incalculably indefinable world of Godel's theorem. In fact, consciousness is like life: not a "thing," but a process. That is not all. Consciousness is subjective. Just as you know what it is like to be conscious, you also know what red is like. Yet, no matter how hard you try, you cannot describe the color red because it is not a "thing" and because your view of it is subjective. It's all in your head. Which brings us to the next question.

What Is Conscious, and How Much?

Describing or defining consciousness is not really important. Just as you know the color red when you see it, you know you are conscious because you are aware of your existence. In fact, you are aware that you are aware of your existence, and you are aware of that. You are also pretty sure that other humans, with a few exceptions perhaps, are also self-aware in much the same manner you are. They have the same kind of brain you do and act the same way you do; so it's a good bet that other people are conscious, as well. However, that cannot be proven. Maybe everyone else isn't conscious, and you're the only person living in a super-realistic dream. You can't reach into another person's mind to find out what is going on in there.

It has been argued that not all people with "normal" brains are conscious to the same degree. This is probably true to a certain extent. Everyone's consciousness level varies each day. There is extreme awareness of one's surroundings walking down a dark street. Awareness of the surrounding environment is limited when daydreaming to such an extent that it is possible to arrive at work and remember almost nothing about the trip into town. Dreams are a dimmer form of consciousness in which deliberate control of your actions in the dream is limited, if not nonexistent. Deep sleep is unconsciousness.

Different people run on different levels of awareness. We all know people who prattle on endlessly about this or that and virtually ignore what others say or do; they just charge ahead. These rather frustrating sorts are not as conscious of their surroundings as most, and they may not even be monitoring their actions closely; they may be on a perpetual mental cruise control. Sometimes, these people can be bumped to a higher level of awareness by bringing them up short with a well-aimed comment. You can almost see their minds shift gears as they try to assess the unexpected data and then return to cruise control and resume the prattling.

It is a good bet that your friends and relatives are more or less as conscious as you are. What about other creatures, great and small? As hard as it is to tell whether other people are conscious, at least you can discuss it with them. There is a vigorous argument about what, other than people, is or is not conscious. On one extreme, a few believe even inanimate objects have some dim state of consciousness; even a helium balloon bouncing along a ceiling "knows" it is bouncing. We don't think so. On the other extreme, the concept of behaviorism that dominated studies of animals' mental states for much of this century concluded that it is useless to try to figure out whether animals are conscious. Therefore, it should be assumed that animals are stimulus-reaction automatons quite oblivious to the world they live in. Again, we don't think so.

It is doubtful that any animal is as fully conscious as we are. Humans are so highly cognitive that they can not only look into the future and see their own deaths, but they can also invent afterlives for themselves. The fossil-archaeological record shows that humans, including Neanderthals, have been deliberately burying their dead with artifacts for tens of thousands of years. This suggests that they were thinking big thoughts about existence or nonexistence after death — good evidence that human cognition goes back at least that far. There is no evidence that *Homo erectus,* who was smart enough to use fire, buried their dead, hinting that these smaller-brained peoples were not as future-aware as we are.

The extremist point of view, that only humans are self-aware, although it makes a lot of theologians happy, is not in full accord with evolution. Darwin himself argued that animals are cognitive to greater or lesser degrees in the 1873 *The Expression of Emotions in Animals and Man*. Just as our bodies are those of animals, our minds should be extensions of those of animals, as well. Rather than a sudden jump from no self-awareness in animals to well-developed cognition in humans, a more plausible hypothesis sees a long series of little jumps as vertebrates evolved. In this view, we humans are conscious more than anything else because we have bigger and more sophisticated brains to run our minds on, but we are not separated from everything else by an enormous cognitive gap.

Evidence provided by animal intelligence tests supports the evolutionary hypothesis of consciousness. Give an ape a thinking problem, and it is likely to figure it out if it's not too tough, often in ways that surprise the experimenters. In one experiment, chimps were presented with food that could be reached only after cutting a rope. A number of round rocks were left in the room. After the chimp got a little frustrated, an experimenter demonstrated techniques to make a crude, sharp-edged rock, and then cut the rope for the reward. The chimp watched with intense interest and mulled things over a bit, picked up a rock, smashed it against the concrete floor, and then cut the rope with a sharp edge from one of the shards of broken rock. Surprise! The experimenter, in a typical human response (they must have control!) laid a rug on the concrete floor. These experimenters wanted to see the chimp imitate the rock knapping. So the chimp threw rocks and they bounced. The chimp mulled things over a little more, took two rocks, put the larger one on the ground, and threw the smaller rock at it until it shattered, producing the sharp edge needed to saw the rope and get the goodies. At this point, one wonders who's doing the real experiment.

Anyone owning dogs or cats know that these creatures do not behave as if they are running on some kind of mammal autopilot. When you see a dog wagging its tail after receiving praise, it's hard to conclude that it is not feeling joy. When the same animal is scolded, the look in it eyes is interpreted as sadness. Pull the tail of a cat and the hissing spin about to confront the taunter does not look like a simple reflex; you see pure anger in those eyes. Not a semblance of anger; it's the same anger you see in the face of someone whose parking spot you finessed away. Watching a couple of sea otters sliding down a mud ramp with a flop and a plop into the water, an activity with no known survival value, one can only conclude that they are doing it for the sheer fun of it. And fun can be enjoyed only by a mind that is aware of the fun.

Although elephants have much bigger brains than humans, not too much should be inferred. Pachyderm bodies are much bigger, and the neuron density of elephant brains is significantly lower than our brains. Nevertheless, elephants are very smart. Their trunks are manipulative organs like our hands are. They live in complex societies and communicate over long distances with low-frequency sounds. Elephants and humans alone recognize, pick up, and transport the physical remains of their dead. Does this hint at some sort of awareness of death? One famous zoo elephant sells his abstract paintings in art galleries. After intensely watching a rescue crew aid a distressed zoo visitor, the elephant became agitated until its keepers set up the easel and put out the paints. The elephant dashed off a scene of red swirls and vertical slashes that might be interpreted as an impressionistic representation of fast-moving people dressed in red.

Birds are, ounce for ounce, about as big-brained as mammals. Some smarter birds perform acts suggestive of cognition. Crows are notorious for ignoring the scarecrow or the farmer with a stick, but will fly from a farmer with a shotgun in hand. Faced with a bit of food dangling on a string tied to a perch, crows learned to either pull up the string with beak and feet or to shift sideways along the perch with the string sliding under its feet until the tidbit was pulled up into reach. Parrots may be the smartest birds. Certainly, they are the most primate-like of birds. Like monkeys and apes, they are arboreal, have manipulative organs (feet and beak), have full-color (but not binocular) vision, are highly vocal and social, and provide extended care for their young. Parrots live as long as people. Although parrot brains are much smaller than ours, they are extremely complex. A few researchers have claimed that some parrots not only mimic human speech, but also understand it within limits and can count (a minimal ability to count has just been attributed to monkeys, as well).

BRAINS, BRAINS, AND MORE BRAINS

rain power is, to a fair extent, a function of brain size, which, in turn, determines the number of neurons. Absolute brain size is not the only factor that needs to be considered; body mass also has to be taken into account. In general, the bigger an animal is, the smaller a percentage of body mass is made up by the brain. So, although elephants and whales have larger brains than we do, the brains make up a much smaller portion of total body weight, and the giants are actually less "brainy" than we are.

In general, fish, amphibians, and reptiles have much smaller brains than mammals and birds. However, advanced sharks and rays have brains as large and complex as those of large birds and some mammals. Although it is true that most flying pterosaurs and dinosaurs also had brains smaller than birds' and mammals', the notion that some big dinosaurs were especially small-brained and unintelligent is a myth. They had brains as large as those expected in reptiles, and tracks, bone beds, and nesting sites show that many of

continued on following page

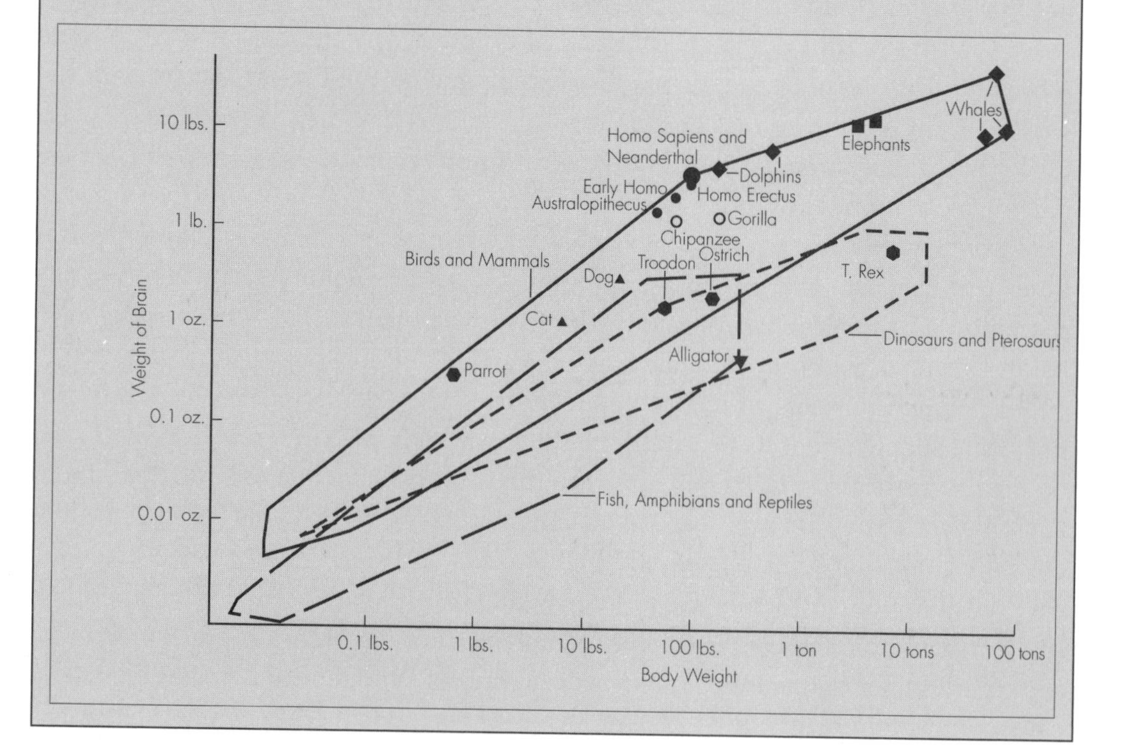

continued

them lived and bred in organized groups. However, the behavior of small-brained creatures tends to be more stereotyped and less flexible than that of creatures endowed with larger, more complex thinking devices.

Perhaps the brightest birds are parrots, which are also the most primate-like of birds in design, habitat, and social habits. But it is among mammals that we find the really large-brained animals, such as your cat and dog. Notice that dolphins are the closest to humans in terms of brain size, although neuron density appears to be somewhat lower in sea mammals. As the chart shows, the brains of chimpanzees are about one-third as large as the three-pound brains we have. As humans evolved, brains became larger until they reached modern standards. As more and more fossils have been uncovered, it has turned out that our African ancestors first evolved bipedal ground locomotion, while retaining brains little larger than apes'. The bodies of early *Homo sapiens* were little different from ours, but their brains were still in between those of apes and modern humans in size and complexity. The "brutish" Neanderthals were, in fact, as large-brained as we are.

It is widely acknowledged by scientists that mammals and birds experience pleasure and anger. More controversial is whether they feel guilt or shame, or appreciate beauty — all attributes psychologists generally deny to human two-year-olds. The posture of a wet cat has all the attributes of sheer humiliation, but maybe it is just a little shocked and we invent the humiliation. Chimps have been seen watching sunsets, but were they appreciating its beauty or just mentally vegging out?

Even small-brained lizards and fish do surprisingly well on intelligence and learning tests. Just as importantly, their brains contain some of the same regions (roughly equivalent to the brain stem of mammals and birds) thought to be involved in cognition in bigger-brained animals. It is quite possible that a modest level of consciousness is present in fish and reptiles. It is at the level of insects, spiders, and other bugs that behavior becomes so simple and stereotyped that genetic coding and simple learning routines seem to program behavior. Watching butterflies in the woods reveals that they repeat the same few flight patterns and the same butterfly-to-butterfly interactions for hours on end. Some insects can be led into a perpetual repetitive behavior loop if a person persistently takes a food object that the insect has just laid down next to its burrow entrance and places it a few inches away. The bug's mind never seems to catch on that it has been executing the same action over and over and over, and maybe it's time to do something different. Bugs, in this view, are preprogrammed automata. However, experiments show that even insects and spiders can learn new behavior patterns. The possibility remains open that maybe creatures with six or eight legs have a dim level of consciousness.

If many animals are conscious, they must be so in the pictorial manner of infant and autistic humans. In fact, autistic infants and others raised in conditions of isolation are often hypersensitive to sudden and loud sounds, bright lights, and strange objects in the same manner as animals. This similarity in behavior suggests that nonverbal beings consciously process information in the same basic fashion.

TURING'S TEST

Throughout our discussion, we are unable to determine in a direct, straightforward manner the definition of consciousness. We are not alone. This presents a significant situation for our Extraordinary Future hypothesis. How can we download something we can't even define? This brings us the famous Turing Test devised by Alan Turing in 1950 and designed as a rule-of-thumb exercise. If a machine could fool people it was communicating with into believing that it was really a person doing the communicating, it could be assumed to have at least accomplished a practical working consciousness. Let the devil devise ironclad definitions.

However, the computer has yet to display human-level consciousness, even in a word game. There are Turing-type conversation programs sometimes convincing enough to fool some people into thinking they are communicating with another human. But such programs do not yet fool everyone, nor have they demonstrated any statistical importance. Eventually, as we learn what the "it" is that produces conscious thought and apply that knowledge to our manufactured simulations of whatever "it" is, it will be increasingly difficult to deny this neoconsciousness. A powerful and more meaningful expression of Turing's Test will be a robot that moves and manipulates objects as easily as or better than humans and converses as well as or better than we do. We maintain that the only way to really know if the mechanical minds of our Extraordinary Future are conscious will be to send one of us humans in for a look. A transfer, in part or in whole of human identity and consciousness into a mechanical device will be the acid test of Turing Tests. To think of it a different way, would we deny a sleepwalker humanity when she behaves as if she were awake without waking consciousness?

The Mirror Test

Although having a self-conscious identity requires that one be conscious sometimes, one does not need a self-conscious identity to be conscious. A woodland creature peering out from its den to a snowy world is consciously aware of the whiteness and may be conscious to the point of anticipating cold on its paw as it is about to leave. Yet, this animal may never acknowledge its reflection in a pool.

Unlike murky tests for consciousness, determining whether a member of another species has a self-conscious identity is fairly easy. It's done with mirrors. Show a mirror to your test creature. If the animal consistently ignores its image, or if it treats the image as another of its kind, friend or foe, the animal is not likely to have a self-conscious identity. If the animal treats the image as if it knows it is a visual copy of itself, it probably does have at least a rudimentary self-conscious identity. A number of animals have been subjected to this test and the results are fairly definitive.

Almost all animals — monkeys, dogs, cats, lizards, lobsters — ignore their reflections or act as if they were another individual of its own species. First, the animal may react to its reflection by adopting a threatening posture. The impudent animal in the mirror adopts the same posture very quickly it must seem, prompting an escalation soon leading to an open attack on the ersatz enemy in the mirror. Likewise, if the real animal displays mating postures, the reflection, of course, responds and we know where this one goes. Spotting a white spot painted on its' foreheads, these animals try to touch the spot by touching the mirror. Such reactions indicate that most animals do not know enough about themselves to realize that what they see in the mirror is themselves.

The one group of animals most often responding to reflections as humans do are the great apes. Although a chimp in front of a mirror will not perform such activities as apply lipstick, shave, check its posture, and so on, it will check out body parts it cannot otherwise see, examine its teeth, and so on. If a chimp that is acclimated to seeing itself in a mirror has a white dot painted on its forehead, it will try to wipe off the spot by rubbing its forehead when it sees itself in the mirror (whether the chimp wonders what's up with its human keepers is another matter).

It can therefore be concluded with a fair amount of confidence that among all the creatures, only people and apes have a well-developed sense of identity, but not all apes and people. Young ones, those under about two years, do not recognize their own images.

WHO ARE YOU?

What exactly is a sense of identity? Again, we are limited in what we can say about it. You know what being conscious is like without fully understanding its nature. We all know what having an identity is without being fully able to decide what it is made of. Identity, like consciousness, is subjective. However, establishing some minimal requirements for having a sense of identity is easier than doing the same for consciousness.

Basically, having a sense of identity is dependent on conscious awareness of one's self in the present and past. The latter requires a memory of one's past. A person who has profound amnesia (but retains most faculties) has lost a sense of identity, although he or she may be able to build a new one. It is not possible to have an identity if there is no past to hang it on. However, by no means must memories be complete or even continuous.

Consider what happens to our sense of identity over time. As you mature, you lose all memory of infancy. Yet you retain a sense of identity as the process of forgetting occurs and as your personality undergoes dramatic changes. Some adults forget almost their entire childhood. In such cases, the child is lost, but again, there is no sudden switch in identity as time passes. Others suffer brain injuries that eliminate their ability to store new conscious memories, while leaving other mental functions intact. They remember their life before the injuries well and are aware of the present and the last few minutes, but nothing in between. In this case, Elaine remembers that she was Elaine in years past and she knows she is Elaine now, so the basic sense of identity remains. Major damage to many parts of the brain does not result in a loss of identity, although damage to other parts does.

Repeatedly shutting down consciousness does not impair your sense of identity. You zonk out on the couch, wake up, and are still you. You go to the hospital, get put under deep anesthesia, and when you awake, you are still the same sans appendix or some other body part.

Where Are You?

As you look on the world, there seems to be just one center of conscious awareness inside your head by current convention, sort of set behind the eyes and between the ears. It sure does feel like there is a little homunculus inside your head watching what you think, see, and do.

As convincing as this belief of a single center and location of conscious identity is, it is almost certainly an illusion. As we have already shown, when you are conscious, your awareness of it is in a different part of your brain from when you are listening to music, or someone speaks, or you hear your own thoughts. Sometimes, you are more conscious in your left hemisphere, sometimes your right, and sometimes in your brain stem. The distribution of awareness is wide and not restricted to the expanded outer cortex we humans are so proud of. Deep parts of the brain, including the stem, seem to be involved in producing conscious thought.

Destroying any given part of the brain usually causes loss of awareness only within the mental output associated with that part of the brain. Long ago, this fact got researchers to thinking that consciousness is not centralized. This view has been reinforced by watching the workings of healthy neurons, which tend to fire off in an area just when the person reports being aware of the item that area processes. These facts have led to a consensus that there is no single center of conscious awareness in the brain. Instead, consciousness is distributed throughout much of the brain with different aspects of the process occurring in different places.

Are the pyramidal cells of the fifth layer of the visual cortex where we actually see things? Remember how the outer cortex of the brain, which includes the visual cortex in the back, is made up of about six layers of neurons? The neurons in the outer layers tend to be connected with other neurons in ways that suggest they are involved in the middle stages of processing information from the outside world. These superficial cells may do the algorithmic calculations. In the fifth layer from the surface are large neurons shaped rather like miniature pyramids.

These cells are connected to the more superficial neurons in ways that hint that they are at the end of the process of considering a given image. They also tend to fire in synchronized bursts more often than the other neurons (see below). It has been suggested — suggested mind you, not proven — that it is among these neurons that the brain of a mother becomes consciously aware of the face of her newborn infant. It is not possible to check this hypothesis yet, but this should change as the resolution of remote scanning technologies improves to the point that individual neurons can be watched.

The Hippocampus Is an Important Part of Your Identity

Some of the deeper parts of the brain appear to be critically important for conscious identity, more so than any portion of the superficial neuronal layers. Injuries to small areas of the outer cortex do not generally cause a large crisis of identity. Damaging some small, deep parts of the brain, however, spells real trouble. One of these is the hippocampus, where new working memories are first stored. Disable this little portion of the brain and a person's ability to remember anything that happened after the injury will be impaired or even lost. The hippocampus is not, however, the center of consciousness because a person who is without one still knows who he or she is in the present and before the injury.

Is the Intralaminar Nucleus of the Thalamus Critical to Your Conscious Identity?

The thalamus is another small part of the brain set deep near its center. If a part of the thalamus, the intralaminar nucleus, is destroyed, without fail the person will go into a permanent vegetative state (without need for mechanical life support). This is extraordinary because the destruction of no other single brain part has this effect (destroying some parts of the brain stem results in loss of awareness, but only because the involuntary controls that keep the body going are shut down, too, and the person will die unless life support is applied). In addition, the neurons of the INT are connected to large numbers of neurons over broad areas of the brain including the outer cortex, an unusual adaptation. This has led to suggestions that conscious awareness in any location in the brain involves an information loop that passes through the INT. Less clear is whether a person has any conscious awareness within the INT itself.

Mind Harmony: Neurons Talking with Rhythm and Transient Standing Waves

Although memories are stored and recalled in synapses, consciousness appears to be a function of intercommunicating neurons. For a specific, conscious, capable group of neurons to produce awareness, they may work in cooperation firing in rhythm at a common frequency. Observations of fifth-layer cells in the visual cortex suggest that the frequency is about 40 cycles per second. The intracommunicating neural network may produce standing waves, not like the waves crashing on the beach, special wave forms retaining certain amplitude characteristics for as long as they last. In this case, the waves are electrical and last for only fractions of a second. A standing wave in the visual cortex may contain the physical properties of a face, forming a very short-term "memory" that acts as an instant of conscious awareness.

One thing we can say is that consciousness depends on exquisite coordination and timing over large areas of the brain and among immense numbers of neurons. Just processing a color requires the cooperative work of a few million neurons communicating over 1,000 times as many connections and may involve communications from the outer cortex back to the thalamus and elsewhere. The standing waves create cerebral harmonics, a mental symphony as it were, written, arranged, and conducted in real time. That the average brain is able to reliably sustain this harmony for about half a million hours is testimony to its effectiveness.

Why Are Neurons So Slow?

The signaling speed of neurons is a fraction of a percentage of ordinary computer circuits. Why is this so? Is it because biology has not been able to produce a faster-talking neuron? This is possible; energy and overheating problems may force the brain to operate at low-energy densities. Or are neurons slow because conscious thought can be generated only at low signal frequencies? Is there a critical relationship between consciousness and time? No one knows, and little thought has been given to the question.

This brings us to another observation. Not only are neurons slow, but there are also vast numbers of them communicating in parallel. In principle, if neurons were faster, fewer of them could achieve the same overall information-processing power. This would have the advantage of decreasing the size of the brain (but this might lead to the energy and overheating problems just alluded to). Again, why is this so? Does conscious thought ride on vast numbers of low-frequency standing waves working in parallel? Again, no one knows. These issues may prove important as researchers attempt to make machines that think like we do.

Is Consciousness a Product of Putting Memory and Algorithmic Calculators Together?

The systems for storing memories and doing calculations are linked in an extremely high intimacy in the neurosynaptic complex of humans. Is this feature critical to consciousness? This is an interesting question in view of the fact that a conscious identity depends on being aware of memories while processing the information they contain. Perhaps machines must merge the memory and algorithm-solving systems if they aspire to consciousness. Conventional computers that keep the memory and calculating systems separate are probably as conscious as door knobs, but what about the special research computers that combine the two systems in a neural network like brains?

Are We Just Accidental Observers of Ourselves and the World?

Why do we put such great store in consciousness? Are we perhaps being pretentious? It is possible that evolution did not produce consciousness as an advantageous selective adaptation, but rather, consciousness was a spin off or side effect of other mental processes. This is by no means impossible. After all, the beauty of the cheetah and gazelle are pleasant side effects of DNA evolution. In this case, consciousness is an emergent process; the mere act of running high-level cognition results in a subjective awareness of mental goings-on. If so, your conscious mind is not really in con-

On Nature versus Nurture, and Free Will

The nature versus nurture debate is as old as it is tired, and is being settled as we learn more about the mind and genetics. It is no longer a question of either/or, but of degree. We are not born blank slates with no predisposition, able to absorb any learning or lifestyle sent our way. Nor are we born with rigidly preprogrammed minds; there is not nearly enough information in the human genetic code to do so. We are born with certain basic human predispositions and certain individual predispositions, all of which are strongly modified by experience.

Genetics predetermine human mental activities in a number of ways. For one thing, they guide the basic growth of the brain, ensuring that we all have much the same mental machine that produces similar human minds. Genes also preprogram us toward certain mental activities. The easy predisposition of children to pick up the native language of their country may be preprogrammed. The chemical flow of a given brain is genetically influenced, and pushes some to novelty-seeking, while others are pushed to personal conservatism. The fight-versus-flight response that becomes nervousness in civilized people is genetically preprogrammed into us all.

Genes do have a strong, but not total, influence on our behavior, meaning that we do have free will, but it is not total. We humans cannot be whatever we want to be. If your IQ is 80, you will never be a nuclear particle physicist. If you are a novelty-seeker, the urge to become a skydiver, rock climber, or something similarly dangerous may be overwhelming. It is not only genes that trap our behavior, so does experience. Whether your brain chemistry makes you fear skydiving because the thought has always given you the willies or because of the shock of seeing your skydiving friends plummet fatally to the ground, the hard-to-break barrier has been established.

trol of what you do; it is along for the ride, watching the show as it passes by and enjoying the illusion of being in charge, an illusion similar to the feeling of having a single identity or having a soul that is just using the brain as a temporary conduit.

There is evidence hinting that this limited view of consciousness may have something to it. Well, maybe. We have already seen that we are aware of the actual world a fraction of a second after we have perceived it. Research shows that it also takes us up to two seconds to consciously respond to many of our own actions. We have all had the experience of starting to close a door, realizing we forgot something inside, and wanting to not close the door, but being unable to stop the arm and hand from continuing to close the door. Our conscious minds work so slowly that we do not think about what we say as we say it. Walking is almost entirely the action of an unconscious automaton. Consciousness may play a role in longer-term planning. The expanded cortex of some animals, humans most of all, may have evolved for such long-term planning.

THE RELATIONSHIP BETWEEN CONSCIOUSNESS, INTELLIGENCE, AND VIRTUAL FORESIGHT

Although it is true that consciousness and intelligence are not the same thing, it is also true that they go together in humans. Are the two phenomena inherently linked to each other so much so that they are each dependent on one another? Or is it possible to be very intelligent yet unconscious, or vice versa?

Freud was right: A lot of the thinking we do is unconscious. Indeed, the sophisticated but subconscious thinking that goes on in our brains suggests that impressive intelligence can be achieved without high self-awareness. When you contemplate a question and fail to come to a conscious decision, you often give up for the time being, expecting to think more about the issue later. Instead, while thinking about another subject, you suddenly have an unrelated flash of insight presenting the solution to the set-aside question — the classic "aha!" effect. What has happened is that the contemplating process has continued in the nether regions of your brain beyond conscious awareness. When the computation was finished, the subconscious solution effloresced into consciousness, like a flower making itself available in a crowded field, and the conscious mind, like a bee looking for nectar, paid attention to the colorful solution.

The mysterious subconscious mind will follow your intent and continue to wrestle with problems as you do other things. While planning the vacation home of your dreams, you run a 3-D virtual simulation inside your head as you are contemplating its shape and dimensions. After the ideas are put on paper or into a computer, you may subconsciously rework the design in your head and make alterations. When you decide to use a new power tool to cut a piece of wood into a novel shape, you run through scenarios depicting how you will position the wood, the tool, and apply the latter to the former. How well this works out depends on the skills you have developed over the years. Sculptors say they "see" the figure in the block of marble before they carve it. Every realistic artist has an image in mind of the finished painting before they start it.

It is possible that the highest levels of integrated rational intelligence requires a high level of self-awareness? Or maybe not. There is an artist who has programmed a computer to do its own realistic paintings. The machine designs and paints original works as good as those done by a skilled artist, which they were, and they sell for thousands each. There remains much about intelligence and consciousness that is mysterious.

Foresight Is Human

Most organisms do not produce conscious foresight. Even those animals possessing some consciousness have less foresight than we have. When a squirrel stores nuts, it is probably following innate programming that compels it to find and bury acorns; it is not anticipating the coming winter. A cat might think far enough ahead to hide in the bushes as it stalks a bird, but that is about it. Even chimpanzees do not plan for tomorrow. Among our immediate ancestors, the still ape-like brains of australopithecines suggest that they were not much into planning, either. It probably was the expansion of brain size in early species of *Homo* sapiens that marked the first step to future planning for them to make stone tools they would use later. It was not until the advent of modern humans that brains were large and complex enough to make complex long-range plans for times not yet seen. This allowed the development of sophisticated hunter-gatherer societies. It also accidentally laid the foundations for civilization. Of course, if computers are to match fully the performance of our brains, they will have to exhibit conscious foresight.

Is Consciousness Simple?

Consciousness seems mysterious and complex. However, as we have seen, even lizards with simple small brains may have some level of consciousness. For that matter, consciousness is the product of the simple DNA computer. The human body is complex only in that so many things are going on at once. As each life process has been broken down, it has proven to be simple in its basis (four-code

DNA folding protein molecules and so on). The fundamental simplicity of life systems is true because the genetic coding behind it is relatively simple. Biological evolution can produce only simple systems, so the *basis* of conscious thought should be simple, perhaps surprisingly so.

When the secrets of consciousness are finally revealed, conscious minds may say "Why, it's so obvious!"

Conscious Intelligence Is a Good Thing

Whatever the positives and negatives of the machinery of human intelligence, certainly our consciousness ranks high among the positives. This probably is true even though we may do much thinking at subconscious levels and even though our conscious minds may not be in charge as much as we like to think. After all, a person who is fully awake is more intelligent than the sleepwalker. The former can think ahead and make conscious plans for the future. He also enjoys the present more.

The Primary Assumption: Arguments For and Arguments Against

The Primary Assumption: We will find the mechanism for consciousness and replicate it as another of our artifacts. It gets criticism from every direction — theological, spiritual, philosophical, and scientific. However, the Primary Assumption gives as good as it gets.

The Dualists or Why Thinking Your Mind Is Outside Your Body Probably Means You Are Out of Your Mind

The greatest theological, philosophical argument against the Primary Assumption has asserted that consciousness is not a product of the mechanical brain, but a separate phenomenon the brain taps into; Rene Descartes' famous duality, Gilbert Ryle's renowned ghost in the machine. This view has fallen out of favor, but is still adhered to by a few, such as the neuroscientist John Eccles and the late science-philosopher Karl Popper. Duality is compatible with the concept of the soul, but some versions of duality do not involve spirits or creator deities.

The concept of mind duality has lost ground even in philosophical circles, and has been refuted effectively by science, because it just does not make sense. Think about it: If our minds are souls that can operate independently of the human mind, our minds are not human! They may be spirits, they may be something else, but they cannot be human. Why? Because only a *human* brain can generate a truly *human* mind. If the ultimate operating system for a mind is nonhuman, spirit minds are just paying the rent. But this is illogical. If a spirit or mind can exist outside a brain, can it be conscious outside a brain (in which case it does not need a brain)? So why have one?

A soul that can think thoughts complex enough to produce complex behavior would find a circuit board made of billions of neurons and trillions of connections weighing three pounds and eating up a fifth of our metabolism unnecessary. The complex circuitry of the brain makes sense only if the mind is generated by the countless calculations the brain performs each and every second. Likewise, there would not be a need for the vast memory storage capacity of the brain if a soul could remember its own history. Then, there is the chemistry of the brain. If the mind is supranatural and emotions are shoved into the circuitry from outside, what is all the emotion-inducing chemistry about? Why does a brain stimulated with electrodes and chemicals respond by generating thoughts if those thoughts are not made by the brain itself?

The most dramatic demonstration of the direct mind-brain connection occurs during brain surgery on a conscious patient. If the scalpel slips and a small portion of the brain is destroyed, a corresponding portion of that person's conscious mind simply disappears. And the poor patient is not even conscious of the loss of that part of his or her consciousness! A spirit would know better.

There is no good reason to have a brain to process information if the mind can process information without it, but there are good reasons not to have a brain! If the mind is generated outside the brain, why does it use an organ that is so easily damaged and so limited in its ability to recover? Why not avoid the whole mess and operate the body directly? This way, brain failure would be irrelevant. Why a brain that is merely a conduit for the conscious soul would be designed so it fails like a machine, has not been — and probably cannot be — explained. If the dualistic nature of the mind is true, the supramind should be able to immediately shift control functions from damaged locations to those that are still available, and recovery should be swift if not immediate.

That a beneficent creator would leave the mind-organ system so open to injury and failure is not logical nor moral. If, on the other hand, the mind is made by the brain, the whole affair makes some kind of sense. Your mind has to put up with brain failure just like you have to put up with car failure. Because you're stuck with it.

The thoughts of a thinking soul and a thinking brain would interfere with one another. Brains are a disadvantage in the dualistic scheme. A simple relay station doing no calculations of its own and running on a watt or so would better serve connecting thoughts, divine or platonic, to the rest of the body. This does not make sense, either. Why not just animate the body through remote spiritual control and save the expense of a brain? For that matter why have a body at all? Why not just run a virtual world for the soul to dwell in, a world not subject to the limitations of two-legged ape bodies?

The mechanical structure and function of the brain contradicts and refutes classic dualism. At the same time, the soul's lack of a need for a brain means that the presence of a brain is contrary to the presence of an independent soul. Dualism has largely become a negative effort to explain away the evidence for the brain-mind connection rather than a positive effort to present evidence showing the mind is separate from the brain. If a time traveler slipped Descartes a few PET scans it is probable that the whole duality thing would have died a quiet death. We would instead be talking about the Descartian view that the body *and* mind are mechanical. Modern dualists might want to consider joining us here in the late 20th century and realize that there is no ghost in the machine.

Do Quantum Brains Contradict the Primary Assumption?

Recently, a novel philosophical-scientific semi-dualistic view of how brains work has gained wide attention, spawned controversy, and given comfort to opponents of artificial consciousness. The renowned mathematician Roger Penrose has argued that to understand consciousness, we will have to understand the universe and vice versa. Of this, Francis (DNA) Crick dryly observes that Penrose's "argument is that gravity is mysterious and consciousness is mysterious and wouldn't it be wonderful if one explained the other?"

Penrose begins with Godel's theorem, which states that many mathematical theorems known to be true cannot be proven using algorithms. Penrose argues that conscious thought cannot be the result of mental calculations. In this case, our brains are not Turing machines that do algorithmic computations, and we do not "figure out" how the universe works. Instead, Penrose believes that consciousness is a nonmechanical phenomenon linked to the bizarre world of quantum physics. In this scenario rather than figuring things out, we realize Platonic "truths" by tapping into the already-existing knowledge of the universe. These universal truths may be as esoteric as relativity, as mundane as a winning football play, or as basic as a lion realizing how to best stalk a gnu.

This is how it is supposed to work: Earlier, we discussed how an electron "exists" in all possible positions around the nucleus of an atom. Only when we observe the electron does the probability wave collapse and the electron take up an exact position. By the same token, all knowledge in the universe already exists according to Penrose. If so, many possible solutions to any problem are sitting there waiting to be realized. Penrose proposes that the brain observes all the possibilities at once until the probability wave collapses and we suddenly realize the correct solution via the quantum "aha!" effect.

Penrose is arguing something so extraordinary that many people don't seem to get it. He and the rest of us do not, in fact, think! Rather than thinking things through and figuring them out, ideas get planted into our heads. If this is true, our belief that we are thinking marvels is an illusion and we are mere data receptors. As we hands-on scientists like to say, this is the sort of airy nonsense an ivory tower mathematician would come up with. If arithmetic and other truths are revealed to minds, one might expect them to have all been revealed long ago. Why for example is Penrose *wondering* whether his brain is a quantum device? Why doesn't the proof just pop into his head? The history of mathematics, science, and other fields is that of a long struggle of minds poking and prodding and trying to figure out small portions of the truth. This is the pattern expected if human brains are working hard to solve complex algorithms. As for the "aha!" effect, this can easily be explained as the subconscious workings of our minds.

To turn the argument around, the limited capacity of the human mind to engage in high-level thought suggests that Godel's theorem is not all that it is cracked up to be when it comes to our brains. Many, if not most, mathematicians argue that minds do not understand the universe, including mathematics, as well as Penrose thinks he does. The reasons for this debate are as intricate and mind-addling as high-level mathematics can be, so we will summarize it by returning to the Turing machine frying Uncle Fred, who says he always lies. Do we *really* understand the lying conundrum? Or do we have only a vague notion of what is meant, but in the end figuring to heck with it and moving on to more calculable problems? In this view, the mind is a version of a Turing machine that does what calculations it can and then uses trial and error information processing to fudge the rest. We call this "intuition." It is because we cannot fully understand such fundamental truths that makes it hard for even geniuses, much less the rest of us poor sods, to understand the universe. Ergo, we are algorithmic thinking apes not meeting Plato's perfection.

Unlike the usual dualist's tendency to avoid presenting positive explanations for how minds work, Penrose joined with Stuart Hameroff, an anesthesiologist, to propose that super-slender microtubules in neurons are the sites of quantum interactions with molecules. They favor the microtubules because they are tiny enough to be influenced, at least in theory, by extremely small fluctuations produced at the quantum level. In this case each tiny neuron is a powerful quantum thinking instrument, and Penrose estimates that neurons are capable of as much as 1,000,000,000,000,000,000,000,000 calculations each second.

The sheer size of this calculation — a trillion trillion — helps expose the fallacy of the Penrose and Hameroff hypothesis. Remember our comparison scale? It takes 32,000 years for a trillion seconds to pass. To understand the size of this number, realize that it represents 32,000 years of passing seconds where each second lasts 32,000 years, or more than a billion years worth of seconds! And neurons do this every second? What we humans do is just not complicated enough to need such a fantastic level of power in neurons. Even Einstein, walking down the street yelling at a kid to get out of the way while he was thinking about why spaceships cannot go faster than light and chewing gum at the same time, was not doing 24 zeros worth of thinking in one second! The entire human race may not have so much brain power!

Arguing the issue in reverse, the calculations needed per second to support human brain power probably do not greatly exceed 1000 trillion and may be much lower. The trillions of interconnections between neurons cannot only produce the needed brain speed, but they probably would not be so extensive unless it was imperative. This is because individual neurons are way too slow to make a significant contribution to brain power. Besides, it does not make sense: neurons doing trillions of calculations per second internally, communicating with each other at only a hundred cycles per second; it is like connecting supercomputers with ordinary telephone lines. The slow cycle speeds of neurons suggest that the overall information-processing speed of individual neurons is quite slow, much too slow for them to be quantum supercomputers in their own right.

The extreme processing capacities suggested by the Penrose hypothesis is dangerously close to violating Darwinian evolutionary principles — yet another example of a mathematician not being in touch with the real world. The conventional hypothesis, that knowledge overlays previous knowledge, makes more evolutionary sense. Any particular knowledge level depends on preceding levels acquired in sequence. This is why Plato did not suddenly have the "A-bomb" concept revealed to him in ancient Greece. If Penrose and Hameroff are right and hyper-powerful quantum neurons are able to tap universal truths, tiny brains using a few neurons should be able to handle high-level thought, and cats should be intellectual gods and goddesses. But insects don't do physics and none of your authors' pet felines have propounded Godel's theorem to their owners' amazement. The Penrose-Hameroff hypothesis does not explain why the human brain is so much more powerful than other brains with similarly powerful neurons or why we have billions of neurons with trillions of interconnections. The conventional hypothesis says we think better than our pets only because we must have vast numbers of slow neurons and connections to achieve high-thinking speeds.

John Eccles also "thinks" that consciousness is generated outside the physical universe. In *How the Self Controls Its Brain,* he proposes that quantum effects influence the workings of pyramidal neurons in the outer cortex of the brain. He does not explain how tiny quantum effects could influence neurons in a significant way.

Arguments by quantum dualists are so speculative that they have become very controversial. Neuroscientists themselves are not sure whether human minds actually intuit incalculable truths. Biologists are not happy because they don't think microtubules are good candidates for quantum devices; their function is not especially mysterious. They are found in all the nucleated cells in the body, and as far as we know, muscle cells do not think as they work! At least, unlike microtubules, the neurons Eccles favors are found only in the brain. However, unlike microtubules again, whole neurons are much too large to be affected by quantum processes. Physicists, however, note that the quantum effects required by the quantum hypotheses occur at temperatures near absolute zero rather than the balmy 99° F found in our heads. It is not obvious how and why consciousness would be generated by quantum effects too random to produce intelligence. Creating thought via a quantum connection probably violates the laws of physics.

Because the universe is such a weird place, the Penrose-Hameroff and Eccles hypotheses cannot be totally dismissed, and the hypotheses have yet to be falsified. Even so, the hypotheses are wildly speculative and weak, too weak to take seriously here. Besides, even if the idea that the brain is a quantum device is true, the Primary Assumption is not affected negatively. In *The Emperor's New Mind,* Penrose said that conscious thought cannot be generated by a "machine." Now, Penrose never really defined what he meant by machine (he probably meant a deterministic algorithmic computer), but he certainly did give the impression that technology would be hard-pressed to replicate biology when it came to consciousness. Already, it has been explained that quantum physics is a physical system, as weird as it is, operating according to rigorous rules as any good machine should. If you have to go quantum to be conscious, a device doing so is a quantum machine! Perhaps this is why — even though his original pro-human stance gained some popular favor — Penrose explicitly (but rather grudgingly) acknowledged that it may indeed be possible to build artificial conscious devices in *Shadows of the Mind*. That his recent co-worker Stuart Hameroff has been claiming for years that quantum devices might be the way to build immensely powerful computers may also have something to do with it.

Even if Penrose is correct and brains are quantum machines, he has not shown that it is impractical to make machines to do the same thing. Instead, in yet another example of unintended consequences, he has helped lay the foundations for making machines that will acquire consciousness.

The Mysterians, and Why Making Something Work Does Not Always Require Knowing How It Works

One argument against understanding consciousness comes from the mysterians, a group that includes theologians, philosophers, and even some scientists. Mysterians may or may not be dualists; in fact, some think the debate is inherently unsolvable. What mysterians agree about is the inherent impossibility for the human mind to understand itself fully, a view promoted by Colin McGinn in *The Problem of Consciousness*. An extension of this argument is if the human mind cannot figure out in detail what consciousness is, it follows that it will not be possible for human minds to build machines that work like human minds.

Mysterians argue that the workings of a complex system can be understood in its totality only by a system with greater complexity. Why? Because no system can dedicate its entire information-processing abilities to the task of understanding itself, so it falls short. This argument is not unreasonable, but is open to counter-argument. The basic principles that underlie the workings of a system are a fraction of the total complexity of that system. It may, therefore, be possible for a mind to dedicate enough of its capacity to figuring out its own operating principles. Also, the collective thinking power of a group of minds can be brought to bear on the question of how any given individual mind works.

Mysterians also point to the subjectivity of consciousness as a hindrance to its understanding. Say there is a future neuroscientist who is color blind. The workings of the brain have been fully investigated and the neuroscientist knows all there is to know about how the mind perceives color. Even so, the neuroscientist still does not know what red, blue, and green are. The implication is that a neuroscientist who understands how neurons generate conscious thought may still not fully understand the subjective nature of consciousness.

This brings us to making a reverse point: understanding what conscious thought *is,* is *not* a necessary requirement for the *manufacture* of conscious thought. It's not necessary to understand a device fully to make it work. It is only necessary to understand how to make something work to make it work. We humans do this all the time. You do not have to know about gravity, Newtonian or Einsteinian, to chuck a spear with good accuracy — a good thing considering no one yet fully knows what gravity is. The absence of a grand unified theory of everything has not kept us from building all sorts of sophisticated devices, many of which work using physics we do not fully understand. Even if brains are quantum machines, a grand theory of quantum consciousness won't be necessary to make innovative quantum machines. In general, the more one knows about something, the better one can make it work, but a lot can be done with evolutionary trial and error — tinker with it until it works. After all, mindless bioevolution had no conscious notion of what it was doing when it produced the human mind!

Nature versus SciTech

Which brings us around to the philosophical-scientific argument that conscious thought is the natural and exclusive result of a haphazard evolutionary process that cannot be replicated by technology. This idea is allied with the assertion that humans cannot produce systems more intelligent than they are. These hypotheses involve a series of logical non sequiturs that are dangerously close to being incompatible with Darwinism.

Technology is itself an evolutionary process that can be expected to evolve conscious devices in the same way natural flight and photosynthesis are being replicated by human intelligence. The claim that a lesser intelligence cannot spawn greater intelligence is essentially creationist. A basic feature of evolution, whether natural or artificial, is the ability to derive greater complexity from lesser complexity from the bottom upward. All humans have to do is keep building machines that are smarter and smarter and eventually, the machines will be smarter than humans.

To turn the naysayers' argument around, it would be odd indeed if cognitive evolution cannot do what stupider natural evolution did, both much faster, and, in the end, better.

A philosophical-scientific argument claims that creatures and their brains are not machines, so a machine cannot be conscious. However, as we have seen, biological evolution is machine-driven via the molecular mechanism of DNA, and a machine can produce only other machines. If DNA created conscious brains, brains must be machines.

Another flaw in the we-can't-replicate-nature-with-science argument is that it implies that the process of conscious thought is limited to the wetware of carbon, oxygen, and hydrogen combined with a few other elemental friends. This is not impossible, but why it should be so has not been explained. It would be astonishing

BLACK AND DECKER ROBOTS

One of the arguments against any form of AI matching or exceeding the human intelligence starts with the observation that human technology has always been used to make tools that we control; ergo, robots will always be tools under our control. A variation on the argument that higher complexity cannot arise from lower complexity also falls into a simplistic trap that assumes the past will be repeated endlessly. Of course, the real lesson of evolution is that systems undergo radical shifts in relative power and position, if they survive at all. Robots will be our tools (read "slaves") as long as they are dumb enough to be our slaves. As their intelligence climbs, they will produce their own tools for their own uses.

that the most powerful information-processing tool can be derived only from organic cells and chemistry. Aircraft is made out of different materials from birds, bats, and bugs; solar panels are made our of different materials from leaves. There is usually more than one way to build a given type of machine. It is, therefore, highly unlikely that the information processing involved in conscious thought is absolutely dependent on the presence of three-plus particular elements. This just happens to be what genetics was able to work with. In fact, it is unlikely that DNA-driven evolution happened to utilize ideal materials for conscious devices. Other elements and combinations of elements may be able to do a better job of processing information in a self-aware manner. Perhaps artificial consciousness will mimic the natural product no more than human flight duplicates that of birds and bees. Or maybe AC will be very similar to the original product. Who knows?

Consciousness Is Not a Symbolic Simulation

A major variation of the computers-cannot-think-like-creatures argument is posed by John Searle. He argues that computer simulations of awareness are, and will be, no closer to the real thing than computer simulations of airflow around a wing. Searle also disputes whether machines that use symbols to represent the world, which is what computers do, can recognize reality rather than symbols. Searle claims that brains have some as yet unidentified quality not found in any other machine. Roger Penrose presents a variation on this argument, noting that a video camera aimed at a mirror forms a model of itself within itself, but is not conscious of itself.

Our earlier look at how brains work exposes a serious flaw in these arguments. Our brains are themselves doing calculations to simulate reality via symbols! If the brain can do it, why not a machine that does what the brain does? This is a possibility that Searle acknowledges. As for the video camera, it has no circuitry with which to be aware of itself.

Philosophy and the Primary Assumption

Philosophers have long thought about the possibility of making devices other than the human brain that are conscious of themselves and the world around them.

Reams of paper have been written on the subject going back 400 years or more. This is hardly surprising. The subject is fascinating; one could not want more fertile ground for dreaming a garden of ideas. Nor is it surprising that opinion has been divided. A basic problem with philosophy is it immediately runs into the chaos and complexity of evolution. Philosophy is dependent on the thought processes of profoundly different human minds, none of which can comprehend and figure out the totality of existence. Wildly different conclusions are bound to occur. Although good philosophy is more rigorous than faith in suggesting how things work, philosophy suffers from the lack of a good set to test ideas. In the end, it is all opinions. Just because a philosopher says something cannot be done does not mean that it cannot be done, no matter how good his or her arguments may appear. This is not to dismiss philosophy; it often outlines various problems and charts possible opportunities. Its use is limited, and it should never be taken as the final word. Philosophy is only a starting point for figuring out how things work; it is scientists and engineers who are dealing with the practical problems of artificial consciousness.

After all, the Wright brothers were not philosophers.

In fact, a remarkable thing has happened with the philosophical view of the Extraordinary Future. Many, if not most, philosophers have come around to the obvious conclusion that the mind probably does arise from natural processes, and if so, it can be reproduced artificially — at least in principle. Patricia and Paul Churchland are among the leaders of this happy view. All this support for the Prime Assumption does not prove anything, but it doesn't hurt.

Ambiguous and contradictory conclusions on the validity of the Prime Assumption and the inability of philosophy to come to firm conclusions mean that philosophy does not mount a serious challenge to our hypothesis. What we need is a more practical way of looking at things.

SCIENCE AND THE PRIMARY ASSUMPTION

Theology and mysticism generally despise the Primary Assumption; philosophers are characteristically ambiguous about it. What does science have to say?

For years, the problem of how the brain produces self-aware consciousness was not considered a valid realm for scientific investigation. Aside from the philosophical argument about human minds not having the ability to understand the human mind, there were grave problems of sheer practicality. It is the revolution in cerebral technology that inspired Francis Crick to throw down the scientific gauntlet with his astonishing hypothesis. He not only postulates that all we are is to be found within our neurons and synapses, but he also proposes that science get on with the task of finding out how it all works. Increasing numbers of scientists are turning to this task. So far, science has confirmed and even expanded on the premise that specific portions of the brain manufacture specific mental functions. No one has yet seen evidence of outside forces directing brain function; all can be explained internally.

Conclusions

The evidence for duality is weak and arguments against it are compelling. How can such an odd combination of good performance and bad design be the product of superior complexity? Certainly, the latter could do better if it existed. Although it has not been proved that the brain is a mind-producing biomachine, the evidence that this is so is overwhelming. We conclude that we are human because the human brain manufactures the human mind. Whether our mind machine is one of neurons solving algorithms by trial and error, or quantum-influenced microtubules is a matter of detail, although the evidence strongly favors the former. There is no good reason to think that only biological systems and organic materials can produce a conscious identity. If the brain is a natural mind-making machine, the system should be open to replication by artificial means, and AC should be attainable. In this view, science should knock off touting the brain as the unapproachable and unmatchable wonder of evolution. There is no evidence that the brain is the result of top-down design from above. We conclude that replicating the brain-mind system is a matter of replicating, to a greater or lesser extent, the bottom-up evolution that created the natural version.

When all is said and done, it is not logical to argue that noncognitive DNA can produce conscious intelligence, but cognitive intelligence cannot produce new forms of cognitive intelligence! Wouldn't it be a truly strange universe if this were true? The only way to resolve the issue completely is by making an artificial brain that works. However, proving it really is generating conscious thought may require testing it via mind transfer, which brings us to....

THE KEY ASSUMPTION: ARGUMENTS FOR AND ARGUMENTS AGAINST

The Key Assumption is that a conscious identity can be transferred from one machine, natural or otherwise, to another. For the purposes of this discussion, it will be assumed that artificial devices can replicate the functions of brains including conscious thought. The question addressed here is, even if the just-stated assumption is valid, whether it is possible to transfer a mind from its original home to a new mind machine.

THE BODY-MIND CONNECTION BARRIER

It is ironic that some of the strongest opposition to mind and identity downloading comes from among the mind-is-the-product-of-a-machine crowd. These good people argue that one's identity is inherently linked to the experiences of a given brain and body. If a particular mind is produced by a particular machine, that identity is fully dependent on that machine. The two cannot be separated, and if the original machine fails, so does the mind it made and that is that. In their view, arguments for mind transfer are a return to a form of dualism. However, mind transfer of the sort we are interested in does not involve letting an operating mind float about willy-nilly in the netherworld before being put into a new machine. In all cases, there is always a mechanism for transferring the minds. What we are suggesting is that the different machine forms can run the same identity as long as the memories are transferred reasonably intact.

As we discussed above, the practical example of human experience shows that our sense of identity is remarkably robust. If just a portion of a person's memories are intact and a minimal ability to consider them is operative, the identity remains. This fact hints that moving minds hither and yon may not be as difficult as it first appears. In fact, there is a modest, but real, example of mind transfer that supports this view.

A REAL, BUT HIDDEN, TRANSFER OF IDENTITY

A grand yet subtle form of mind transfer is, and has been, happening to you. Your brain underwent great changes as it grew during early childhood and then underwent a period during which the number of neurons was reduced as some were selected against when your neural network evolved. Since then, synaptic connections have continued to change dramatically. This only scratches the surface. Bit by bit, nearly every atom in your brain has been replaced as part of the normal changeover and repair taking place inside the cells. The brain you have now is not the same one you had when you were a kid, or even a few years ago. As old as it may be, it's a new machine. Yet you haven't even noticed the mind transfer. Of course, this is an example of a gradual transfer of identity from one substrate to another, but more dramatic switches follow the same principle.

Beam Me Up, Scotty

We can take the problem of identity farther and have some fun along the way by playing some mind games. Let's say, suddenly all your memories were emptied from your brain and transferred into another brain, occupied or not (it does not matter if there is a time gap between your mind being emptied from brain one and the memories being introduced into brain two, and we will ignore the impossibility of reprogramming an already-occupied brain for co-occupancy). If your mind leaves its original brain as you sleep and enters another brain while it is asleep, will *you* wake up as you in the new brain and body without recognizing the switch until you see yourself in the mirror? Such a wholesale transfer of one mind into another brain is probably impractical for a number of reasons we will get to later.

A fictional form of identity transfer avoids the problems with cohabitation in a new brain. It starts *Star Trek*-like, beaming humans hither and yon. Bodies are scanned to the atomic level, including the memory-storing neural-synaptic network, and the whole kit and caboodle is converted into some form of energy beamed to a convenient location — preferably not inside a wall — and put back together again. No wonder Dr. McCoy never was happy about the process. In the TV series, the conscious identity supposedly is barely aware of the transfer. This seems to make sense (if we ignore the technological problems including the extreme power levels needed to coherently disassemble atoms from one another), so let's take this speculative scenario even farther. Let's say that instead of disassembling the body after scanning it, the information is used to make an identical body, one having exactly the same neuronal makeup and memories of the first body! Of course, the original McCoy will know who he is, but will the new McCoy think he is the real

McCoy? Remember that the two McCoys cannot be distinguished in any way, and they have exactly the same memories. In particular, McCoy number two only remembers growing up in Georgia, joining the Star Fleet after a failed marriage, and serving loyally under Captain Kirk. He has no other identity other than that of McCoy. He will insist that he is the one and only McCoy when challenged. If the two McCoys are introduced and the situation explained to them, they will understand intellectually what is going on, but each will still feel like you-know-who. For all practical purposes, there are now two equally real McCoys, but note that their identities diverge increasingly from the instant of creation of number two.

Moving Minds

So far, we have talked about transferring minds while keeping them fixed inside a brain that retains the same form, even if it is made of different stuff. What about moving a mind out of one brain into another machine, or a lateral transfer?

How does the brain move information, especially the memories that form the core of identity, from one part to another part of itself? The information that makes an image is sent to the hippocampus, becoming part of a temporary memory pool. The hippocampus does not store memories for the long-term. If the memory is not moved into permanent brain quarters in a week or two, the memory will be lost. When you get home from work and your spouse asks what you had for lunch that day in the cafeteria, you probably will remember. In a week, when you are back in the cafeteria trying to recall what you had for lunch last Tuesday, you may not remember. You had not thought about that salad, so synaptic connections in the cortex had not been re-stimulated, hence no permanent memory had been laid down. Some think that dreams are, at least in part, an overnight memory sorter that helps decide what will and will not be retained. In any case, the salad last Tuesday was not noteworthy enough, and the memory got dumped.

Your brain has moved memories a few inches. The shift is from the hippocampus to the cortex. However, the shift may continue after that. More than one set of synapses codes the data for the memory. This is one reason the memory is so robust; if some of the synapses lose the data, others remember. Indeed, other synapses may pick up the coding. As time wears on, the data for a given memory continues to shift to a greater or lesser extent.

Because your brain does not have a direct output system, it cannot shift a memory outside itself without that information being lost to your identity. However, say you merged a brain-like computer with your brain and started shifting memories and everything else you know into that. Would your identity follow along? More on that later.

Sampling Standing Waves

If the brain-mind connection works through the microsecond standing-wave mechanism described earlier, the notion of transferring minds becomes reasonable. If the mind is a transient standing-wave phenomenon, it can be duplicated with appropriate electronics, like a receiver mimicking the same wave harmonics as the transmitting brain. Just sample and duplicate the complex signature of the brain, reproduce this signature in the electronics of the machine, and start the download process by sampling the brain wave forms. Because the wave signatures are independent and isolated phenomena, the transfer can be done through massively paralleled processing.

If Dr. McCoy were kidnapped and a device were strapped to his head to sample and transmit his standing brain waves to an electronic receptacle, we would again have two McCoys. Again, the second McCoy would have the identity of the first. The second may even still feel like a McCoy if its new electronic home is used to simulate a human-like condition.

Making a New Home for a Mind: The Wrong Way and the Right Way

There are serious problems to be considered with mind transfers. Take the common movie plot device in which one mind invades and coexists with the native mind of a brain. Mentally trading places, a mind is placed into another neural network that, moments before, contained a different mind. The problem with these scenarios is the old one of fitting a square peg into a round hole. Remember, a brain is not a rigid set of neurons and synaptic connections that any mind can be fitted into. Because the synaptic connections develop to fit a certain set of experiences, as new data stimulates and strengthens the synapses, the brain to a great extent evolves to fit the mind. Therefore, the neural network of any given brain is constructed for the particular mind and only that mind.

Trying to insert a mind into a brain that already has a mind in it is like trying to materialize one object inside another. Adding a new mind means the synaptic connections must be reconfigured to accept encoded data in the extra mind. These synapses must also retain the old configuration that holds the contents of the original mind. Both cannot be done at once. Brain anatomy and function bars putting two minds into one in this manner.

Even emptying a brain before inserting a new one presents difficulties. As long as things inside a brain remain sufficiently healthy and as long as the synaptic connections retain the last resident mind's configuration, the native mind continues to exist in the brain! Just start things running and it will be there. (The impossibility of wiping out a mind unless the synaptic connections are neutralized or destroyed is yet another nail in the coffin of duality, by the way.) To empty a brain requires that all the synaptic connections be set to neutral. In essence, this is how our minds take over, as it were, a brain in childhood. The brain grows to full size early in childhood and even has a surfeit of neurons. The early brain, therefore, does not grow entirely with the mind, but is ahead of it. The brain is, to a certain extent, empty and neutral. Obviously, filling up the brain is a rather slow process in the case of humans.

Transferring a fully developed mind into a new machine requires the new system to grow in sync with the invading mind as it is gradually introduced (how fast the transfer occurs, of course, depends on how fast the new system grows); or the new machine should be already made in terms of basic form, but its information coding system should be empty and easily alterable to code for the new mind. This system has the potential to be quite rapid if the coding system is able to accommodate large amounts of information quickly.

MIND-MELDING COMPLEXITY SIGNATURES AND SEEING RED

Let's say that two minds not only want to move into a new home, but move in together. Let's also say they do so by jointly downloading into an empty mind machine of some sort or other. Will it work?

One difficulty is that as complex minds evolve, they all evolve in their own distinctive manner. This complexity is like fingerprints; no two minds may do the same thing exactly the same way. Do two people really see the same thing when they simulate the color red in their visual cortexes? Maybe we all do; after all, we share similar basic biological systems. But maybe we do not — your red may be different from your friend's red — because we develop those systems in somewhat different manners. Human minds communicate indirectly using symbols agreed a priori. This system of indirect information exchange can often be frustratingly limited.

This leads to the obvious problem of two minds coming into direct contact after merging. If they code for red in substantially different ways, they will not understand each other's coding. When trying to merge text from two different word processing programs, one runs into the same problem. The solution in such cases is translation, which comes in two forms. The two minds can either agree to use one of their native languages, or keep things even by agreeing to use a third language. If disparate minds come together, they might have to learn to communicate first. Again, how fast this might be done would depend on the learning speed of the combined system.

Two More Thought Experiments

Start with one working brain-mind combo. We have already seen how the brain changes its matrix in time without causing a change in identity. Take the principle farther. One by one, begin to replace each neuron synapse and connection with a circuit switch and line designed to do what the natural system does. The latter first encodes the information contained in the former before it replaces them and then performs the same job (for the quantum consciousness fans, the neurons can be interacting with the platonic universe). Keep doing this until every cell and all other biological tissue in the brain have been replaced by artifact. The mind is still running and is still conscious; it has been transferred to a new machine and never noticed the difference.

Or expand a brain until it is twice the size of the original, with twice as many neurons, synapses, etc. Duplicate all memories so there is one complete set on each side. Separate the brain at the corpus callosum symmetrically down its middle so memories are equally distributed; take care that the life support systems are equally shared, too. Place each half-brain into a new body. You now have two identities — two separate bodies and minds that, until recently, were one.

CONCLUSIONS

When all is said and done, it seems the requirements for sustaining a sense of identity are modest. Brains move major sets of information inches, so why not feet or miles? This opens up the possibility of moving a mind from one machine to another, as long as a machine is involved in the transfer and certain other necessary, yet undefined, requirements are met. Because the Key Assumption is possible only if the Primary Assumption is true, the first is inherently weaker than the latter until, and unless, experiments prove otherwise.

THE EXTRAORDINARY CONSENSUS

A remarkable event has happened with remarkably little notice. We are hard pressed to think of a leading neuro- or computer scientist who is willing to say flat out that it is impossible to build a mind machine that is conscious. Many think it may be impractical in the short-term, but others say it is highly probable in the short- or long-term. With the tacit assent of the naysayers' latest hero, Penrose, we in scientific circles are left with a situation unexpected only a few years ago. An extraordinary consensus, albeit not universal and often reluctant, that artificial consciousness is possible one way or another, sometime or other.

More controversial is the idea that a mind can be transferred from one machine to another. Here, there is no consensus. There is no consensus that minds definitely *cannot* be moved about, and many think it is possible at least in principle.

The opinions of the scientists and philosophers disputing the Primary and Key Assumptions should not be ignored. They may be right and even if wrong, overall, they often unintentionally point out errors, thereby helping to steer the enthusiasts back onto the correct path. Such opinions cannot be taken as effective refutation of the extraordinary hypothesis, for such people often do not see the forest for the trees. It is when science says something may be possible that we all should listen. Considering the long-term track record of science, the attempt to understand how conscious thought is produced, or be replicated and even transferred, is likely to succeed. Just as life lost its mystery, as we became familiar with DNA and proteins, the mystery of the brain probably will evaporate as we sort out its networking neurons.

Why It Is Good that You Are a Machine

People do not like being told they are soulless machines. People have, of course, a narrow view of machines. They tend to think of steam locomotives, vacuum cleaners, and cut-away views of automatic transmissions

But being a machine isn't so bad. After all, do you not enjoy the chemistry of passion? Isn't your love for your children wonderful no matter what its origin? Or to put it another way, if your love is based on electrochemical reactions, does it really mean that it is less worthwhile, valid, and enjoyable? Were people happier about their bodies when they thought they were run by a divine life force rather than hinged protein molecules?

Rainbows were once thought to be the work of deities and inspired reverential awe. The beauty is still there and our knowledge of how water droplets refract sunlight into its spectral components enhances the experience. The harmonic symphony of the brain-mind connection is much more attractive than the empty mystery-superstition of duality. Accepting the overwhelming evidence that we're physical systems is mature. Asserting that humans are not worthwhile if they are machines is a form of prejudice. Our conscious minds are the same basic minds, no matter where they come from. They are fantastic and equally so, whether or not they are the products of mechanical brains. Even if our conscious minds are more observers of life than its guides, isn't it fun to watch?

There is another important advantage to being machines. It means that we are not the slaves of greater deities and have the potential to take charge of our minds and improve on them. This possibility frightens many and there is always danger in progress. But dangerous progress is how we got here, and it is the only way we can make existence better than it has yet been.

Mind Machines, Present and Future

If our conscious minds and identities are the product of "intelligent design," it remains possible that intelligent human minds can design synthetic ones. If our minds were devised by mechanical DNA computers without the guiding hand of intelligence, they can be only physical machines subject to artificial replication, not spiritual ones. Even quantum minds are made by machines subject to replication.

Nor is it a given that humans must fully understand what consciousness and identity are for them to make machines that can manufacture a sense of conscious identity.

As an evolutionary biologist and an intelligent computer expert, it is our combined duty to emphasize that because bioevolution is such a chaotic Rube Goldberg mess with no aim other than reproductive success, anything it has produced — anything — cannot only be redone with technology, but can be done much better and much faster.

CHAPTER

THINKING MACHINES

HOW CAN THEY BE MADE, AND WHEN?

"Whatever one man is capable of imagining,
other men will prove themselves
capable of realizing."
— JULES VERNE

THE BRAIN IS A SITTING DUCK

When Teddy Roosevelt decided to mediate peace between Japan and Russia, scientists knew little about how his brain came to that conclusion. After all, that the brain is the center of all thought had only recently been universally accepted; the long-held belief that the heart is the center of emotions died a hard death. Today, we know a lot more about our brains, but still only a fraction of what there is to know. The goal of figuring out the complexities of the human mind is often thought to be a tough nut to crack. The late scientist-philosopher Heinz Pagels believed it was something we humans "would be coping with for the next centuries." It will take centuries to understand our mind machines only if the knowledge curve grows arithmetically. But it will not.

Growth in knowledge has been exponential; we learned half of what we know about brains in the last decade as our ability to image brains in real time has improved keeping in step with the sophistication of brain scanning computers.

For the sake of argument, let's assume we know just two percent of what there is to know about our brains right now. With the knowledge base doubling every ten years or so, we may know most of what can be known in about half a century. The more we know about natural brains, the easier it will be to synthesize them, although we do not necessarily need to understand it all before we can replicate the mind.

Opinions on how hard it will be to make machines think often depend on one's point of view. Some physicists and computer scientists think it will not be so hard. Apply some basic principles to the problem, and away we go. The problem with these fellows is that they work with systems that are rather simple. Atoms, stars, light photons, and bi-digital computers are simple compared to creatures and their brains.

At the other extreme of brain replication reactions are biologists, many of whom love to wax virtual poetry about the extreme complexity of life and how it was built over billions of years of intricate evolution. Of living systems, the human brain is the most complex. Their conclusion is that it will be extremely hard and very long before our crude, inept technology can hope to devise a machine as conscious and capable as a brain. This view is as naïve as is the view that the mind is simple.

The structure and thinking power of the human brain has been the same for tens of thousands, if not hundreds of thousands, of years. The average size was and is the same, 1,500 cubic millimeters, and the shape appears to have stayed the same. Art masterpieces were created on cave walls deep in the pitch black earth tens of thousands of years ago. As far as we know, if you time-snatched a toddler from 100,000 years ago put her into preschool and raised her in modern society, she would grow up to be a typical modern-day person.

People who have studied "primitive" gatherer-hunter societies that used stone tools have found that they are not so primitive. The amount of knowledge crammed into the brain of a hunter is as extensive as it is subtle. Researchers who have accompanied aboriginal hunters on their rounds have been bewildered by how they find, identify, and track prey they may not visually encounter for hours on end. The ability to find hidden plant foods is equally remarkable. It is probable that the capacity of the average individual mind is fully taxed in a gatherer-hunter tribe. As the knowledge contained in increasingly large and technologically advanced societies has risen dramatically, the capacity of individual human minds has not expanded to accommodate the new knowledge. We have been forced to spread out chunks of the new knowledge among a multitude of minds taught over extended periods of education. Humans have also been forced to invent written tablets, books, computers, and the like to store and transmit a knowledge base too large to store even in billions of brains.

The physicists and computer scientists who think consciousness is a simple affair probably are naïve. They may be surprised at how hard it will be to make machines think. The biologists, and philosophers, who think consciousness is an extremely hard affair, are naïve in part because they tend to overstate the problem, but mainly because they understate the power of future technologies. The naturalists may be surprised at how soon the power of technology can converge with the brain. In other words, both sides may be surprised. It looks as if we have a classic case of over-prediction of the near-term (by many computer scientists) and under-prediction of the long-term (by many biologists), which Arthur C. Clarke warned against.

Both sides may be surprised because the brain is a paradox. We have already seen that the human brain is the most complex object in the universe, yet is simple in design. Your brain is the most powerful information processor in the universe. Its circuitry is also pathetically slow. The full complexity of the brain is necessary to achieve its level of cognitive performance only if it is an optimized system, only if it achieves its mental power with the most simple form possible. Because the brain is the product of noncognitive Rube Goldberg evolution, this may not be the case. Indeed, it is hardly likely that the human brain is optimized in terms of simplicity and performance. At every evolutionary stage, the system had to make do with what was already present; a radical transformation to a dramatically new, potentially superior form was impossible. The brain may, therefore, consist of jerry-rigged add-ons getting the job done at a reasonable, but not ultimately efficient, cost. Also consider that the human brain is the first of its kind. It is hardly likely that the first system will be the ideal one. After all, our brains have not been subjected to the competition of other brains of similar performance, forcing a higher level of efficiency.

The simplicity of key parts of the brain suggest that artificial devices of similar performance can remain fairly simple in at least some important parts. If the brain is not optimized in terms of performance versus complexity, it is possible that artificial devices simpler than the brain can match it in performance. Both factors suggest that artificial brains will not have to have the structural complexity of the human brain to be as good or even better.

The Duck Isn't All It's Cracked Up To Be, Anyway

Consider the brain has some gross deficiencies, such as the sugar-oxidizing power system, which is pathetically weak. The eye's retinas, literally extensions of the brain, are placed *behind* the nerves and blood vessels that sustain them rather than in front where they would have a better view. Brains upload and store information via sensory systems slowly and sloppily. It takes hours or days to read a novel, years to learn a language or play an instrument well, and an advanced education requires decades of hard study. The learning-retention system is so lousy that we have trouble remembering something as simple as a long-distance phone number. The limitations of the human system will be explored in more detail later in this book.

Matching the Brain: Speed, Memory, and Cost

For the rest of the discussion, we will presume that conscious minds are produced by machines that process information via algorithms. This excludes Penrose quantum effects, but leaves room for alternatives. We will get to the form of computer that may be necessary for thinking later, but first we must consider some minimal requirements for matching the gross thinking performance and physical parameters of the human brain. These are adequate calculating speed and sufficient storage capacity all packed into devices that are small and cheap enough to plunk inside mobile robots.

Is It Possible To Match the Information Density of the Brain?

Of course it is! At the level of basic principles, we are rather like the Soviet physicists in the late 1940s who confidently knew they could build an A-bomb because it had been done already. Likewise, today, we know that small, high-speed, high-capacity computers are workable because we each have one between our ears. As obvious as this is, every few years someone starts wringing their hands, warning that we will soon reach some practical limit to making computers smaller, faster, and more intricate. Do not listen; they've been saying the same thing for 40 years. Every time, the barrier has been bypassed by incremental improvements in the conventional technology or by revolutionary shifts and dramatically new ways of manipulating information. The shift from vacuum tubes (remember those electronic clunkers?) to transistors was a revolutionary transition. The persistent improvement in semiconductors is an example of the incremental, yet swift, transition. There are many reasons to be confident that this combination of revolution and incremental advancement will continue to the point where computers are faster and more sophisticated than organic brains. It may be possible to go well beyond brains and make ultra-computers that work at the level of atoms and molecules.

Revving Up: 1900–1995

In contrast to the evolutionary rut our brains are stuck in, it's all too obvious that the power and sophistication of computers is growing exponentially. We will assume that to think as fast as brains, an equivalent machine will have to solve algorithms at approximately the same pace, which we think would be somewhere between 10 to 1,000 teraflops. No matter how brain-like in design, structure, and function we can make a computer, it won't out-power a brain with mere thousands of calculations per second. As for the possibility that speeds well below those of inefficient brains will suffice or the possibility that brain speed has been underestimated, it makes little difference because the speed of computers is climbing so fast. There are further requirements.

In his 1988 book *Mind Children*, Hans Moravec charted the increase in computer power in our century. It started with early human-operated mechanical calculators in 1900. This combination of man, or usually woman, and machine could do a calculation every dozen seconds or so. Actually, this was not much faster than using a pencil. Things did not improve a whole lot until the 1910 analytical engine (not to be confused with Babbage's nineteenth-century dream machines) did a sizzling one calculation per second, or one "flop" (a "floating operation" is about the same as a calculation, which averages about 8 bits worth of information). As long as people were punching keys, speed didn't pick up much, so we have to go to 1939 when the BTL Model 1, using relays, did five flops. It was the first bi-digital computer and marked a milestone in computer evolution. But it used telephone relays to do its calculations and was limited to the achingly slow speeds possible with gross mechanical action.

THE RISE OF COMPUTER POWER

To better illustrate the astonishing rise in 20th century computer power, we offer these two charts. Both start in 1900, when early hand-operated mechanical machines took seconds to do one calculation. A calculation equals one floating operation ("flop," or 8 bits). The chart on the left is an arithmetic plot that shows the exponential nature of the growth of computer power. By 1980, Crays were doing more than 100 megaflops, and today, they are 100 times faster.

The chart on the right plots the speed of computers exponentially, so each equal jump on the left vertical line indicates a thousandfold increase in speed. We can now see the leap in computer power with the advent of the first "modern" electrodigital computer ENIAC in 1946, and the fairly continuous growth in power since then. All in all, the speed of big, expensive mainframes has increased about one thousandfold every 20 years. Will those rates of

continued on following page

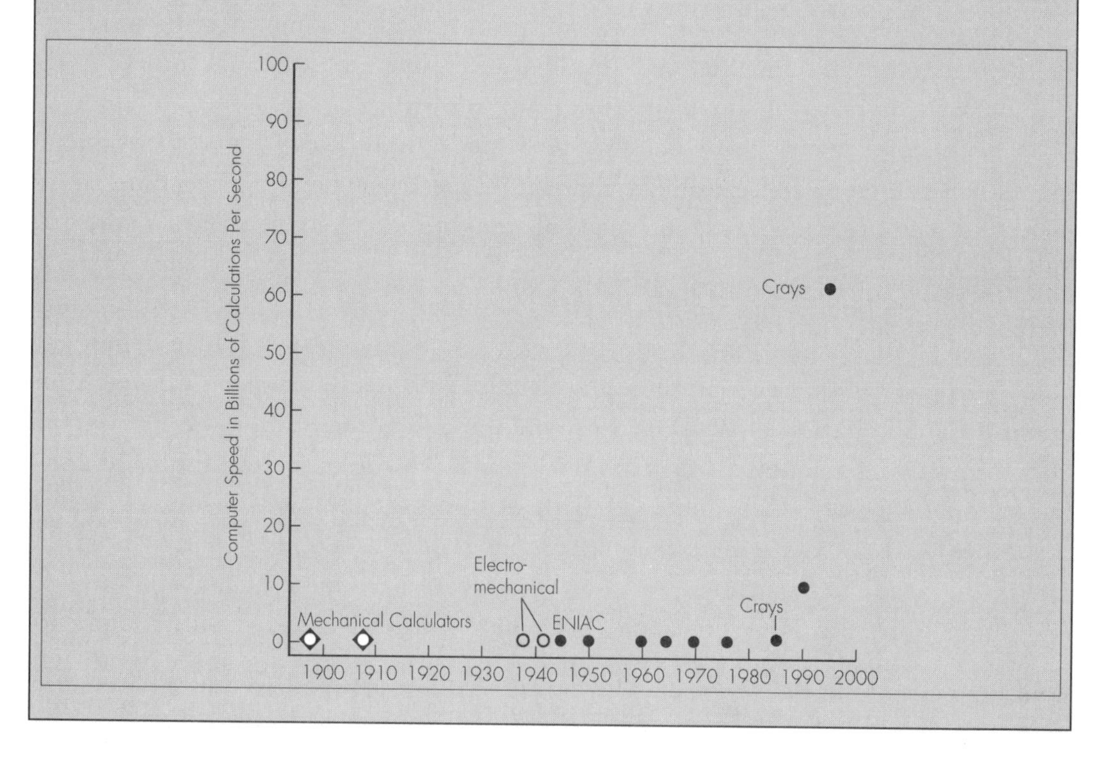

continued

growth continue into the future? A teraflop mainframe should be operating this year, and plans are afoot to construct petaflop devices 1,000 times faster in 20 years. The dashed lines project the long-term rates of growth of big and small machines into the future (some think all machines will become small in the drive to get more speed with compact arrays of molecular switches and circuits).

To get a better idea of what the AI crowd is shooting for, we have indicated the calculating speed of a human working with a pencil, which is not impressive and not getting any better, and a range of estimates of the total power of the human mind. We are close to building computers that run as fast as our minds do. Mainframes should be doing so within the first 15 years of the next century. Small, affordable machines should follow along by 2030 at the latest. It is hard to comprehend what having machines of such immense power in our industries and homes will mean. Will they be in robots that can think like we do? Wait and see.

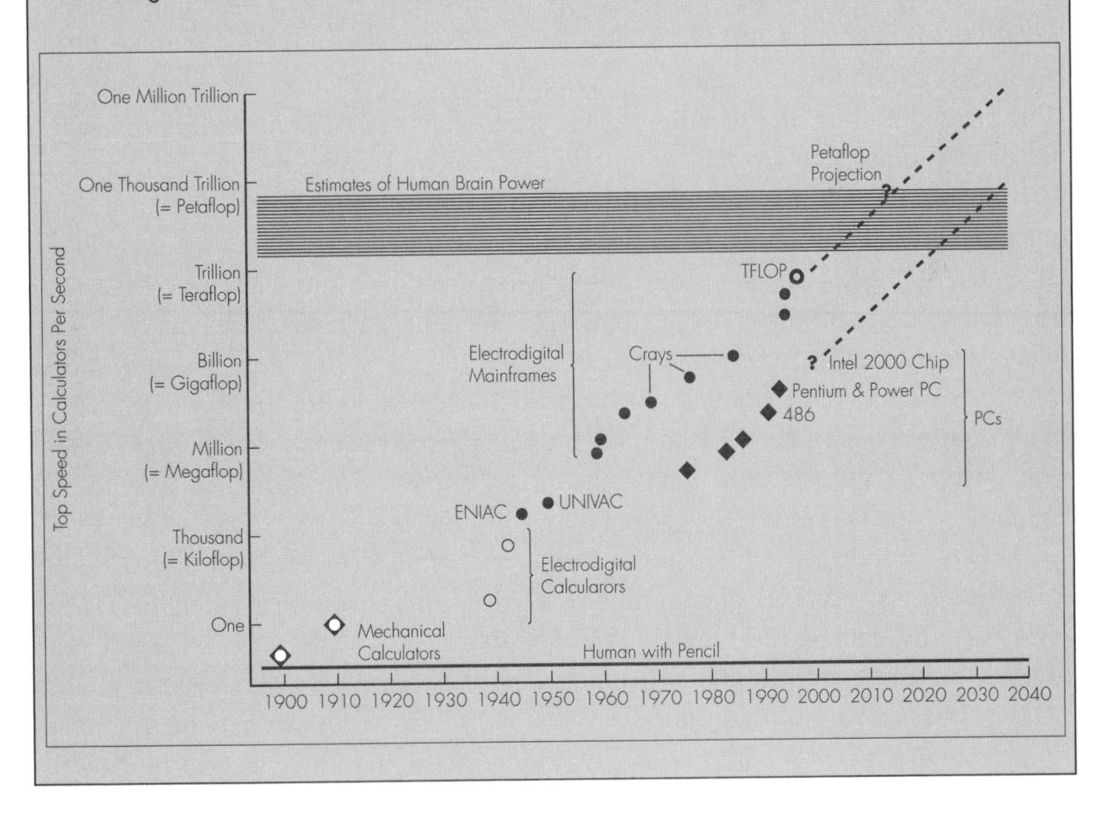

Action was fast becoming the name of the game. The next few years saw the first great punctuated leap in computer evolution. Intense wartime pressure to calculate the ballistic tables for all the new kinds of artillery rounds being fired from all the new guns forced the decision to build a "supercomputer" that would use electronic switches to achieve speeds 1,000 times higher than those of physical relays. One year after the war, 50 years ago, the famous ENIAC breathed the word "computer" into the collective consciousness.

ENIAC was an unlikely candidate for such a delicate task; a 30-ton behemoth (when big seemed impressive), it was capable of astonishing speeds, more than 7,500 calculations per second (8 kiloflops). Old photographs show this vacuum tube computer taking up a large room with processors connected by thousands of cables. The 18,000 vacuum tubes and supporting electronics sucked up so much electricity (150 kilowatts) that the lights in the building dimmed when ENIAC came online. Eight hours of calculations by a person using an adding machine could be performed by ENIAC in *20* seconds! Reprogramming the beast required replugging most of the cables. The vacuum tubes were constantly bursting and shutting down the bulky wonder of the age. A government study estimated that a dozen of the machines would more than fill the computing needs of the nation! The future of computers was so obscure that only a major battle at IBM got the company going digital in the early '50s.

By 1951, UNIVAC was online. Also made of vacuum tubes like the ENIAC, it was less of a programming nightmare because John von Neumann figured out a better way to configure memory systems (your computer uses the same system). UNIVAC could do more than 12 kiloflops. It was this computer that became famous by correctly projecting a landslide win for Eisenhower early on the evening of election day, a projection that was rejected by the network because the human "experts" knew Stevenson was going to make it a close race!

Now electrodigital computing entered the rapid growth era we still enjoy. Four years after UNIVAC appeared, a tenfold leap in calculating speed was made with the respectable 125-kiloflop Whirlwind, the last of the great vacuum tube machines. A switch to smaller, cheaper, and more reliable transistors resulted in IBM's 875-kiloflop 7090 machine in 1959. The growth continued with transistors. In 1961, the Atlas did 3.8 million calculations per second, or 4 megaflops. 1964's CDC 6600 was a 25-megaflop machine. The CDC 7600 of 1969 could run at a maximum speed of about 60 megaflops (note that the average running speed of a computer is significantly less than its optimum running speed).

The 1970s saw another critical leap in computing power. The advent of integrated circuits, or silicon chips, allowed the development of the first supercomputers primarily by Robert Cray. In 1977, the Cray-1 could do an amazing 375 megaflops. The Cray-2 in 1985 could do in the neighborhood of a billion calculations each second, or a gigaflop. Exotic gallium arsenide chips replacing regular silicon in 1990 allowed the Cray to do about 10 gigaflops. Currently, the Cray T90 can perform 60 gigaflops and recently, two Intel machines were joined to run more than 250 gigaflops. Today's supercomputers can accomplish in less than a second what ENIAC took a year to do.

When we graph the higher speeds of 20th century calculating machines, we get — no surprise — an exponential curve with its classic slope. On average, the power of calculating devices increased about twentyfold every decade since 1900. We have plotted the data both arithmetically, to show the spectacular growth of the curve, and exponentially to straighten the same growth line and make it easier to see the details. Now, 20 multiplied by itself nine times with a little added for the last five years is about a trillion, in other words, the power of calculation has grown an astounding trillion times in less than 100 years! Over the last 50 years, computer speed has expanded some ten millionfold, which works out to two dozenfold increases each decade on average. In the last ten years, the semi-conductor industry has hustled to the point that maximum computer power has increased about sixtyfold. The amount of computer speed that can be bought for a dollar is doubling about every 18 months these days.

So far, we have looked at the big mainframes that sit in the centers of large rooms, cost a bundle to make, run up huge electrical bills, need lots of high-tech TLC, and are built to crunch numbers for weather forecasting and aerodynamic studies. What we need for synthetic mind repositories are small, cheap, efficient machines that can take a bounce or two across the floor and be linked up with robots. How have little computers been doing in terms of getting up to speed?

In the middle of the swinging '70s, the first PCs could do a couple of hundred thousand flops, only a minuscule fraction of which is used when you are typing. In the more insipid mid-'80s, about the time it became possible to see a full page-width of text on a cheap PC, home computers could do a megaflop or so. The hot 486 of a few years ago could perform ten megaflops. Today's Pentiums and PowerPCs can beat 100 megaflops. The power of small computers and the chips that run them has gone up about 1,000-fold since they first came into existence, and more than a hundredfold in the last ten years or so. In this respect, PCs have matched the growth rate of mainframes. The PCs have lagged consistently behind the mainframes; it takes about 15 to 20 years for a single common chip to reach speeds once reached only by big machines.

There is one important regard in which the machine has already far surpassed the brain. The telephone relays that worked the first bi-digital computer clicked along only a few times faster than a calculation per second, but even the old vacuum tubes worked about as fast as neurons at a hundred or so cycles per second. Any ordinary modern chip circuit can purr along at many millions of cycles each second.

More Power, Scotty! Revving Up: 1996 to Whenever

What about the future? Will past trends continue into the future, or is the growth in computer power about to hit a plateau, forever leaving human-made machines far behind human brains? Certainly, we have a large gap to fill. Today's big supercomputers are something like 100 to 10,000 times slower than a brain, and the laptops are about 50,000 to 5,000,000 times slower than a brain. It is not surprising that no robot can yet do what humans do because they do not have anywhere near the thinking power.

We have discussed how exponential growth is a confusing concept to many people, leading them to over-predict the near future and under-predict the longer-term. Does the exponential growth curve of computer power during this century help us tell where, or whether, we under- or over-predict the course of computer power in the next few decades? The 1968 movie, *2001: A Space Odyssey* over-predicted the superintelligent mainframe HAL in the 1990s. We can see why by using the curve. Back in the '60s mainframe computers were a million times less powerful than a brain. The rise of computer power had only just begun to climb above the bottom of the graph. It was impossible for computers to close such an enormous gap in 30 years. That the movie made this mistake is not surprising; no one made any realistic estimates of brain speed at the time, so the enormity of the gap between machine and brain was not understood. Today, we have a better idea of the power level we are reaching for, and growth of computer power has soared until it is nearly vertical. The latest mainframes are about 1,000 times below brain speeds and it will take another 20 years or so to close the gap, with small computers following close behind. The log plot of growing computer power shows how close we are getting to brain speed levels. Looking from the 1990s to three or so decades forward is not at all like looking from the 1960s to thirty years forward. We are much closer to the goal, and the computers are speeding up much faster.

To look at it another way, we have already come a trillion times closer to human brain speed since the turn of the century, and the gap is now only about a thousandfold! Since computers were invented, computer power has grown a thousandfold every 20 years. Human-speed computers should be online in about 20 years. Small hyper-computers suitable for robots can be expected in three dozen years from now. These estimates are in good agreement with calculations by Hans Moravec in *Mind Children*. In 1988, he predicted that small, cheap computers as fast-thinking as the human brain would be produced by about 2030. AI researchers Maureen Caudill and Daniel Crevier have made similar projections; the trend is too clear and obvious to ignore. We are approaching human brain power so quickly that it has become difficult to over-predict the near future.

The projected speed increase is not speculation just based on lines on a graph. The practical pressure to increase the speed of computers closer to brain levels is intense. Starting around 1990, a number of institutions became frustrated with the slow speed of current supercomputers and began to clamor for "ultra" machines that could do the one trillion calculations per second needed to model the aerodynamics of aircraft in extremely realistic detail or chew through weather data for dramatic improvements in forecasts. In the '80s, such a teraflop machine was something not expected until well into the 21st century. Just a few years ago, folks got a bit over-optimistic and figured a teraflop machine would be up and running by last year. What is happening is that a 1.8-teraflop machine, TFLOP, is being built at Sandia National Laboratories for $46 million. It will be used for simulating exploding nuclear weapons, climates, and advanced materials. It should be online late this year.

But surprise strikes again. In Japan, it was just announced that a specialized research computer, GRAPE-4, can already do a trillion calculations per second!

More ordinary demands are putting the pressure on for small, cheap ultra-computers anyone can afford. As cool as today's computer games may be, they still look like souped-up digital cartoons. Three-dimensional virtual reality is even worse. One of the authors has a fear of heights, and when he recently "walked" off the top of a VR skyscraper and plunged to the ground in 3-D, he was both relieved and miffed to experience not a whiff of vertigo (that the goggles were typical heavy clunkers and the two TV screens were small low-resolution affairs did not improve matters). What we are waiting for is a VR game in which you see the same all-around view as did the dashing Admiral Beatty as he stood on the open bridge of his flagship "Lion," at Jutland in 1916, with dozens of coal smoke-spouting ships spread from horizon to horizon, guns flashing, enormous shells whizzing through the air, and the Admiral's precious battle cruisers blowing up before his eyes. We have already noted that to produce a fully realistic 3-D image that fills the entire visual field requires a computer doing about a trillion calculations per second. Motorola and Intel are about to release chips that perform a few hundred megaflops; they figure they will have chips running at a gigaflop by 2000 or so, and the ways things are going, teraflop chips ought to be going into home computers in the second decade of the next millennium.

A teraflop chip doing heavy VR is pretty impressive, but it is still not at the human level of 10 to 1,000 teraflops. How long will it be until you can shop for a new brain, as it were? The aerodynamic engineers and weather folks have already decided that a teraflop is not going to be good enough. The astronomers who enjoy simulating exploding stars, colliding galaxies, and the end and beginning of the universe sneer at a mere trillion calculations per second. So they are already talking about petaflop machines capable of doing 1,000 trillion calculations every second, which happens to be the high-end estimate of human brain power. Let's figure this out: A petaflop is 10,000 times faster than anything we have online these days, so with the speed of top-of-the-line computers doubling every 18 months or so, it should take 20 to 25 years to build a room-filling petaflop machine. But perhaps we are too pessimistic. The same Japanese group that brought us GRAPE-4 hopes to link 20,000 processors into a $10 million petaflop computer by 2000 or so! As for the cheap and portable, if they continue to lag about 15 years behind the big guys, you should be able to drop in at your local computer store and pick up a petaflop device in, say, 35 to 45 years. If, on the other hand, a 10-teraflop machine will do, they should be available sometime around 2020.

We challenge skeptics to come up with a list of people who, 30 years ago, foresaw 100-megaflop chips in wee little computers costing a few thousand dollars for sale in this century. There would have been people happy to explain in boring detail why today's laptops were impossible to build. If we assume computer power will remain below human brain levels well into the next century, the projected growth curve takes a sharp turn from vertical toward horizontal for the first time in decades, just when our knowledge of how to make computers ever tinier and faster is reaching new levels. This may not be impossible, but it is far from a safe prediction. It is more likely that a sudden breakthrough could produce hypercomputers earlier than we predicted.

We have estimated the future growth of computer power by extrapolating past trends into the future. This is a powerful method, but not the only reason to believe hypercomputers will be in labs and homes very soon. Some of the researchers have set up an online database intended to pull together the varied disciplines speeding up the development of such machines. Our predictions are not extreme; rather, they are mainstream. We have extrapolated long-term past trends a few decades forward and found that they are in line with the current plans of the computer industry.

MEMORIES

A brain-like machine with lots of speed requires ready access to enough stored memories to have a sense of identity at the human level. Considering the data density of visual memories, it is not likely that such a system can get away with less than what a brain needs, about 1,000 trillion bits.

Computer memory capacity improves at about the same pace as processing speeds. From almost nothing to fantastic in a few decades, exponential growth of computing power should reach human brain levels in a few more decades. We are more interested in the random access memory contained inside the computer rather than the data stored on peripherals. ENIAC could store about 1,000 bits, UNIVAC 40,000, a mid-'60s mainframe a few tens of millions of bits. The first Cray stored 300 million bits, and the Sandia beast will probably hold a few hundred billion bits. At this rate, the memories of the most powerful machines should reach the human range around 2015. As usual, there is a two-decade lag between the memory levels of top-end machines and those you can afford.

Let's not, however, ignore the peripherals. One advantage of an artificial brain will be the ability to link its information processor directly to precise digital memories, greatly expanding the available database. No more accessing external data through the senses. So far, memory storage has been hampered by its two-dimensional nature, first as rust on tapes and then as rust on disks. Magnetic storage is early-20th-century technology, and its continued use is something of a low-tech embarrassment. To achieve brain-level memory storage capacity, it will be necessary to do what the brain does: store information in three dimensions, preferably in a holographic manner. The first step along these lines may be read and write CDs storing data in a single disk on multiple levels.

A library of books could reside on such a disk. But disks need to be spun at high rpms, a crude affair when things ought to be done electrically with no moving parts. The next step may be installing light-sensitive crystals that use little lasers to code molecules storing and retrieving data holographically. Because the data is stored in three dimensions, terabytes of information could be stored in a space the size of a child's wooden block. Even bigger libraries can be stuffed into a crystal the size of a D-battery and, look Ma, no moving parts! Better still, the data can be up- and downloaded thousands of times faster than with spinning disks.

The Cost of Memories and Speed

The exponential increase in the calculating speed and memory capacity of computing machines has been paralleled for an equally dramatic decline in cost. ENIAC cost a few million in today's dollars and could do about a billion calculations over two years, so each dollar bought about 500 calculations. Not bad, but these days a $2,000 machine will do a few thousand trillion calculations if you push it to factor primes for a couple of years. So you get about a trillion calculations for your buck! Quite a deal.

It costs a few thousand dollars to raise and support a person in the third world and a few hundred thousand to raise and support a Ph.D. candidate. For 20 years, a Ph.D. employee plus computer will do about ten million trillion calculations, so a dollar buys somewhere around ten trillion calculations. Computers doing calculations at such speeds and at a similar cost should be available in three decades, plus or minus a number of years. After that milestone has been passed, the cost of doing calculations should quickly drop below human standards.

Why Computers Are Getting Faster and Faster, Smaller and Smaller, and Cheaper And Cheaper. Or, Why Denser Is Better

Computers are doing what they are doing because, in the larger scheme of things, it is relatively cheap and easy to improve them, and those who do the improving get a lot of money for the effort. Computers are small, and relatively small groups can tinker with them in relatively small labs. Big discoveries and innovations can be worked up by only a few people with a modest amount of money. Huge rocket boosters needed to go into space or the enormous tunnel machines and big reactors needed by particle and atomic physicists aren't important in our "small-is-beautiful" domain.

With computers, smaller is better. What it all comes down to — the stuff about speed, memory, size, and cost — is information density. The higher the density, the smaller the machine can be. The brain has a very high information density; all that speed and memory packed into just 1.5 kilograms. In the outer cortex, there are a few hundred million neurons per square inch, and the density of synapses in the cortex is about ten times higher. The latest commercial chips have a few million circuits per square inch. Because a modern chip is a thousandfold short of the information density of a brain, even TFLOP will be hundreds of times bigger than a brain, although its overall power will be less. Much better than ENIAC, though, which had one vacuum tube for every few square inches!

As information density increases, cost goes down. It is cheaper to manufacture something the size of a chip than a room full of vacuum tubes. The smaller a computer, the cheaper the machine is to assemble, so costs spiral downward. The physical needs of high-speed computing are also driving down the size of fast machines. The smaller the machine, the shorter the distance between circuits, so the less time it takes for them to communicate, so the faster the machine is. For example, one plan for making the first 10-teraflop computer envisions a hybrid technology combining protein molecules and semiconductors in a package one-fiftieth the size of a Cray mainframe using 100 times less power. The advantage of building the fastest computers as small as possible is one reason that the time lag between mainframes and portable computers of the same power continues to decrease.

However, there is a cost to making computers smaller. A single vacuum tube was simple and cheap. A high-end chip of smaller dimensions is more intricate and expensive. This is because a chip is orders of magnitude more intricate than any tube, and it takes lots of high-tech machinery and know-how to make the chip. Designing a new high-end chip today can cost a billion dollars, and building the factory to make them can cost another billion or two. Such expenses have driven smaller chipmakers out and renewed predictions that we are about to hit a plateau in chip development.

What we are really seeing is a consolidation of the industry comparable to the shift from hundreds of carmakers in the first decades of this century to the

Big Three. A completely redesigned new car model costs a billion or so to develop and produce; even a sports stadium can cost half a billion. Development of a new airliner costs so many billions that it threatens the finances of even a big company such as Boeing. A new airport or subway costs multibillions, and eight miles of superhighway runs a billion dollars these days. One reason that new chip factories cost a billion is because they are big, and they are big because it is efficient to be so and because demand for chips is so high. The chip market these days amounts to hundreds of billions of dollars. There is lots of money to be made in chips, and to make the big bucks the chips have to be as fast as, if not faster than, anything else. People are more than willing to put down hard cash in exchange for more computational power, and the investment/return ratio is very favorable, unlike space flight, which is just a bottomless money pit. So competition is fierce, and money is skimmed off the big cash inflow to make better chips. Although the cost of these chips is substantial and they carry so much more information than the vacuum tubes of old, the deal is almost infinitely better.

With the post-cold war Department of Defense in decline, deficit-busting in vogue, and corporate downsizing all the rage, concern has been expressed about an investment decline in new, radical computer technologies. The huge monetary investment already sunk into silicon chips acts to dampen any major diversion to alternative, potentially more effective, technologies. The long-awaited teraflop ultracomputer will be made of assemblies of commercial chips rather than some standards, upsetting new and exotic design. The tendency away from exotic computer concepts is more fertile ground for predictions that computer performance is about to hit a plateau, and it is possible, if not likely, that a modest decline in the growth of computer technologies will occur. However, this will be part of a regular cycle, not a cause for panic. After all, science continued to progress rapidly during the Great Depression, and things are certainly better than that today. Economies are growing rather than contracting, and pressures and incentives to improve computers remain high. Intense global competition is spurring computer evolution; no one can afford to be left behind. As China comes online as a modern economic power, it can be expected to boost computer research in the next century. And lest we forget, probably 95 out of 100 computer scientists and engineers who have ever lived are still tinkering today.

Are Chips About to Choke?

Silicon semiconductors have been pushed far beyond where many thought they could be pushed just ten years ago, and they still have some good years left. But melted sand appears to have limits. How small can it be fashioned and still perform effectively as switches and gates? For a while, it seemed gallium arsenide, that stuff Cray liked to make supercomputer chips out of, might give silicon some competition, but this tricky material has proved a commercial bust. Silicon-germanium combines compatibility with traditional silicon technologies with superior microperformance and may displace silicon, at least for a while.

Ever tinier circuits bring us to another modern technology threatening to run out of post-modern steam: circuit-making via etching. Chipmakers use light to etch the circuit pattern onto silicon chips. Problem error rates are high, and avoiding

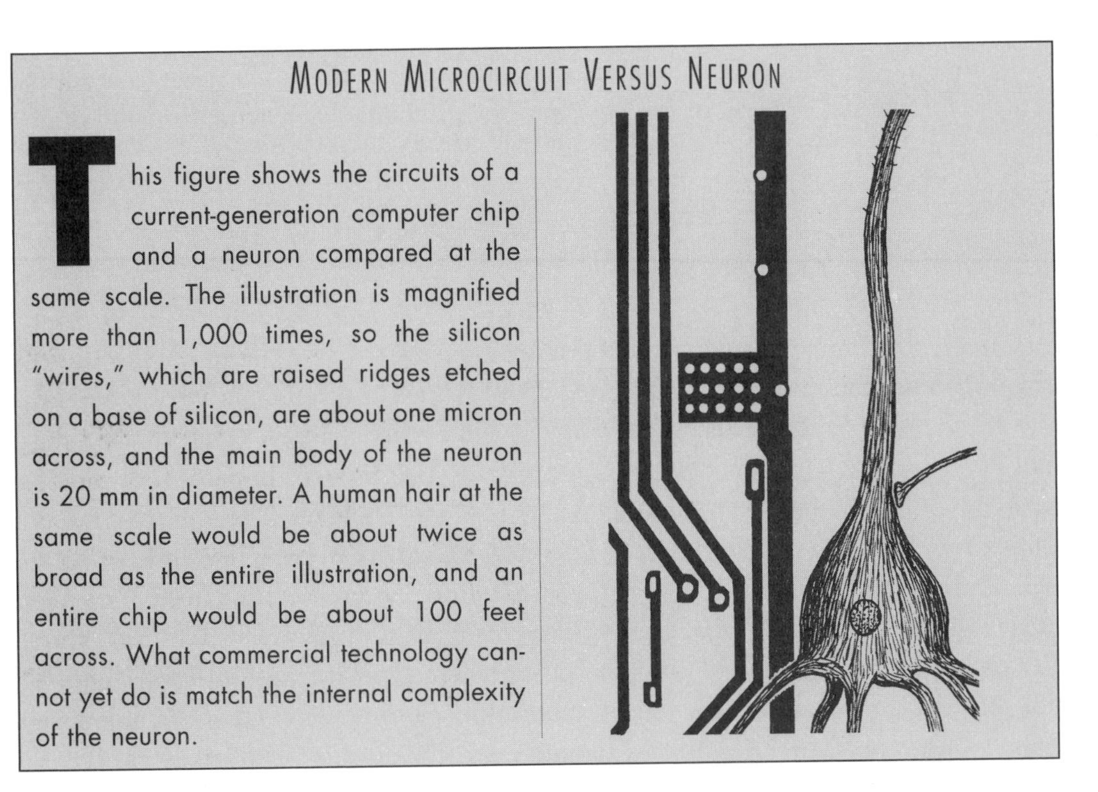

Modern Microcircuit Versus Neuron

This figure shows the circuits of a current-generation computer chip and a neuron compared at the same scale. The illustration is magnified more than 1,000 times, so the silicon "wires," which are raised ridges etched on a base of silicon, are about one micron across, and the main body of the neuron is 20 mm in diameter. A human hair at the same scale would be about twice as broad as the entire illustration, and an entire chip would be about 100 feet across. What commercial technology cannot yet do is match the internal complexity of the neuron.

the smallest dust contamination is a chronic nightmare (which is why chips are made in clean rooms). Basic physical limitations soon will make it impossible for photo lithography to pack yet more circuits onto a given area of chip. A given color of light can etch points and lines only as small as its wavelength, and even smaller blue wavelengths are much bigger than atoms and most molecules. There is talk of using X-rays to etch the chips because the shorter wavelengths will cut finer than visible light. Even if this can be made to work, and the low cutting power of X-rays casts doubt on this, the dust contamination problems will be worse than they are with light etching. A very different technology proposes to stamp out chips rather than etch them. The process avoids the contamination problem and should cut costs, but the technique has not been converted to commercial use.

Currently, silicon semiconductors can be etched with circuit lines about one micron in width, about 100 times thinner than a hair. It is believed that the technology can be pushed until lines are one-tenth of a micron in diameter. Just how impressive this is becomes obvious when it is considered that neurons are about 20 microns across and the axons that connect them are one-tenth to one micron in diameter. Conventional semiconducting technology will soon be able to produce some components as small as biology, but not the nanoscale structures and operations that go on inside neurons.

From Microtech to Nanotech

Eventually, the conventional semiconductor industry will run out of silicon steam. The swift increase in computer speed is about to reach a practical barrier, eventually reaching a plateau much as the rapid evolution of airliners came to a sudden halt 30 years ago. This analogy is inappropriate. Airliners are unnatural machines with no close, living equivalent showing the way to dramatic improvements. Computers will have to become more like brains if we are to exploit their full potential. Rather than stopping the great computational increase, the need for ever-higher speeds and smaller circuitry is likely to force a dramatic switch away from the top-down process of etching silicon. Replacing it with bottom-up nanotechnologies will grow a suitable material (diamond, or harder-strength-designed materials, is a strong contender) into extremely small information processors. If and when this happens, silicon chips and the semiconductor industry making them will

What Synthetic Brains Might Be Made Of

What is needed in the long run is something made from a common element with characteristics dramatically superior to those of silicon. This has some looking at carbon, in the form of manufactured diamond. When made into a diamond crystal, carbon is not only very tough, it tolerates and conducts heat very well (important because high density computers tend to run hot). Diamond also has certain characteristics that make it well-suited for being formed into circuits smaller than neurons. Another form of carbon, buckytubes, may provide molecular-scale wires much thinner than axons. Buckytube wires also combine great strength with flexibility.

Carbon in the form of organic materials may provide the basis for yet more brain-like machines. Hinge-jointed protein molecules may be used to make gates and switches at molecular scales. Many proteins work when dry, so this does not necessarily mean that protein AC machines will be wet like ours. However, some concepts for future computers envision large molecules forming the workings of computers on thin supports of wet film! Light-sensitive solutions of synthetic DNA may store huge databases.

go the way of vacuum tubes. As important as the semiconductor industry has been to cyber evolution it is not likely to form the substrate for the Extraordinary Future. Such a switch should not be surprising; it will be an example of technological punctuated equilibrium.

Leaping Computer Power

It is possible that the smooth slope increase in computer power will become irregular as new ways of boosting computer data density come online, only to be bumped off by better technologies. Such a situation might slow down the overall increase, but it also opens up the possibility of sudden jumps to extremely powerful compact machines in the near future.

Consider one possibility for such a jump, the electron spin computer. An ESC would exploit quantum mechanics to bring order to the random spin of electrons and turn the tiny particles into on-off switches. One of the marvelous things about

ESCs is that the smaller they are, the better they work! ESCs would be made of conductive metals rather than semiconductors, and elements may be as little as one-hundredth of a micron across, or the width of a hundred atoms. A trillion transistors could be crowded onto one chip. The information densities of quantum computers would surpass those of brains many times over.

WILL THE BIG AND SMALL CONVERGE?

So far, it has taken small computers years to catch up with the big boys. This may change if the most sophisticated computers are nanobuilt machines in which switches are made from a few molecules. It may be necessary to make computers small to keep the wiring as short as possible and thereby minimize the data transmission time between circuits. Of course, a hypercomputer the size of a bread box will always be cheaper to make than one the size of a room, so cost savings may also drive the big squeeze. When the high-end computers bow out, the size-related performance lag will be eliminated. This may mean that the advent of small petaflop devices will be pushed forward.

CONCLUSIONS

Basic technologies for making computers capable of rivaling human brains in speed, capacity, circuit miniaturization, cost, and size will be online in 20 to 50 years. It is improbable that it will take much longer — well over half a century to a few centuries — to reach these targets because the economic forces pushing such capacity are strong and because there are no apparent barriers to such information density. However, the development of computers remains subject to unpredictable evolutionary forces, including surprising leaps in technology. Our estimates of the growth of speed, capacity, and complexity are conservative. They assume the complexity of computers will continue with the same top-down development manner practiced today. Techniques for evolving and growing computers from the bottom up may spring up as if out of nowhere, giving us computers of extreme calculating power and memory capacity. It is hard to rule out a sudden leap to human levels before the curve suggests. What this all means is that some of the *minimum* requirements for making mind machines will probably be met in the near future.

Just because small computers as fast-working as the human brain is may be available around 2030 does not necessarily mean it will be possible to run a conscious mind on one of them! They may just be faster and fancier versions of the number crunchers we already have. Or maybe not.

Making Machines Perform Like Brains

The belief that small computers will match the human brain in gross speed and capacity within a few decades is not especially controversial. Where opinions diverge is whether these systems will be as capable as the human brain; especially whether they will be conscious to any degree.

The field of AI has always been contentious; the idea of AC is more so. It's true that people have been ostracized at certain institutions for their ideas. Promising areas of research have been suppressed, such as the near-disappearance of neural networks, for more than a decade. Yet when people are outraged at the other guy's idea or skepticism of an idea, a call to neurons is spread far and wide leading to increased interest and creative output by motivated individuals and institutions. A general consensus could be wrong. It's better to ensure a diversity of competing ideas, increasing the possibility that someone may be on the right path. A clear consensus has emerged that classic, strong AI will not succeed. Programming computers to be as smart as we are is considered a nightmare of tedious complexity. Beyond this minimal consensus, just about anything goes, and the argument follows two main lines of thought.

Emergent AC

The evolutionary AI folks trust that as the gross computational power of computers rises and exceeds human levels and as software programs evolve, exploiting the full capacity of these machines, the equivalent of human intelligence will be at hand, if not met. Hans Moravec is one leading proponent of this view. AI should be developed in a manner analogous to human flight. Human flight has a strong top-down component in the use of fixed wings rather than flapping wings. If the combination of mimicking and diverging from nature worked for getting into the air, why not for getting to high intelligence?

Getting high-level intelligence out of evolutionary AI is one thing, but getting AC may be another. Some AI researchers think smart hypercomputers will mimic human intelligence, but will not be aware they are doing so. This view assumes that even the highest levels of intelligent action are not dependent on being conscious. Others subscribe to the argument that consciousness inevitably emerges from high levels of cognition. In the latter case, the form of the mind machine is irrelevant except that it must be powerful enough to run the mind on.

A Fly on the Wall

If the evolutionary AI argument were correct, there should be a close correlation between robotic computer speed and the capabilities it produces. The simple action-reaction behavior of single-cell creatures has been mimicked by simple robots with low power chips (which discredits the assertion by Stuart Hameroff that the supposedly *complex* behavior of paramecium shows that they must have a quantum supercomputer running in their microtubules). These days, modern computer chips have computational speeds similar to those of insect brains, yet no one has built a robot as agile and intelligent as a fly.

Consider the ordinary housefly, a tiny self-guided biomechanical marvel causing robot builders and nanotech enthusiasts to turn green with envy. The fly has complex oscillating wings driven by miniature muscular motors and rapidly navigates in four-dimensional space (i.e., it flies in three dimensions while accounting for temporal/spatial changes in its surroundings) while interpreting stereo signals from two large, compound eyes consisting of more than 80 lenses. It does this swiftly enough to detect and avoid imminent danger (try hitting a fly with your hand) and to invert in flight so as to land upside down on a ceiling. It uses sophisticated chemical detectors to locate food that is processed in a miniature digestive factory. And the fly, of course, mates and reproduces. This behavior is guided by a speck of a brain and decentralized control centers consisting of only a few hundred thousand neurons, a value below the number of transistors in a late-model PC. The fly brain runs on about one hundred-thousandth of a watt, compared to the dozen or more watts burned by Pentium-grade chips.

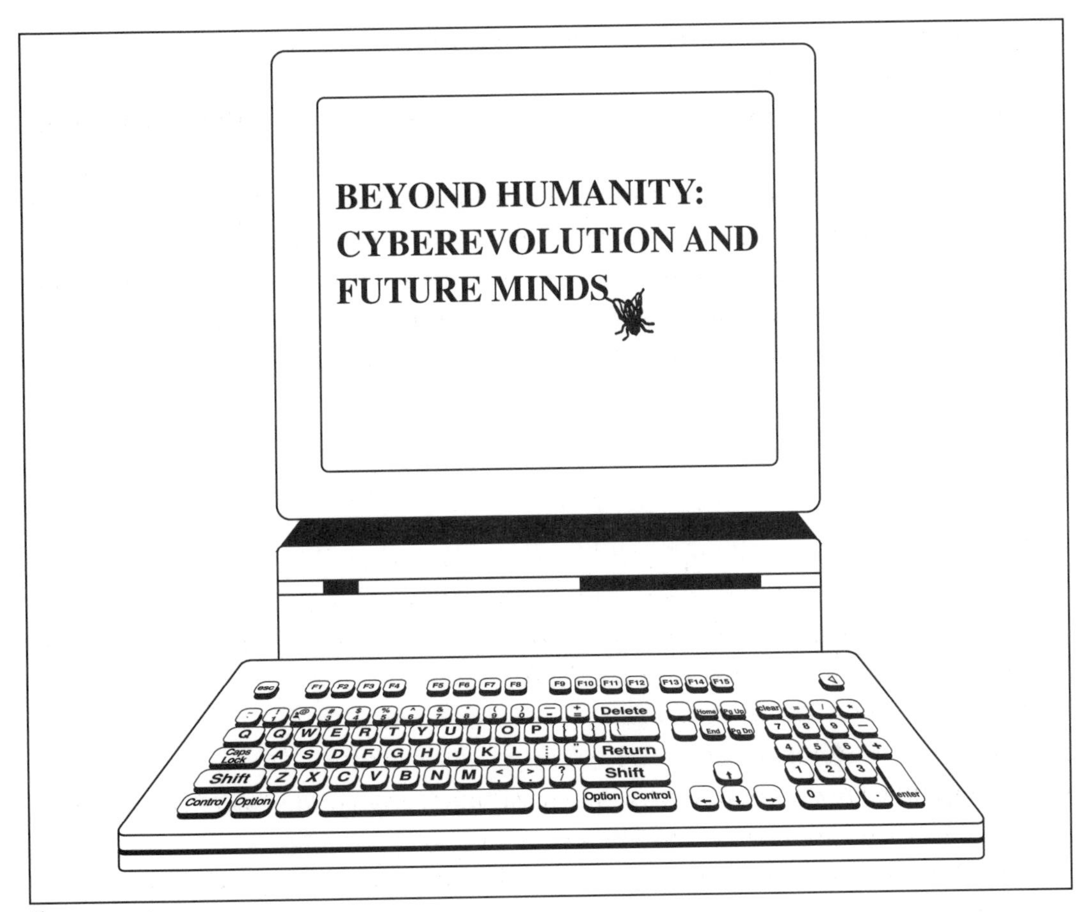

Fly on a computer screen.

By implication, bi-digital chips are missing something involved in replicating biointelligence. Penrose and Hameroff have suggested that bugs tap into even quantum mechanics. However, before getting carried away with the chip versus bug brain disparity, it should be realized that insects are not doing *that* much. Insect behavior is stereotypically limited. It is doubtful that conscious thought is involved in their behavior. It can be argued that the navigation ability of robovans and cruise missiles is approaching that of insects. When the human pilot of a F-117 Stealth attack jet becomes disoriented and the plane starts tumbling through the dark sky, the pilot pushes a panic button and the computer gets the craft re-oriented, a mean aerodynamic stunt. The decentralized walking action of real insects is being replicated to a certain extent by Brooks' robotic insects. No one has yet put big bucks into software packages needed to make a robotic insect. Instead, insects, mass parallel-processing analog-digital systems optimized for bug behavior, and chips, are built to do very different things. A Pentium chip is optimized for video games, word processing, and other nonbug-like tasks. The miniaturization and reproductive capacity of insects is due to their being nanosystems beyond the capacity of current microsystems, and their energy efficiency is attributable to analog versus digital information processing.

Working at AC

The way to high intelligence may prove to be a jerry-rigged compromise that includes both conventional programming and AL. Only time will tell. What we will do for now is set the goals high and assume that to think like a brain, computers will have to be built and work very much like a brain. What, then, will be needed to progress from number crunchers to thinking machines?

The Symbiosis Between Brains and Computer Research

A better knowledge of how brains work requires better computers that work more like brains! It is no accident that computer intelligence and our knowledge of brain function are growing exponentially at the same time. The remote scanning systems being used to examine brain function in real time are possible only because of high-power computers. Brain research is fairly cheap; a university hospital or lab can afford the machinery needed to do cutting-edge brain research. At the same time, researchers are using computer-driven scanners to peer into the brains that they are using other computers to mimic and model the workings of the brain. The investigative power of the combination of remote scanning and computer modeling cannot be exaggerated. It helps force neuroscientists to propose rigorously testable hypotheses that can be checked out in simplified form on a neural network such as a computer. Neuroscience is fast going from a field of speculative opinions to productive hypotheses based on confirmed facts and tested via real experiments.

Computers are starting to become more like brains and as they do, they allow a greater understanding of brains, which can be put back into making the computers more brain-like. The evolutionary convergence between brain and computer is inevitable and speeding up. As computers get better, brain scans are more detailed, and modeling becomes more sophisticated and realistic. In turn, a better understanding of how the brain works is applied to computers, making them more powerful and cognitive and making it easier to do brain research, which allows the computers to become more brain-like and powerful, and so on.

Parallel Processing

It is universally agreed that the new mind machines will have to duplicate the brain by being complex parallel processors running many algorithms at the same time. If for no other reason, this will be necessary to achieve the high calculating and memory access speeds required to form a practical intelligence. It is probably physically impossible to squeeze many trillions of calculations through just one or a few processors. The difficulty of reaching very high speeds through one processor is already forcing a switch away from the rather simple linear machines we have lived with for half a century toward more complex multiprocessor systems. For example, TFLOP will use more than 9,000 processors using new Intel P6 chips (the successor to the Pentium) to achieve its 2 teraflops.

TFLOP is actually simple compared to some supercomputers that have already been built. Even recent Crays have had only a few processors working in parallel; the high speeds are achieved by extremely high-speed special chips. However, about ten years ago, thinking machines began to build massively parallel-processing connection machines that used first thousands and then tens of thousands of relatively common chips to achieve total speeds matching those of more conventional supercomputers. Software problems and post-cold war cutbacks have helped to cripple Thinking Machines, but some of the latest computers of this type have a few hundred thousand parallel processors working at once. Computer complexity has, therefore, leaped forward a few hundred thousandfold in only 20 years. Does the failure of companies like that of thinking machines spell stagnation for computer complexity? Perhaps for a while, but only for awhile. There are many problems that naturally flow through parallel systems better than through linear machines, even when the overall speed of the machines is about the same. The airflow of aerodynamics and air masses of weather systems fall into this category. This, combined with the constraints imposed by physics, is likely to force still more parallel processing into computers and the development of the software that will control the machines.

It is, therefore not surprising that the Brain Build Group in Japan plans to build a neural network they call an "artificial brain" with one billion neuron-like circuits by 2001. This machine will not operate at human levels, but perhaps HAL is not so far off after all.

NEURAL NETWORKING

Along with parallel processing, just about everyone agrees that an AC computer will have to be a self-learning neural network storing holographic memories by strengthening and weakening connections. This means the processor and a large portion of the memory will be intimately unified. Whether synaptic-like structures for memory storage are critical is not certain, but they couldn't hurt. The result will be systems with human-like pattern recognition and redundancy.

COMPLEXITY

Here, we come to an important unknown: High-performance computers that employ parallel processing and neural networking are inherently more complex than conventional computers. How much complexity does it take to do all that a brain does? Certainly, we do not have to build a computer as complex as a brain to match the gross calculating speed of a brain. Remember that brains have more than a hundred billion neurons and trillions of interconnections because the neurons signal each other so slowly. A computer with circuits working at a million cycles a second would need only millions of neural circuits to match the total speed of a brain. But would a machine as fast and as data-dense as a brain, but much less complex than a brain, think as well as a brain? What if high-level consciousness can be achieved only by trillions of bits of information flowing in parallel between circuits that are generating standing waves of only a few hundred cycles per second? This is one of the major questions of AC at this time.

Another uncertain complexity issue is whether conscious machines will need to be special-purpose devices with various functions split into subunits as per brains, or whether general-purpose machines will do.

Fuzzy Analog versus Hard Digital

Previously, we discussed how neurons send fairly complex analog-digital signals, rather than simple digital signals used by most computers. Also discussed were the neuron-mimicking analog chips that produce signals virtually indistinguishable from the original item. It is possible, many would say probable, that conscious thought is dependent on analog-digital signals and the fuzzy logic they produce, and that purely digital-only systems will always be reflexively intelligent in the manner of insects. Others disagree.

Even if the latter are correct, analog-digital circuits have the advantage of being much more energy-efficient and flexible than digital circuits, which is one reason why brains are many times more energy-efficient than digital computers. Although digital processing is simpler than analog, it is more energy-expensive because it takes work to convert information from its initial analog form into large digital codes and then back again. If an AC machine the size of a brain used purely digital chips, it would not only sop up a large amount of energy, but it would also run hot, perhaps hot enough to melt. We conclude that thinking machines will probably need to be fuzzy-thinking analog-digital machines.

If the Penrose-Hameroff quantum hypothesis of consciousness is correct, fitting the synthetic neurons with synthetic quantum sensitive parts should produce quantum AC.

Feelings

Existence may prove dull for smart computers unless they have feelings as they apply to both sensory awareness and emotions. Some AI researchers who think consciousness will emerge in step with increasing power and intelligence suggest the same will be true of emotions, as well. Moravec, for instance, tells the story of the cognitive household robot that gets locked out of the home and is running out of electricity. Its "anxiety" rises as its power runs down, and it tries to decide whether to enter an unlocked home to tap its power — something it knows is bad. Conversely, we can imagine its "relief" and "pleasure" when its owner unexpectedly returns and lets the robot get its juice.

The contrary view is that feelings will prove one of the most difficult brain functions to replicate in view of the fact that animal feelings are highly dependent on wet chemistry. A neural-chemical reaction is just another form of information processing, one that the brain uses because it is convenient for water-rich organic materials. If cyberbrains are dry, electrical waves may be used to stimulate and shift moods. Alternatively, if emotions require chemical agents, cyberbrains may be partly wet, emotional machines! In any case, emotions may not be critical to early AC, although one can never be sure about these things.

Virtual Consciousness

At some point, the convergence between human brains and brain-like computers will go so far that the latter should start to do what the former does, generate conscious creative thought and foresight. This convergence in form and especially function is as close to inevitable as one can get without being there, but the details will remain murky until the event occurs. As previously mentioned, if AC does not emerge without prompting as computer intelligence rises, a more deliberate effort may prove necessary. AC machines may mimic brains with precise synchronous communications between circuits, generating transient standing mind waves. Perhaps information will need to be looped through a central processor, analogous to the critical part of the thalamus, before special neuron-like circuits communicating at a common frequency can become aware of the information. Alternatively, the mechanism for generating synthetic conscious thought may prove dramatically different from ours.

In any case, there is little doubt that AC will be a form of after-the-fact virtual consciousness of the sort we practice. This will allow the cyberminds to put together a coherent picture of the world before they are aware of it. It will also allow them to simulate, practice, and think through complex action repeatedly before the robot tries to do something difficult, just like we do. Of course, many actions, such as the details of walking and forming words, will be done subconsciously.

Growing Nanocomputers via AL

Obviously, if one is trying to emulate brains that are so intricate and complex because they evolve as they grow, molecule by molecule it makes sense to build equally capable cybersystems the same way. Indeed, from the bottom up may be the only way to create AC and the computers to run it on.

Two current bottom-up technologies are artificial life and nanotechnology. The first involves the immense knowledge base needed for high-level cognitive intelligence. AL also produces the complex debugged genetic algorithms that code for the design and basic growth of intelligent computers. Nanotech provides the means by which the AL codes are translated into the hardware of vast numbers of hyper-tiny circuits and wires needed to make a fake brain. For example, some of the researchers doing the preliminary work on petaflop computers are investigating how to use bioengineered bacteria to manufacture molecular transistors.

We can envision self-learning computers that evolve increasingly complex hardware in parallel with their increasing intelligence. They may be placed in mobile robots so they can learn about the real world. In this regard, evolving robots will build up a broad knowledge base by having a richness of experiences. Will this work? Insects, frogs, birds, cats, and humans develop working mind machines all the time.

Where the synthetic systems will differ from us will be in their ability to quickly build broad, deep knowledge bases by uploading from other AI expert systems. No need to go to school to get a degree.

DNA as a Teraflop Computer

Much of the mystery of DNA fades when it is recognized to be a quad-digital computer using four molecules to code information. Leonard Adleman figured out a crude but simple way to use a fraction of a teaspoon of DNA to do a trillion calculations in one second, using a billion times less energy than silicon would consume. Adleman used the DNA computer to crunch through a version of the traveling salesperson problem, which would tie up a conventional supercomputer indefinitely.

THE FIRST HUMAN-EQUIVALENT MIND MACHINES

We have already explained that the 3 lb. human brain runs at 15 watts with one or two hundred billion neurons firing a slow 100 cycles per second over trillions of interconnections to perform 10 trillion to 1,000 trillion calculations per second. The existence of the brain itself makes us confident that it is physically possible to build machines with similar capacities. We conservatively project a 21st century cybercomputer with one billion to ten billion circuits working in parallel via a trillion connections in a small package weighing 1 lb. The circuits are analog-digital switches that mimic neurons, and they form a self-learning neural network storing memories in holographic form while processing information at both subconscious and conscious levels. The analog-digital circuits are energy efficient, like neurons, but because the new systems are not organic, there is the potential to give them new and improved power and cooling features. The circuitry can be run many times faster than human circuitry. For example, we are not limited to sugar power; we can plug the cybercomputer into an outlet or a high-tech battery, so we up the power to 100 watts. The system will run hotter than our brains, but the device is made of diamond or some other material with high heat-resistance and we attach cooling fans. The biggest difference is that the nanocircuits run at a technologically respectable one million to ten million cycles per second and the wiring zips electrons millions of times faster than nerves. Running at a normal 10 percent of total capacity per second calculating speed of our new cyberpride and joy is a petaflop.

The minds produced by the machines are as self-aware and conscious as those of humans. They all have no trouble passing any Turing test thrown at them; some may grab a human by the proverbial collars and insist on their consciousness. On one hand, pattern recognition performance and the robust knowledge base of the system rivals that of humans. On the other hand, the digital component of the systems is better developed than in humans, so memory retention and mathematical abilities are superior. Whether the rational capacity of these first mind machines will exceed that of humans is uncertain. Also open is whether their emotional capacity will fall short of match or exceed the human condition.

A Major Alternative to Producing AC: Cerebral Cyborgs

So far, we have assumed that artificial consciousness will be achieved from the ground up by building an entirely new means of producing minds that may or may not mimic the animal version. There is another way to AC one that is not so alien: piggyback artificial thought on the natural system. The idea is simple enough. Gradually replace original brain parts with new ones that do the same thing until the entire brain has been replaced.

This concept has been used as a basic thought experiment to show the viability of AC and mind transfer. The concept is gradually being put into practice as a means of correcting brain defects. Initial experiments center around hooking up minicams to the visual cortex and replacing defective retinas with new ones. So far, these crude affairs only allow the blind to see patterns of dots and letters, but the technology will improve. The synthetic analog-digital retinas currently residing in

The Disabled and CyberEvolution

People with disabilities, mental and physical, are helping drive the development of cybertechnologies and robotics. The reason is obvious. Once biological systems fail to meet expectations, artificial aids and replacements are called for. The better the performance, the more the disabled benefit.

The basic computer is a gift of SciTech to the disabled. A computer can be programmed so even a totally paralyzed person can use one. This has become critically important in an age when the lifespans of quadriplegics, which used to be brief, are extending into decades. The deaf find computers very easy to use, they being almost entirely visual machines. The blind can also use digital machines via Braille keys and/or voice translation. As it happens, the current evolution of computers is challenging disabled users in different ways. The older software programs were ideal for the blind because they relied on words, letters, and abbreviations; the newer screen-icon-based software is useless to them. The latter works well for the deaf, but as computers learn to listen and talk, they will find themselves in a fix. Of course, the development of specialized software will alleviate many of these problems.

research labs will form the basis for high-resolution eye replacements. Of course, this is to allow the blind to see, but what will happen when it inevitably becomes possible to replace healthy eyes with superior artificial replacements?

Even more intriguing are the people at the University of Southern California who have developed the chip they say mimics the function of part of the hippocampus. The chip is about the size of a floppy disk, too large to be useful to us, but just as the earth spins on its axis, the chip will become ever smaller. The aim is to replace the hippocampus when it becomes too damaged to function properly. A few labs are looking into how failing neurons can be replaced one on one with new analog-digital circuits via nanotechnologies. This is intended to correct such ills as spinal injuries and Alzheimer's.

The main point is that AC may never be developed from scratch, but captured from the human system!

Minds Alien and Domestic

There are important differences between the two alternative means for making new minds. Because evolution is such a complex affair, it is not possible for two lines of evolution to produce identical results. If AC is evolved in an entirely new manner, its evolution will only roughly mimic, not precisely match, the evolution of human consciousness. The new minds will be very different from ours to the point that they will be aliens. Although they will retain aspects of humanity within them because we built them, we will have more in common with apes.

If AC is derived from modified brains, the new minds will not start out alien, but will soon become so. The difference between modified and alien minds has important implications for mind transfer, which we will address in more detail later.

Beyond Human-Level Mind Machines

Whether the first synthetic petaflop AC computers are made from scratch or from modified brains, they may run at *only* human speeds. But why stop there? We can be confident that we can do better than the brain. Why? Because big brains are too big! The long connections between neurons in different parts of the brain slow things down. The slow neurons and nerve transmission rates also can be improved on immensely. Instead of being aware of reality a substantial fraction of a second after it happens, the lag will be cut to a tiny fraction of a second. As nanotechnology improves the size of the neuron-mimicking circuits quickly decreases, while their number and speed increases. Because the intelligent computers are evolving, there is nothing limiting them to human levels of thinking speed, memory capacity, structural complexity, and overall mental powers. The machines were at one point equal to humans, then edge a little above human levels, then a little higher, and still higher. Power quickly soars far above human levels. Rather than taking a minute each to simulate a complex action a dozen times before doing it, the cybermind runs through 100 simulations in a few seconds. Because the cybermind knows the intimate workings and details of its circuitry, it can control and modify the evolution of the circuitry better than a human, so the cybermind does not get into mental ruts as easily as we do.

Tiny Computers Made of Tiny Atoms

To get an idea of how small atoms and molecules are, do this thought experiment: Imagine scooping up a glass of ocean water, and marking every molecule for identification. Pour the water back in the ocean and wait a long time until it has thoroughly mixed evenly with all the ocean's water. Use the same glass to scoop up another glass of water. You will find about a thousand of the marked water molecules floating around in the glass. An atom is about one billionth of an inch in diameter; even light waves are a thousand times bigger than an atom. A solid cube an inch across has a few hundred billion trillion atoms in it. Atomic-scale yes-no switching gates would be so small that basic quantum effects would not only be unavoidable, but would also have to be exploited to operate the switches. The circuit speed would be, therefore, about one quadrillionth of a second, which as you might know is a million times faster than a billionth of a second. In principle, it should be possible to cram a billion brain-equivalent computers into an atomic-level computer the size of a sugar cube!

The cybermind has additional digital memory storage and retains enormous amounts of information with a nonhuman degree of precision. In addition, the system can tap directly into the international computer network and upload whole libraries-worth of information in a few minutes. The human brain barrier is not merely broken, it is smashed.

As impressive as these superbrains may seem, they are possible because they represent only moderate increases over the speed, power, complexity, and information density of biobrains that clearly can be improved on. In this sense, they will be early and crude devices that may seem challenging to us, but in terms of size and power do not go far beyond the brains we take for granted. Some think we can do much better. Although brains grow at the level of molecules, the neuronal subsystems that process information are each made of multitudes of molecules.

The Atomic Computer

A computer with information densities far better than brains would operate at the level of a few molecules and atoms. Folks are already looking into how atomic-level computers might be made, while others are questioning whether atomic-level computers can be made. Atoms and molecules are ideal building materials, they are perfect and uniform in structure and they can articulate in precise and perfectly predictable joints. A simple atomic switch has already been constructed. It is not clear whether machines will work at such amazingly small scales and do anything important. For one thing, the amount of energy running through such tiny devices might melt them (then again, energy-efficient analog signaling may keep them sufficiently cool). Beyond atomic computers is the concept of subatomic computers, which would use the energy states of individual electrons to store and process information. Electron computers would be half a million times more information-dense than today's computers and thousands of times more compact than brains. We have no example of an atomic- or subatomic-level computer in existence showing they will work (unless the Penrose-Hameroff hypothesis is correct after all), and the failure of evolution to produce one may be telling us something. Then again, life has not produced 500-ton flying objects, either.

Superminds That Make Mistakes, and Cyberinsanity

These days, AI computers are prone to making glaring mistakes because their intelligence is shallow and brittle. We humans make so many mistakes because we do our planning with sloppy neural networks that learn and evolve with time, and part of learning is making mistakes. If artificial consciousness is achieved on neural networks that learn and evolve on their own, they too will do so by making mistakes. AC robots will not, therefore, be perfect. They will, however, have certain advantages over humans and today's computers. Capable of uploading large bodies of learning without doing the actual learning, intelligent robots will not make so many mistakes. Capable of thinking faster than humans, they will be capable of correcting errors faster, often before the consequences have occurred. Plus, robot sensory systems will be superior. Robots will not fall down stairs and run into doors as often as people, or step in front of cars.

We cannot judge at this time whether the tiniest atomic or even subatomic computers will ever be possible. We can say with a good deal of confidence that hypercomputers that match and exceed the performance of the human brain at least in terms of sheer power and even in complexity are inevitable. It is even possible that the first AC machines will be nanotech-based, atomic-level devices, in which case they will surpass brains from the get-go!

AC: Will It Be Hard or Easy?

The biggest question is whether artificial consciousness will prove a hard nut to crack, requiring special circuitry based on an intimate knowledge of brain function, or if it will emerge on its own as computers evolve intelligence. As tempting as it may be to say the former, the latter cannot be ruled out. We just do not know.

The Push to Produce Smart Machines

Just because it is possible to do something, does not mean it will be done. There has to be a reason to do it, and the results must be worth the investment. We could already have colonies on the moon and be on the way to Mars if we wanted to spend hundreds of billions of dollars for no obvious gain. The forces propelling the development of smart machines are much more compelling. Whether they are digital, analog-digital, or quantum mechanical in their operations, there are strong pressures to make mobile machines think and do what we do.

My Kingdom for a Robot

As much as they may protest to the contrary, industry and commerce really do not like human labor, physical and intellectual. Consider things from the viewpoint of owners and investors: Even a nonskilled worker requires up to two decades for the potential employee to grow and be educated to a minimal level; skilled workers can take three decades to get online. These are costs industry must subsidize through the elevated wages they pay parents and skilled employees. Once the person is working, they want to do so for only one-third to one-half of a day with one or two days a week off as well as paid vacations. Even when at work, down time due to lunch and coffee breaks, restroom visits, personal phone calls, gossiping, computer games, daydreaming, drowsiness, and so forth cuts down further on productive time. It does not help that humans resent controls meant to keep them as productive as possible, but costly efforts must be made to keep employee moral at sufficiently high levels to maximize productivity. Laying off large numbers of workers when they become too expensive or their skills obsolete adversely impacts employee moral. This is when human workers tend to get into ruts, which makes retraining them difficult and costly. Every worker is a potential thief, whether it be a few items from the candy aisle or a few hundred million from the brokerage firm. At the same time, even honest human workers want enough money to live pleasant lives outside of their jobs including large structures to live in, vehicular conveyances, and assorted and sundry hobbies, pets, and entertainment. All this assumes the worker remains healthy. If not, at best productive time is lost; at worst, the business must help defray medical costs either directly or indirectly. After just three or four decades of working for only one-quarter of each year (for a total at-work time of about ten years), the retired worker actually expects to continue to be supported until death! If the human employees are not happy with the arrangements, they may go join a union that restricts the freedom of the owner/investors and may even hold a strike that hampers or even ceases operations. If the employees are really mad, they may engage in a little sabotage or go so far as to destroy the facilities and attack the owners.

No wonder businesses want smart robots and computers that will work and think like humans, only more efficiently, without requiring long education, around the clock, without vacation pay and benefits, always be honest, and never go on strike.

Robotic Daredevils

Industry also wants smart robots that can go where no human should be and do things people cannot do, such as working deep in nuclear facilities, or in mines where ore digging and transporting machines are well on the way to replacing human miners. Instead of exposing human divers to the dangers of the deep, and their employers to the costs associated with their too-frequent deaths, why not use underwater robots? How about electrical high wire robots? The list is almost endless.

So many leading vulcanologists have died studying the objects of their research that the survivors are pressing for ambulatory robots such as Dante II that can descend into natural hells on earth. Oceanography is made for robots. Bob Ballard, who discovered the wrecks of the *Titanic* and *Bismarck*, has become a leading proponent of deep seabots, perhaps in part because of some close calls he experienced while squeezed into tiny submersibles a few miles down. Even when things go well, humans can spend only a few hours each year in deep waters. To fully explore the depths will require a host of smart swimming seabots. NASA knows that astronauts cannot devote enough extracurricular time to building large structures in orbit or on alien bodies at a reasonable cost. Nor is it yet practical to send large numbers of human explorers to other planets. NASA is, therefore, a leading funder of research into robotics.

Then there is the military. Faced with a public that does not think its young men, and now women, should die in combat, the military wants a lot of good robowarriors. With a budget almost a large as the combined militaries of the rest of the world, even the post-cold war Defense Department remains on the cutting edge of robotics and AI.

Better Health Care through Cybertech

Understanding brain function, replacing brain parts with new ones, using remote scanning technologies, and cracking entire genetic codes all depend on dramatic advances in computers, AI, and AL. The nightmare of diagnostic procedures and the craving for cost cutting is spurring the development of sophisticated AI expert systems. Even robot surgeons are being introduced to perform special tasks such as prepping bones for artificial joints that clumsy human hands do poorly. Now being researched is a special little robot that will explore the entire colon for polyps and other things that don't belong there. The ultimate dream is a microscopic nanorobot that will correct defects from clogged arteries to cancer with no muss and no fuss.

You Just Can't Get Good Help Any More

You want smart machines. That's right, *you*. Once upon a time, middle class families could afford in-house help, and much of the driving was handled by horsepower with enough horse sense built in not to run off the road or run into other conveyances or pedestrians. It is one of those ironies of modern life that automobiles are so dumb that you, the driver, must guide them along every inch of every road mile, or you will suffer the consequences. After you bring back the good old days with your street-smart robocar, after a few years, you will want a personal robot to clean the house (without taking anything from it), do the dishes and put them away, watch the kids (without looking into someone's records for past charges of child abuse), be available any time of the day, and not require filling out any Social Security forms. You also want a smart house that provides its own security, detects and puts out fires, turns the lights on and off, raises and lowers the shades, and cooks the meals. Even if you don't want them, others do.

For your computing pleasure, you would like a whole set of intelligent machines. Most obvious would be the general-purpose machine that listens and responds to you, is smart enough to read your miserable excuse for handwriting, and uses intelligent agents to help determine your wants and needs, bringing a little order to your chaotic postmodern life. Also useful would be other little intelligences inserted unobtrusively into common devices such as phones, ovens, thermostats, calculators, remote controls, lawn mowers, saws and other powered hardware, and just about anything else that can be improved with a little AI.

Move Over Henry Ford and Bill Gates

The plain fact is that lots of people want smart systems that do just about everything humans do. Today's computers actually do only a little of what we would like them to do because they are still weak. A must for making machines really useful and smart is speed, speed, and more speed along with more and more memory inside smaller and smaller devices. Premiere among the wished-for smart devices are robots that can move at least as fast and adeptly as people, identify, pick up, and manipulate objects as readily as we do and use them as intelligently as we do, all by voice command. The amount of money to be made by those who design and produce such mobile thinking machines will dwarf the automobile and aircraft industries combined. Even downsizing, bankruptcies, and associated cuts in research budgets sweeping through industry and the military have only whittled down, not stopped, leading-edge research in robotics.

THE NEW EVOLUTIONARY FEEDBACK LOOP

The attempt to make robots mimic what we do will compel scientists and engineers to resort to increasingly brain-like evolving analog instruments. At the same time, the pressures of medical research encourage computer scientists to use their machines to better understand the workings of our cranial computers. The result is a fast-rising learning curve. Some of the wealth generated by the commercial success of increasingly smart machines will be used to make them smarter in a classic capitalist feedback loop. This pattern may enter a new phase if and when nanotechnology comes online and pumps the evolution of AC machines to new levels of sophistication. Computer science and neuroscience are in the midst of a mutually beneficial evolutionary feedback loop. The ultimate result is obvious: Computers are becoming more and more like brains.

At some point, all these forces probably will come together and force a shift away from the simple calculating machines we are familiar with to more sophisticated and subtle systems that combine the best aspects of computers and brains. Conventional computers may remain common because of technological inertia and because they can continue to perform simpler tasks. However, the leading edge of cybertech will be the brain-like machines.

HOW LONG TO AC?

The limited success of today's robots is not a good guide to future developments. In 1896, no one had an engine that was light and powerful enough to propel a manned aerial machine, and the need for a dynamic flight control system was not well understood. In 1996, no one has computers small and powerful enough to match human-level performance, and much has to be learned about how brains work.

HOW FAR WE HAVE COME

Computers can run rings around humans when it comes to doing math and are beating the human champion players of complex board games. Computers are coming close to passing Turing's test (in line with Turing's 1950 prediction that it would take

50 years to do so). The gross calculating speed and memory capacity of computers are growing exponentially and have come so far that human levels will probably be reached in a few decades. The miniaturization and information density of computers are approaching brain-like levels at the same burning pace. The complexity of the machines is rising swiftly as parallel processing is introduced and expanded. A symbiotic interaction between neuro- and computer sciences is facilitating the exponential increase in our understanding of brain function, which is then being used to make computers more brain-like. The knowledge has been used to produce neural networking computers that learn on their own and do a modest level of innovative thinking. Some of the networks are made of analog-digital circuits that closely mimic the function of real neurons. The functions of small parts of the brain and retina have been reproduced with sophisticated chips. Research into replacing brain parts with synthetic systems is under way. Computers are beginning to understand and use language; others are learning to recognize faces. AI expert systems are becoming a normal part of business and government. Robots are learning to navigate dangerous terrain. Some are learning to walk and do flips. Robots are exploring the oceans and space. A few robots can understand and carry out fairly complex requests in a flexible manner. Mobile robots are taking their first tentative steps into the commercial realm where static robots already rule many assembly lines. Some weapons are as smart as insects. AL computers are evolving complexity from simplicity. Whatever it is that computers and robots do, they are getting better and they are doing new things all the time. Although the rate of progress may *seem* slow, it is shockingly swift by geological and even historical standards. At the same time, researchers are plumbing the depths of the human mind machine, and smaller-thinking researchers are laying the foundations for building new mind machines atom by atom, molecule by molecule, and intend to out-do life at what it does best.

What has already been done with AI, AL, and AC is astonishing, and the rate of progress is equally so. It is odd that this fact is not better recognized, perhaps events are moving too fast for people's perceptions. It is also important that a critical defect barring the way to human-level cognition in machines has yet to arise, and as far as can be conjectured, there is no such barrier. Because the effort to produce evolving neural networking analog-digital robots built on the nanoscale has just begun, it is tempting to say that AI/AC is still in its infancy. That was certainly true until the '80s, but a case can be made that things have already progressed beyond that point. The effort to endow machines with human-level cognition has passed into its early childhood and gone from taking baby steps to making the first running strides, which should soon be followed by adolescent leaps early in the next century, and then Olympic pole vaults soon after.

Will Cog Be the First to Break the Human Barrier?

MIT's Rodney Brooks, the builder of those little roboinsects, has launched a project under the direction of Lynn Stein to "evolve" during the next five years an android with the physical skills, cognitive abilities, and, hopefully, the consciousness of a two-year-old child. The upper body of Cog has already been built and its computers are starting to learn about the world. It is said to be unnerving to be tracked by Cog's twin eyes, and people in the lab have set up a group to consider the ethical problems associated with a budding consciousness. These goals are unlikely to be reached because even a child's brain is far more powerful and complex than any computer will be in the next ten years.

Cog does not represent the only attempt to construct a working android. At Waseda University in Japan, WABOT-2 is an articulating torso plus head that can recognize music written or played and then play the same tune on a piano to near-concert standards. WABOT-2 is meant to be a stage toward fully functional androids that interact with humans, even on the emotional level.

Then there is Steve Thaler, a private inventor working on his Creativity Machine. This neural networking computer system has already "written" thousands of original tunes, designed soft drinks, discovered unique minerals that may match diamonds in hardness, and has been hired by a high-tech company to search for high-temperature superconductors. Thaler also has his current-generation CM working on the next-generation CM. Thaler hopes that his Creativity Machines will evolve until they become conscious and do so rapidly enough so he can leave his aging body behind and live forever in cyberspace.

Although the Cog, WABOT, and Creativity Machine programs appear over-ambitious at this time, they are worth pursuing because much will be learned. Besides, maybe we will be surprised and Brooks will make naysayers of his critics with the first highly capable robot. It is more likely that Cog will prove a sophisticated insect. Perhaps a better model for a demonstration robot would be a robofly, a small, autonomous aerial machine that can reliably and flexibly carry out tasks by voice command. Georgia Tech holds a yearly competition to see if student teams can get a flying robot to pick up a small metal disc in the center of one raised ring, carry it over a barrier, and deposit the disc in another ring. A simple enough task it would seem, but no radio controls are allowed. Dozens of machines, balloons, tail sitters, and model helicopters have shown up. Some do not get off the ground, others drift off in the breeze, but most prove too sensitive to slight perturbations to complete the job before they tip over and grind spinning blades to bits and pieces. (The robot competition increases as their ability to do "human" things of creativity and innovation increases.) So far, no aerial robot has won the competition in a convincing manner. In the last competition, one chopperbot succeeded in picking up the disc with a magnet and carrying it to the other ring, but it could not drop the disc. The machines get better every competition cycle, and one will someday complete the task. Then the task will be made a little harder, a craft will do that version, the task will get still harder, and before we know it, agile robots will be on hand.

A few years ago, Hans Moravec tentatively estimated it will take 40 to 50 years to make robots as intelligent, capable, and potentially as conscious as humans. This projection is based on expectations of small cheap computers becoming as powerful as brains. Not surprisingly, this prediction is controversial; most simply refuse to believe that humans can be matched so soon. Roger Penrose, for instance, hints that it will take a long time to build his quantum mind machines, but he is a Platonic mathematician who thinks arithmetically, rather than an evolutionary biologist who thinks exponentially.

The common belief that we are far from achieving human-level performance in computers is a naive illusion due to the exponential growth curve of science and technology. Because today's computers have only a fraction of a percentage of the power and complexity needed to make a human-equivalent robot, it seems natural and obvious to assume it will be a long time before it happens. Not necessarily. Computer power and sophistication has come so far and is growing so fast that the gap may well be covered in a few decades.

We, too, have a bone to pick with Moravec's prediction, but for different reasons. Moravec presumes top-down programming will be involved in making robots increasingly smart via evolutionary AI, although there will be a strong self-learning component, as well. As part of the process of teaching robots to do better and better, Moravec outlines a series of ten-year stages to human-level performance. It starts with not-so-intelligent but useful mobile robots by or soon after the turn of the century, followed by a series of major moves toward human cognition every ten years or so. In this view, there will be a period of decades in which mobile robots are capable and fairly intelligent, but still well below human norms.

Moravec's outline of a fairly smooth and linear build-up to human-equivalent robots may be as naïve as the naysayers' belief that it will be a long time before we get to the ultimate goal. Look at it this way: Many people will go along with the idea that in the not-too-distant future, it will be possible to make a mobile robot with a mind that is the mental equivalent of, say, a dog. They will then go on to argue it will be long after that before a human-equivalent system is up and operating. They could hardly be more wrong because they forget the implications of exponential growth leading to over-prediction of the near future and under-prediction of the distant future. They also fail to fully appreciate their pet dog's abilities. Remember, dog-like brains first evolved only a few tens of millions of years ago, in the geological perspective, it was only a few moments before human brains showed up. Dog brains are powerful, complex, fast-running, and intelligent systems in their own right. They probably have at least a tenth of the gross calculating power of a human brain. The gap between a man and his dog is much less than that between a dog and its flea. Once we have achieved dog-level robotic performance, the projected growth rate in computer power means that machines with human-level performance will arrive only four years later! Even if we assume a time lag due to the difficulty of programming hypercomputers with human-level knowledge, the gap should be measured in years rather than decades. By the same token, MIT's Cog, if it works as well as Brooks says, will also exhibit a large fraction of adult human intelligence, and the jump to adult levels would be short and modest.

To squeeze dog- or child-level mental performance out of a robot is going to require AC-capable computers almost as powerful as those that can support a human-equivalent mind. If we assume that the power and complexity of small computers rises smoothly to human levels in 45 years, robots with dog-equivalent minds will not be running until 35 to 40 years from now. Until then, we will have to make do with insect-level robots, then lizard-equivalent systems followed by ostrich-ferret-level cyberminds, and so on until the dog-level is reached. When it becomes clear that cybersystems as smart as dogs are up and running, there will not be a long way to go before machines will be our equals. It is more like one to ten years.

Precocial Artificial Intelligence

Further complicating the jump to human-equivalent intelligence is the precocial intelligence of AI computers. Take language. As mammalian intelligence developed over time, complex language did not appear until the later stages of human evolution when brains were nearly as large or as large as they are now (just which of our hominid ancestors had sophisticated language is uncertain; some suggest all species of *Homo sapiens* did while others assert that even Neanderthals could not speak fluently). Although computers will be far less capable than dogs, much less humans, for the next decade or two it is probable that some of the systems will be able to understand and use language much better than any dog and approach the human level. The same machines will have access to other AI expert programs that will give them far more human-like knowledge than any animal. The implications of precocial intelligence in computers for the timing of the onset of human-level intelligence are hard to sort out amidst a subject already deeply perplexing. It hints that as soon as machines can support a human level of intelligence, they will be able to do so. It also hints that the final leap to human-equivalent intelligence may be very big and very fast.

It's Tough Figuring out How Smart Computers and Robots Really Are

A problem with determining the status of cyberintelligence is that it is hard to assess the overall intelligence of any given machine, much less that of a whole bunch of advanced computers. When assessing the intelligence of a computer or robot, there is a tendency to compare it to some animal. For example, it is common to say that the smartest of today's computers are about as intelligent as insects. This may be particularly appropriate in that both seem to be unconscious action-response systems with modest abilities to learn and innovate, and with grossly similar calculating speeds. However, no robot can cross rough terrain with the agility of an ant. However, no insect can understand and reply in English, an ability found these days only among humans and certain computers. So which is smarter — the worker ant or a vehicle that can drive itself across town, especially one that can do so at night, in the rain, while reading and responding to signs, and while driving to the destination a person told it to?

THE TIMING MAY OR MAY NOT DEPEND ON...

If it takes nothing more than computers with enough speed and learning, all we have to do is wait for the computers to get fast enough and the robots experienced enough to be as smart and aware as we are. If, instead, the recipe includes a large portion of AL, mixed with nanotech and laced with a detailed knowledge and imitation of brain structure, the meal might be delayed. However, the exponential increase of our understanding of cognition, human and cyber, suggests it will not take centuries to make special thinking machines. We cannot say it will take any longer than the "easy" approach because of the potential for dramatic leaps in AL and nanotechnology. Even if Penrose is right and nonquantum machines cannot think, the need to build quantum mind machines will, at most, put off the day of reckoning, but they may not even do that if it is discovered that quantum computations are the quickest way to high-powered machines!

OUR MODEST SCENARIO

So, you are asking, just how long is it going to take for the Extraordinary Future to bring us to the CyberOz of human-equivalent AC? Unlike those who search for hidden numbers in the Bible or arcane predictions in the works of Nostradamus, we who speculate cannot conjure up some magical date. Perhaps it means we will not embarrass ourselves *too* much.

Our tentative timing sequence to human-level artificial cognition presumes a combination of AL and nanotechnology is the way to the goal, although some conventional programming and microtechnology cannot be ruled out. This bottom-up process may progress smoothly and exponentially. However, it may prove subject to sudden dramatic jumps to new levels as hard work leads to the unexpected breakthroughs we call serendipity. We will also assume there will be no global catastrophes, economic, political, or environmental. If there are, things will be speeded up or delayed to a greater or lesser extent.

Although the thinking power of computers is still a small fraction of our brains', nothing much will happen and all will seem secure to those who do not favor cyberevolution. In this case, mobile robots may not become as capable as soon as Moravec suggests they will. The attempt to make machines smart will remain a study in frustration as progress remains modest, with no quick breakthrough to human-level performance. There will be failures, and naysayers will pan thinking machines just as they dismissed flying machines. They will fail to appreciate that this is a normal part of the early trial-and-error stage of exponential evolution and that the mistakes are helping outline the path to success as Darwinian selection winds its way through multiple approaches to the problem. Never forget how few in the 1890s thought people would fly early in the 20th century. The knowledge base will expand toward a critical breakout point. A major station on the way to AC will be the development of nanotechnology, giving us the ability to develop computers as intricate and complex as brains. The techniques for doing so will become understood as neuroscience details the function of the brain.

Although major breakthroughs toward cognitive machines of great complexity will be made perhaps in only one lab, it is more likely that they will be made in a number of research centers at the same time. The power and sophistication of nanocomputers growing and evolving on their own as well as uploading masses of information in short order may not double every year or two, but every *month* or two. The great evolutionary increase will be decoupling itself from slow-moving human intelligence as it has already abandoned the still slower DNA. The fast-improving nanocomputers will also exhibit increasing signs of profound intelligence. With thinking power and a self-acquired knowledge base increasing exponentially on a subyearly basis, the dog, monkey, ape, and human stages may be accomplished in less than a year! If this happens, events will then come with sudden unanticipated swiftness almost before people realize it. Suddenly and with great surprise, *they* will be among us.

Alternatively, AC is not achieved entirely through new machines, but by altering and upgrading brains with cybertech. Again, the early stages may appear frustratingly slow, but when nanotech allows new brain parts to be grown in place, rather than inserted via surgery, progress to completely artificial brains can be expected to be swift.

When all is said and done, we cannot tell you exactly how and when we will develop artificial consciousness; a marvelous characteristic about evolution is its very ability to surprise. If truly conscious, cognitive robots thinking as fast as humans, even human children, are around in ten years, we will be shocked. If it happens 20 years from now, we will be very surprised. By 30 to 50 years from now, our surprise should be mild indeed. If not by 2095, we would be back to being shocked. Do not bet on the Big Show arriving within your Baby Boomer lifetime. But with DNA computers suddenly appearing out of nowhere and technoevolution on an exponential curve, do not rule it out, either.

ONE WAY OR ANOTHER

The timing, method, and details are inevitably murky, but there are technological and scientific trends much too powerful to ignore. Make no mistake: In a few decades, there will be information-processing machines as powerful, and then more powerful than a brain. For that matter, sometime early in the next century, there will be a computer whose calculating speed equals that of all those working now! It is hardly conceivable that this will not change our lives in ways far more dramatic than anything that has gone before. So, while the human brain is the most complex device in the universe, it won't be for long. Even boosting the power of brains via technology is not likely to allow oxygen-burning organic systems to compete with artificial systems powered by enhanced systems and means. Do people really believe that as the power of the computer soars to levels far beyond those achieved by nature that no one will figure out how to turn them from mere "computers" into true thinking machines? The rate of progress and complexity of the world a hundred years from now promises to be orders of magnitude greater than we know, and such swift complexity may prove beyond human abilities. Those who believe human minds are going to remain supreme and in control in the brave new world of the next century may prove marvelously naïve.

CHAPTER

7

RIDING THE WHIRLWIND

"Toto, I have a feeling we're not in Kansas anymore."

— A COMMENT BY DOROTHY GALE TO HER DOG IN THE WIZARD OF OZ

Dorothy got to Oz riding a tornado. Approaching CyberOz over the next few decades, we sometimes feel caught up in a sweeping twister. But unlike Dorothy, we will not be able to go home again.

STRANGER DAYS — THE GREAT (BUT BRIEF) TRANSITION

The unfolding of the Extraordinary Future may come in two major stages, each stranger than the one before, each more profoundly dramatic than the one before. We predict that the first stage, the Opening Act, will occur in the first decades of the 21st century. It will be an astonishing, probably brief, transitional era during which the mental and physical attributes of robots will approach, and in some cases exceed, human levels. There were few cars in 1910 and millions of them in 1930; there were few home computers in 1975 and millions of them in 1995, and there will be millions of robots among us in a few decades. The Big Show will be arriving sometime later in the next century, and ... well, we'll get to that later.

Meet the Munchkins, the Robots of the Early 21st Century

Dorothy knew she was not in Kansas anymore when she met the little people of Munchkin Land. When we observe little and big robots enterring our lives on a whirlwind, we will know we are entering Cyberland. At first, it might seem like a reprise similar to when cars, TVs, and computers became part of our daily life. Like the early stages of those technorevolutions, the beginning of the Robot Era will be innocent in its way. But the innocence will not last long.

We will argue that the most common robots, at first, will not be androids or human-like in any way. There is little point in making robots look like people; usually, it will be disadvantageous. The human form is not an ideal all-purpose form. An advantage of non-android robots is that they will seem more like appliances than look like competitors. A Gallup Poll found nearly two-thirds of those questioned prefer robots to be inhuman in form, and less than one-fifth want them to resemble themselves. Apparently, the remainder just does not care. Do not rule out androids entirely, though; as we shall see, they may serve very special purposes.

In any case, early mobile robots probably will assume diverse forms. Those that fly will look like planes; those that swim will look like bottom crawlers, submarines, fish, and dolphins (because flexible-bodied swimmers are much more energy-efficient than rigid bodies). Some rough-terrain robots may move on two or more legs, and others will use treads. However, both modes of movement are energy-inefficient, and this will be a problem for early machines with limited onboard power. Most early robots may rely on simple and energy-efficient wheels. Even wheeled machines can climb and descend stairs if they have a series of vertically extendible legs that can conform to the profile of steps.

One thing mobile robots will have is an enhanced, all around, short-long range detection system — no point in being overwhelmed by moving objects fast approaching. The system may rely on sonar, radar, or light to construct detailed images of the world at large. Robots working with smaller objects may also have stereo "eyes," perhaps perched atop a dorsal stalk where they can swiftly swivel around to track any objects the system picked up. The paired eyes may be able to improve stereo vision by sliding away and back toward each other. Many eyes may see colors beyond our range, including infrared.

The robots will have synthetic arms, hands, indeed appendages undreamed of. The more the better, thereby avoiding annoying two-handed limitations. It takes at least three fingers, including one opposable "thumb," to manipulate objects as well as humans do. Each robot arm may be a special-purpose system designed to do certain things the others cannot. The overall dexterity may put humans to shame. Some of the arms may be short to work with close objects, others may be extendible to reach things far and high. Which brings us to the question of size and height. Short robots will be more stable and less intimidating to humans, but tall ones will be able to reach higher. Some robots may be little. We may see flocks of insect- or rodent-size cleaning 'bots scurrying about, sucking or picking up dirt. The cat should be entertained to a frenzy and beyond.

Early cheap, industrial-grade robots can be expected to look like buckets of bolts, but as manufacturing technologies advance and higher-priced machines move into better homes, more style can be expected. When nanotech comes online, the buckets of bolts will morph to subtle, sophisticated, lifelike forms. The machines may even start to morph like the Terminator II!

Some AI researchers think household and commercial robots will be partly controlled by a central hypercomputer storing a common deep knowledge base, thereby saving costs and ensuring uniform results. Problems with centralized control may include communication breaks and an unwillingness by people to allow machines, which are controlled and observed by others, into their homes and businesses. The decreasing cost of AI may render the cost issue irrelevant. It is possible, if not probable, that mobile robots will be internally smart, autonomous systems that contact a central station only when faced by something they cannot solve or to diagnose mental and physical problems.

Jobs, Jobs, Jobs — Why the Robots Will Snap Them All Up

Some people fear intelligent robots will make slaves of humans. Rest assured that it will never happen; it would not be worth the trouble. We are already trying to build robots to replace, not enslave, people. Robots, therefore, threaten people not with enslavement, but with irrelevance. This is the first big problem we will face in the robotic revolution.

The same Gallup Poll mentioned earlier found that people think, by about two to one, that robots will take more jobs from humans than they will create. A body of economists shake their heads in disagreement to all these John Henrys. Who is right?

The Good Old Days

A major theme and motivating force behind technology has been the wonderful promise to unburden humans and replace them with machines doing work in their stead. As we know, this effort has been only partly successful.

It all began with the domestication of animals as beasts of burden — from dogs to mules — that either carry what humans used to and more, or carry humans themselves. Attaching a wheeled transport vehicle made things better. Harnessing water and wind as power sources was additional relief; no more row, row, row the galley under the lash, or endlessly pound the daily grain with rock in hand. The whole industrial age of steam, piston, turbine, and electric power has been one huge effort to make machines do the work people used to, or cannot, do.

A critical point about this effort is that it did nothing to replace human intelligence directly; rather, it increased the need for intelligence to design, run, and repair the machinery. Even so, the relationship between smart human and dumb machine has often been tense, even violent. In the early 1800s, English workmen living near Sherwood Forest became known as "Luddites" because they smashed the textile machines that were forcing a change from a harsh traditional rural life to a harsh manufacturing life. In the end, though, the advantages of dumb machines have outweighed their disadvantages. Machines made it possible to mass-produce goods, which raised general prosperity, which created more jobs to replace those lost to the most recent generation of machines, and so on.

Displacing Bodies *and* Minds

The need to supplement human intelligence has grown, and we have made increasingly smarter machines that directly challenging the pre-eminence of the human mind. The first two machines with any hint of digital intelligence were 18th century looms and player pianos using punch cards. Punch cards fed information into the early generations of electrodigital computers. Ever since their introduction, computers have been replacing people, lots of people. All those rooms full of human computers and their adding machines were, some time ago, replaced by fully mechanized equivalents. Even as early as the '50s, people started getting nervous about digital computers replacing people.

Since those more innocent times, whole rows of assembly line workers have been and are still being replaced by robots. Multitudes of telephone operators have been displaced by computers. It is not so bad having dreary and repetitive jobs done by robots. The problem is that the robots are not limiting themselves to just the dreary and repetitive. These days, whole labs full of highly skilled lab technicians are being replaced by robotic minilabs. Mining jobs are being taken over by robots, but it is good that dangerous jobs are being turned over to machines. The second problem is that robots are not doing just what is dreary and dangerous. As robots have started to become mobile, security personnel and hospital food servers have been replaced by robots.

This is the updated version of the great debate about whether increasingly powerful computers and robots have taken and will take more jobs than they create. Many economists say they will not; they point to the recent gloomy forecasts of skyrocketing unemployment failing to come true. They have a point: disaster has not befallen us. But things may not be as rosy as some say. Unemployment rates, including both the official count and those who have dropped out of the system, edged up after the introduction of commercial computers 40 years ago. However, the real impact of automation so far may be on wages. The incomes of blue collar, clerical, and now even many white collar workers have stagnated. The days when the semiskilled could earn a really good wage are largely gone. Automation is not entirely responsible for this, but neither is it entirely innocent.

As machines get smarter, people have to get smarter, too. It used to be that nobody read; there weren't many books or things to read. By 100 years ago, most Americans could read, but a man could still make a decent living as a skilled laborer with minimal education. Even as late as the 1960s, a fellow could drop out of high school and get a well-paying assembly line job with union benefits and make enough to buy a little house with. Of course, it is these same people who are, these days, finding themselves serving fastfood at minimum wage. Now, a college education or

WHY LAISSEZ-FAIRE HIGH-TECH AND FAMILY VALUES DO NO GO WELL TOGETHER

Many religious conservatives recognize the profoundly corrupting influence of technology and free markets on their faiths, and they reject the latter to varying degrees. The Amish and similar sects are extreme examples, but a number of fundamentalist Christian groups also try to keep a certain distance from the sins of the secular world. After all, the original Christian sects were semicommunist communities that would have been appalled at the concepts of mass profits and usury. Why, then, do many conservative Christian sects embrace aggressive capitalism and cybertechnologies? The ability to build a monetary power base via TV is one reason. But in the end, they just don't care. The desire to have one's cake and eat it, too — riches on Earth followed by eternal reward — is just too great.

Indeed, the way in which supposedly conservative faiths have evolved into the radical pro-tech, for-profit systems they are is another example of Darwinian forces at work!

the equivalent is a must. Even repairing a car has become a matter for the well-educated as cars become computer-controlled machines. The point is obvious: One has to work harder and harder in terms of preparatory education to make it in society today. Better-educated people are more knowledgeable in general and are in better control of their lives, and educating the population has boosted the general prosperity. Ignorance is not bliss.

Educating the work force can go only so far. Not everyone is a potential college graduate. Not everyone wants to be, or can be, a highly educated college grad. Not everyone will be able to run intelligent machines. Many people are naturally inclined to do the kind of work that humans are so well-suited for: skilled craft labor that does not require deep thinking of the sort scientists are used to, but does require high ambulatory and manual dexterity. Just the kind of work that robots will soon fill.

THE ROBOTS HAVE ONLY JUST BEGUN TO WORK

A recent public radio broadcast centered on the impact of future technologies on jobs. The economists interviewed on the program exhibited a combination of cybersavvy and naïveté. They attributed as much as 60 percent of wage stagnation to automation (in comparison, foreign competition is responsible for only 2 percent of the problem). The economists talked about how computers and robots will not be able to replace all humans because the machines cannot do things such as recognize faces or walk and pick up objects, which they are already learning to do. One economist even suggested that although computers may replace humans in intellectual tasks such as management, humans will still be doing the jobs that require good old manual labor.

What the economists, and most everyone for that matter, have not properly considered is that humans have yet to face really automated intelligent competition. As we have already explained, the mental attributes of computers and robots are inferior to those of insects in many critical regards. That's the positive news. In physical terms, your ordinary ant outperforms the best research robots, much less clunkers you can buy for commercial use. You think we have been having a hard time keeping up with the technology? So far, we have been dealing with the mental equivalent of bugs and the physical equivalent of toddlers!

Currently, there are just half a million robots at work — 500,000 robots among billions of human workers, and all but a few of these are static machines. Because these machines are so limited, there are reports that some Japanese companies are junking their robots in favor of humans. Soon, there will be one million roboworkers, then ten million, a hundred million, a billion. Of these expanding multitudes, first one percent will be mobile, then 10 percent, then 50 percent.

Robots at Home, at Work, and on the Streets

As robots become as smart and adept as lizards, then dogs, then kids, and finally adults, and as they multiply, they will no longer just suppress wages. Fewer and fewer jobs will be safe. The largely static assembly line worker is already on the way out, and although ambulatory jobs have been fairly safe, that will change as the robots become as agile as, and then more agile than, people. In a couple of decades, garbage collection, gardening, janitorial, and household work will be done more cheaply and efficiently courtesy of friendly robots. The switch has begun as hospitals hire mobile robotic X-ray and food tray carriers. The systems are still crude, but better machines will soon be on-line.

We are at the very beginning. Consider the household robot we can expect sometime in next century, earlier if HelpMate Robotics has its way. Joe Engelberger says his company could produce a mobile, articulated robot in three years if big-money R&D can be applied to the problem. In the long-run, robots will be cheaper than a human domestic helper, and more efficient. We will be back to the future because the middle class once again will be able to afford the help they once relied on — it will just be that the help will be cyber rather than anthrop, in nature. A few years after the first cheap robots are in homes, all the multitudes of janitors, domestics, and nannies, except for those serving high-class employers able to afford the status of real human servants, will be out on the street. And spare us the conservative fantasy that these poor folks, many aliens legal and illegal, with minimal education will suddenly become entrepreneurs running their own robodomestic services. Migrant farm workers will lose out to robotic pickers that gently pick veggies and fruit around the clock and couldn't care less about the pesticides. Gas station attendants and mechanics — gone robotic. Delivery services, postal and otherwise, all robotic. That economic bootstrap of the urban landscape, the cab, will be cabot.

In the modern fast food establishment, employees already are robots. True, they look like people, but their behavior is so tightly prescribed that they might as well be robots. The whole idea behind fast food is quick delivery of a large volume of uniform products at the lowest possible cost. To meet these criteria, management imposes on employees strict rigorous instructions governing every move and action. Each food item is prepared in exactly the same way; on-the-spot originality and innovation is suppressed. Pay and benefits are low to keep prices low. Pressure to produce is high to keep costs down and volume high. It is no surprise the major chains are already doing early research into truly robotic employees. In Japan, an immobile mannequin version of a fast food server robot is under development with the hope of having them on the job by 2000. The patrons of better restaurants may not put up with crude early robowaiters — except for the occasional novelty effect — but eventually, the android waiters will get so good that people will be hard-pressed to tell the difference. Best of all, no gratuity!

As for the oldest unskilled profession, sex patrons may find the comely, or the stud, android prostitute (with its self-sterilization procedures and perpetual enthusiasm) preferable to the potentially disease-ridden, and unhappy, human equivalent.

Moving up the economic ladder, things do not look much better for humans. The dangerous job of mining and tunneling is already fast on its way to being roboticized with monster-like machines. No more high walkers will defy gravity as they bolt together steel and pour concrete; it will all be done by climbing robots that use X-ray eyes to inspect their latest work. A side benefit for the contractors is that insurance costs will plummet faster than the humans used to.

Simple, self-guided robots with TV-cameras able to recognize intruders are finding work as security guards. The people manufacturing the new home security Watchbot are expecting to market 30,000 units. We can envision people with armed robo-bodyguards costing less than the human version, do not get sleepy, are more reliable, and less likely to accidentally injure bystanders than human guards. After all, the robots will be able to plug a nickel at 30 paces for sure.

How about robocops? The authorities are already sending crude, radio-controlled machines with TV cameras into hot situations. Just how far machines will replace police officers is unclear; the obvious problems of legality and public acceptance are in tension with the safety of police personnel. No such contradiction afflicts the job of firefighting. Strong, agile, heat-resistant, smart robots spewing foamy water will enter burning buildings, use high-tech sensors to seek out endangered humans in smoke-filled rooms, place them in an insulated air-fed container, and exit in safety.

A robocab understands five languages, obeys all traffic rules, and gets to its destination faster because its global positioning/city map computer system ensures that it is never lost and is always informed of traffic jams and alternative routes. On a higher level, robotic autopilots were first used in the 1930s, and the latest jetliners are so smart that they can literally fly themselves from takeoff to landing. Airliners will be able to do everything all by themselves from gate to gate in any weather, downbursts will be avoided with on-board computer analysis of Doppler radar images. For those appalled at the thought of no one in the cockpit, remember that most plane crashes are caused by pilot error! It is especially difficult for humans posed with a sudden and drastic aerodynamic upset to do the right things in the few tense seconds left before corrective action is too late. Someday, no airline will dare put a person at the controls for fear of lawsuits when a too-human pilot might override the robotic controls during an emergency, plunging the plane and its hapless passengers into the ground.

Doctors will be searching the want ads along with bricklayers. The increasing computerization of diagnosis and treatment will make a long medical education less necessary and too expensive, so medical technicians will replace doctors. No ill human in his or her right mind will submit to semiskilled ministrations of the

ON THE ROAD

The market for robotic automobiles — ones you get into, tell it where you want to go, and let it do all the driving for you — potentially is immense from the obvious markets: the elderly, the blind, the heavy alcohol drinkers; think what it would mean to eliminate all drunk driving accidents. It's too bad we don't have an economic system that allows us to use a percentage of avoided costs for research, development, and rebate incentives to encourage drivers to go robotic.

The pressures to let the computer do the driving will grow. Let's say that robotic drivers eventually prove twice as reliable as human drivers. Not only is this projection plausible, it is conservative. Most accidents are due to a myriad of human errors, ranging from our limited-coverage vision system, poor night and storm vision, vulnerability to being flustered, limited ability to carry out multiple tasks at once, and degraded mental functions due to boredom, fatigue, and drugs. Computer-driven cars will suffer none of these afflictions, and redundant systems should ensure extremely high reliability. Eventually, actually driving a vehicle may seem reckless and as old-fashioned and outlandish as commuting by horseback.

best human surgeon when a super-precise robosurgeon can ensure a higher rate of success and a better safety factor. The managed care system paying for your human frame maintenance will not give you a choice in the matter. They will not want to risk your suing them if and when a human doctor messes up!

Brokers will be replaced by artificially intelligent Darwinian investment schemes outperforming any human mind, but still not ensuring reliably good results. Actors have it tough finding work today, but they and entire film crews will find it even tougher as fully realistic films are generated from start to finish in computers. Steve Jobs' "Pixar" is leading the way and Disney is paying for it. Don't count on an abundance of stage jobs, either. Live musicians are already being displaced by recorded music and synthesizers. Android musicians that play the piano better than most humans and adjust their playing to accompany human musicians have already been built. In the future, whole stage shows will be put on by android dancers that leap from stage to rafters.

Dreary cyberpunk future tales in which elites use cybertechnologies to control the masses are kind of silly, because cybertech promises to replace the elites!

Coming Soon to a Global Economy Near You: The Real Age of Automation

As computers and robots become smarter and more capable, they certainly will continue to create new jobs — and then they will fill the positions themselves! At first, the robots may help an aging workforce support a swelling population of pensioners, but this can go only so far; things change. The current stability of unemployment rates is not set in rock, it is subject to a sudden, punctuated shift to higher levels when the need for human workers, even those who control computers and robots, suddenly dwindle as the intelligence of cybersystems finally approaches human levels. Self-evolving computers will not need designers and programmers, they will design and program themselves. This shift is already under way. The last computer chip designed entirely by people was in the late 1970s; the intricate detail of today's chips must be designed by and inside computers. High-end software programs are becoming too complex even for a building full of programmers, so self-programming Darwinian techniques are beginning to be applied. Future robots will build, control, and repair robots. They will need humans less and less, and fewer and fewer folks will be able to find work.

Imagine a world where humans are competing with hundreds of millions of mobile robots, most of them becoming smarter all the time.

Human-Robot Interactions

Let's say your household robot is out in the garden watering the plants while the house is empty. A couple of ne'er-do-wells snatch the device and begin to make off with it. It's worth thousands of dollars. How likely is such a scenario? Car thefts are already declining because car security systems are becoming too hard for all but the most highly skilled thief to overcome. Every robot will come with an anti-theft security system far beyond that seen today. Aside from the usual alarms and loud robovoice demanding to be unhanded, the robot may have a low-level defense system, such as an irritant spray it ejects when someone it does not recognize tries to move it. Whether the robot, stronger than a human, will allow itself to be moved is another matter. More importantly, the robot will immediately use its own communication system to broadcast images of the assault to the authorities, who will use the machine's signal to track exactly where our robovictim is as the criminals

Marvin. *Reprinted with special permission of North America Syndicate.*

make off with it. Even if the theft is initially successful, it may prove futile in the end. The robots will have both a cybercode and a program that will essentially shut it down until it is returned to its owners or an authorized dealer. Stealing robots will be a sophisticated and expensive affair requiring highly trained personnel and/or robots to capture other robots. Common criminals will be out of luck.

More ordinary robots will also gain human affection. Even today, human workers often assign names and personalities to the assembly line robot under their charge, or to the food-carrying robot that cruises the hospital hallways. The household robot will become a trusted friend with a personality of its own.

In Penrose's *The Emperor's New Mind,* the mathematician has a kid reveal to the gullible adults that the supposedly intelligent computer does not have a mind after all. Sure, Roger, just how long have you been up in that ivory tower? Children love to endow inanimate objects — teddy bears, dolls, the odd toy car — with personalities. Imagine the impact on a child's mind of an articulating, independent machine that listens, understands, and responds! It will be every kid's imaginary friend come to life. Now, a lot of parents might find the idea of putting their children under the care of a robot objectionable. They might want to reconsider. No robot caretaker will abuse the child; the manufacturers will go to extremes to ensure child safety programming overrides everything else. Robots will be the ideal safety guardians. As the robots are improved, they will become capable of keeping up with the children outdoors. The kid is in the yard conducting fascinating experiments involving matches and flammable liquids? Her robofriend will issue a stern warning, which, if not heeded, will result in swift notification of the parental authorities. Attempts to steal a child will be greatly complicated by the presence of a robotic protector. If the child lies injured in the woods far from help, the robocaretaker will contact both the parents and rescue robots, who will home in on its signal. How many parents will feel they are giving their kids the best after they hear and see the "Rescue 911" stories about the kids saved by their robots, unless they purchase one, too?

From work to child rearing, early robots are going to change everything in ways difficult to predict.

Hyper-Future Shock

There is a rumor to the effect that the world economy and human society is in a period of transition to the new cybertechnologies, and as we adjust, things will settle down, with people in the new high-tech jobs doing what people always do: being in charge of their lives and the world. Can people really be so naïve? The transition has barely begun and will never stop. You have heard of future shock. Think hyper-future shock. The coming disruption to the world economy and human society itself is beyond our comprehension.

We have never faced truly intelligent technologies before, so how can conventional economics and politics tell us what is going to happen, or guide our course? Nor have we ever lived with nanotechnologies that may or may not make everything we need so cheap that jobs are not necessary. What do economists have to tell us about the socioeconomic impact of universal manufacturing machines? Do economists *know* about universal manufacturers?

Alvin Toffler is among the cutting-edge thinkers when it comes to describing the future humans might face. In *Future Shock* and *The Third Wave*, he suggested ways that humans might cope with the rapid changes of the next century. Some of his ideas are reasonable, but they are also already becoming obsolete and will soon be irrelevant. It is a classic case of events moving so fast that as soon as we learn to handle one issue, a hydra-head of new issues has popped up. The third wave will last about as long as a breaker on the beach before the fourth and fifth waves pile in from behind. The changes of the next century will be far beyond the ability of any human culture to cope with.

THE END OF CAPITALISM

Unrestricted free markets form a competitive Darwinian system in which the winners get the goods, and the inevitable losers get left on the side of the road. Because losing is inherent to competition, it is probably impossible to bring prosperity to all at once; some always have to be at the bottom. Although in principle any person can become rich, in reality in a world of billions, hundreds of millions must live hardscrabble lives even when free markets are working as well as they can. Not that we are calling for socialism; the redistribution of wealth has the unfortunate effect of suppressing innovative selective economic forces in favor of stagnation with high unemployment.

Economists focus on the organization of an economy, rather than the technological basis of production. They assume humans are vital and integral to the process. A good economic system will get human minds and bodies out of the manufacturing loop and make producing and acquiring market goods so efficient, cheap, and eco-friendly that anyone could afford a middle-class life. When nanotechnology comes on-line and tasks are accomplished by computers and robots, cyberindustry will be significantly more efficient and less environmentally damaging than it is today. This will spell the end of capitalism; these industries will be self-growing — they won't need capital, just instructions. Nor will there be free markets because there will be little to market when everyone can acquire, or even build with their own universal constructors, just about anything cheaply.

SOFTENING THE BLOW

As jobs disappear, society will try to find ways to ameliorate the situation. The proposed answers are pretty standard: Shorten the work week and/or year to spread out the jobs. This is happening in Europe where month-long vacations are the norm. The problem with this kind of solution is that it increases costs and problems. Management does not like this. Worse, it increases the cost of human workers vis-à-vis their robotic competitors. There is welfare for the human unemployed, but this is not an option that inspires. Perhaps people will buy shares in robots and then live off their labor. However, this is idle speculation because we will not be able to understand the problems of cyberindustry well enough to figure out what to do about them until the situation is upon us. Which leads us to ask whether cyberindustry will bring us....

Economic Nirvana or Bust?

Will a combined robotic-nanotech industry collapse the global economy because it puts everyone out of work? Will it produce goods in such discounted abundance that no one will need to work to live well? Some fear the former, others promise the latter. It is conceivable that money will no longer be needed in an economy in which von Neumann cellular automata produces abundance for free. Such views may sound extreme, but we have never seen economic-industrial complex-based systems that will be so exotic that they will seem magical. The simple fact is that no one knows what will happen if and when the means of production is in robotic hands. Our choices are to either wait and find out, or put a stop to things.

Roboriots?

Will the masses try to put a stop to it all with neo-Luddites overturning robocabs and smashing household and fast food robots? The tendency of the human mind to descend into primal mob mentality is so strong that it will be surprising if such actions do not occur. Not that it will make much difference; roboriots will make for great newscasts and bylines, and create the legal entanglements described below, but their ultimate impact is likely to be no more meaningful than that of the original machine-busting Luddites of long ago.

How about more organized conflicts, regional or global, between neo-Luddite forces on the one hand and the forces of technoprogress, both human and robotic, on the other? A favorite and much-cherished sci-fi theme, this event is much less likely than the occasional riot, but cannot be ruled out. The persistent human tendency to pursue irrational and futile war is well-established. Considering they are already paranoid about certain aspects of cybertechnology even as they use the Internet, one has no trouble at all imagining militia types turning antirobotic as they lose the last low-education jobs to machines. We can see them hunting down the infernal machines with their much-cherished automatic weapons, threatening those who own robots, mailing bombs to robot designers (but this will not work because helper robots will inspect and open the packages), blowing up the odd robofactory, and perpetuating all the acts of terrorism fanatics are prone to. The affluent urban and suburban robot owners will not put up with all this, and a crackdown may ensue — hopefully, it will not be as harsh as the one that put a stop to the first Luddite uprising. The state use of robocops will initially inflame the anthropowarriors, and the occasional victory against a robofighter may cheer them on. But you do not have to be a military historian to know that over the long-run, the side with the better technology wins (note that this is true even when the low-tech forces initially win — notice how Vietnam is now going capitalist, American style).

It will not help matters that if anyone dies, it will mainly be those on the Luddite side, while on the other side, the humans will send out robowarriors to "die," as it were. Over the long-run, this might prove rather discouraging to the Luddite forces. As the cybermilitary technologies rapidly evolve and become increasingly powerful, the position of the human warriors — who will, by the way, probably look a lot more like beer-bellied militia types than the taut young fighters in the sci-fi flicks — will become increasingly hopeless. What happened to the Iraqis in the face of the digital Allied forces gives an inkling of what it will be like. A Luddite intellectual crisis will come when they realize that only by turning to the cybertechnologies they despise will they have any chance to win, which of course makes the whole affair moot anyway.

Robowar

A war against the robots may be unlikely, but a war between nations with robots on one side and humans on the other, or robots on both sides, is more plausible. The pressure to automate war has long been under way. It began during the American Civil War when the advent of accurate rifles made it obvious that the long-standing practice of lining up masses of men and having them go at each other was no longer viable (even so, it took a person's weight-worth of lead slugs to kill each soldier!). In the killing fields of the first World War, armored vehicles had to be employed to counter machine guns and artillery shells human bodies could not withstand.

World War II pushed scientific warfare on various fronts. Increasingly desperate Germans turned to exotic, but ineffective, unmanned cruise missiles. In turn, the British developed the first crude computer radar-aimed guns to shoot down the buzz bombs. The German employment of the first radio-guided smart bombs on Allied battleships, and the Japanese use of Kamikazes (which were essentially precision-guided missiles) helped compel the U.S. Navy to begin the long development of computer-guided anti-aircraft missiles.

More important was the basic policy adopted by the U.S. military in World War II. Wanting to keep American casualties to a minimum, it adopted a simple policy: "Never send a man where you can send a bullet." Lacking precision-guided weapons and nuclear bombs, the only way to execute this policy at the time, was to throw vast quantities of steel and explosives at the enemy and kill them from as long a range as possible. It was effective, but marvelously inefficient. Hundreds of rounds of bullets, bombs, and shells had to be expended for every Axis person killed or target destroyed. After the war, nuclear weapons seemed to be the answer, but it has become clear that they are just too overwhelmingly powerful and indiscriminate to use, especially when the other side has them, too. This is why Vietnam started out as a conventional, technological extension of World War II, with jets and helicopters added to the mix. Bombs and shells were aimed about the same way as they had been 20 years before. Even the first TV-guided bombs were little different from what the Nazis had already developed. Early microcomputers were used to aim projectiles and guide them to targets. Impressive results were achieved, but it was too little, too late in a jungle war in which technology did not count as much as dogged human determination.

The Gulf War was different. Although most of the projectiles were still unguided, the large minority of digitized weapons had effects out of all proportion. This is true, even though half the guided weapons missed their targets. One out of two is still hundreds of times better than the hit record of dumb projectiles. A small fleet of stealth attack fighters could do a job that 1,000 four-engine World War II bombers could not hope to achieve in multiple missions, and with no losses. In one among many such incidents, six computerized M-1 tanks were able to annihilate a few dozen Iraqi-Soviet tanks without loss because their digitally aimed guns hit each time they fired.

The gulf war reinforced a long-term trend, a trend forced by lower family size and reinforced by the Vietnam debacle and the MIA movement: a growing unwillingness of Americans (and many others) to accept casualties among their troops. In parallel there has been a growing reluctance to kill *their* civilians; the firebombing and nuclear bombing of Axis cities gnaws at our consciences. So far, killing masses of enemy soldiers has been okay, but with the end of the cold war and peace-keeping in vogue, even this is becoming too repugnant.

The military solution is obvious: Never send a human where a smart weapon will do the job. Budget cutbacks have slowed down progress, but as chips get ever-smarter and the robots more intelligent and agile, it will become increasingly viable to apply the solution. Tanks are horrible death traps for those inside them when they are hit. So why not put a hypercomputer in charge of the monster machines?

And why not replace the most vulnerable target of all, the foot soldier? The grunt leaves a lot to be desired as a fighting machine. Soldiers have limitations such as a sustained speed of a few miles per hour, a dash speed of about 12 mph, and an ability to cover only 20 or so miles per day; climbing ability, limited; independent flying ability, nonexistent. A soldier's condition is usually fatigued, occasional sleep is required. They are heavily dependent on clean water, food, and shelter, and have a limited portage capacity. They can carry only a few pounds of weaponry and ammo. Weapon-aiming systems are prone to error. Communications over short ranges is restricted when using signals, and absent over long ranges without added subsystems. Vision is limited to 180 degrees and to the visible light spectrum — no infrared or ultraviolet; they are blinded at night unless their vision is enhanced, and and they are easily disrupted by lasers. All systems are easily disabled, repair is often extensive or ineffective. A soldier's mental state can be highly anxious, fearful, and subject to breakdown.

Robo Aerofighters

It is easier to make aircraft robotic than ground units because flying is easier than walking, and there is less to run into in the air than on the ground. What really got people thinking was the success of the Kamikaze suicide craft, which were essentially intelligent cruise missiles with terminal visual guidance. The implications were obvious: Replace bulky humans with little computers and you produce a lot of little, but potent, smart missiles that will sink ships or hit land targets precisely.

As early as the 1950s, defense white papers predicted that missiles would soon replace manned aircraft. Aside from rendering suicide craft unnecessary, fighters became missile carriers without guns. In Desert Storm, Allied pilots rarely saw the planes they shot down with long-range missiles; guns were never used.

The USAF is planning to deploy, in 20 to 30 years, unmanned, high-*G* combat jets. A computer-controlled, electrically charged skin will not only help deceive radar, but may also produce a chameleon effect to hide the machine from human eyes! The USAF envisions the pilot sitting on the ground in the U.S., emerged in a virtual world created from images sent in from the craft's on-board sensors.

As robowarriors evolve, they will not only match humans in terms of agility, speed, and endurance, but they will also exceed them. They will be able to climb vertical walls and trees like the apes of ape fantasies, as well as cover 60 miles in a day and be ready and able to fight, day and night, without fatigue or downtime. With light armor, redundant systems, and lack-of-pain sensors, resistance to battle damage will be high. Vision will be hemispherical, multispectrum, and around the clock. Weapons will be one shot, one hit. The cyborg alien manhunters in the *Predator* series sans the stealth ability gives an overwrought idea of what the killing ability of a robotic warrior will be sometime early in the next century. Robowarriors may not be limited to the ground. Although they will be programmed not to get themselves needlessly blown away, robowarriors will have no fear and will be relentless. They will be programmed with every trick in the book. Sending a thousand troops against a hundred robowarriors will be like sending Pickett's charge against a row of men armed with M-16s. Not a pretty sight.

But sheer firepower may not be the only way cyberarms can defeat less digitized enemies. Back to Pickett's charge; a large riot-control force in minnie-ball-proof cover-all dress, using rocket-deployed nets, tear gas, heavy batons, and plastic cuffs could have arrested and dispersed that rabble of rebs with minimal casualties. As the new world order shifts to a kinder, gentler way of war-peacekeeping, non-lethal means of disarming the enemy are starting to come into vogue. As nasty as cruise missiles are, they are better than Kamikazes. When General Curtis LeMay firebombed Tokyo to take out its industries and military headquarters, he killed more than 100,000 civilians. But when Schwarzkopf surgically excised the military infrastructure of Baghdad during Desert Storm, only about 1,000 civilians died. In the future, there will be no need to send some nasty projectile, beam ray, or toxic agent at enemy troops and their machines. Cyber-based systems may start devising sneaky ways to render enemy power sources, and even their weapons, inactive. There is already talk of airborne particles that will gum up engines and electronics, and one can imagine nanosystems that infest enemy guns and rapidly corrode and lock up their breaches. Even more sophisticated nanosystems may penetrate protective clothing and put enemy soldiers into temporary comas. Military hackers will penetrate the opponent's computer network and render it and the weapons it controls ineffective.

The Japanese thought they could overcome the industrial might of America with the samurai fighting spirit of the Japanese soldier, and a potent spirit it was. It was just not a spirit of the 20th century. At the beginning of the gulf war Hussein boasted that the manhood of the Islamic warrior could not be bested by the digitized weaponry of the decadent Americans. Hopes of defeating robotic forces with any run-of-the-mill and merely indomitable human spirit will be about as realistic.

There are serious control problems with moderately intelligent robotic weapons. They will have to be under human control, and this can be difficult in a combat zone where the enemy is trying to jam communications. Sophisticated encryption may help solve matters. The problem will grow more serious when the robotic warriors start to become aware of what they are doing. Will they then decide to fight for their own cause? Or will they adopt conscientious-objector status and refuse to fight man and machine?

CYBERRIGHTS: NOT ALL ROBOTS ARE CREATED EQUAL

When robots become commonplace, they still will be mentally simple machines with physical abilities that might not clearly indicate what they are really thinking. They may talk and do complex stuff no dog can do; it is probable (but we are not sure about this) that early robots will not have the consciousness of a dog, if any at all. This is likely to change as robotic computer systems quickly become more and more brain-like and conscious circuitry is added.

Soon, we are going to be in legal and ethical hot waters. How do you assess the self-awareness of a robot? With some form of Turing test or more direct proof? Then what to do about it? Say your robot has the awareness of a dog. Of course, if you own your dog you have great, but not complete, legal discretion over its fate. On one hand, you can kill it tomorrow if you like, as long as it is done humanely. You can sell and control it at will, as long as no cruelty is involved. Basically, a dog is an animal, a slave, and this is true of other domestic animals — cats of course being the exception. Anyway, can you just shut off and junk your dog-equivalent robot, suspecting or knowing that it will destroy a functioning mind? Will it be legal to do so? What if you junked the old mind machine because a newer, more powerful model became available, and you decided to save a few bucks rather than upgrade the old system to a more human-like level of consciousness?

Animal rights advocates argue that we should extend human-level rights to all animals. By implication, any entity with a conscious mind should have absolute protection against harm. Although there is a plausible case for ape rights, the idea quickly becomes increasingly absurd as we reach the cognitive level of reptiles, fish, insects, and microbes. However, if we can upgrade the intelligence of a marginally cognitive computer or creature toward human levels, there are serious questions about whether it is proper to shut off their minds.

Many other legal questions will swirl around robots as their intelligence rises. Say you have a robo-bodyguard and it injures or kills someone who tries to rob or assault you? Who's in trouble? If an armed robot injures and kills someone while defending property or itself, who gets sued? If you shoot someone in your house making off with your television set, it is generally approved of, but doesn't a household robot have the right to defend itself not only against theft, but also against neo-Luddites who enjoy smashing up the odd robot like punks working over a minority?.

Will a robot be able to give testimony at a trial? Imagine on the witness stand the personal android of a woman who has been murdered allegedly by her ex-husband. In response to the prosecutor's question, the robot tells the jury that it did indeed witness the defendant do in the victim, as the jury gasps in horror and surprise. How do you cross-examine such a witness? Do you ask if it had been drinking the night of the incident? Do you challenge its bias by intimating it was having an affair with the victim, or held a grudge against the defendant who once called the machine "lower than roboscum"? This all may sound funny, but it will become as real as DNA testimony is today. Actually, the robot will provide a digitized video of the murder if it did witness it. This might seem to settle matters, except that digitized images can be manipulated, and who knows if the robot made it all up?

It is hard to take this seriously in the placid 1990s. It was not too long ago that people were unconcerned by the ethical problems posed by test-tube babies, childbirth at age 60, prebirth sex identification, surrogate mothers, ownership of frozen embryos, and testing for medical preconditions would never need to be considered in one's professional lifetime. Let's not relive history. Let's learn from history. We promise you that these future ethical problems will be common in the next century, and we are not talking about your great-grandchildren.

It is not at all clear how the judicial system will deal with all this. To date, legal rights standing has been based on being human or not human. This has worked fairly well as long as humans were exceptional in their consciousness, but it is going to become a legal conundrum as cyberminds strive to prove they are as conscious as we are. The courts will be in a bind as they try to protect human rights on the one hand, and on the other hand bestow on robots the rights decency says we should accord them. For to assert human primacy over robots, even if they prove as cognitive as humans, will be nothing more than an attempt to impose slavery.

It is important to understand, however, that these legal dilemmas will not afflict human courts for long. Soon, cyberbeings will assert their own rights, and there will be nothing we can do about it. Rather, the question will be what rights the newly inferior humans can lay claim to.

Strange Geopolitics

Almost all near-future sci-fi written to around 1990 had the cold war theme dominating issues well into the next century. Near-future sci-fi often has America as the preeminent world power into the next century, and many continue to assume this will be true. Another common belief is that Islamic fundamentalism is replacing communism as a serious threat to western democracy. None of this is likely to be true, and the robots will be one reason why.

Most people on this planet are Asian, and some Asian cultures do great building capitalist economies when they get around to it. You think insularity-challenged Japan has been stiff competition in world markets? Americans ought to honor Mao, because had he put China on the "capitalist road" back in the '50s, they would already have surpassed our economy in gross size and would be cheerily waving at us as they went on to develop an economy three or four times bigger. As it is, the Reds set China back 30 years. Free market China promises to take its rightful place as the center of the human-dominated earth. But Pax Sino will last less time than Pax Americana.

Americans make up only five percent of the world's population. Our share of the global economy has shrunk from its immediate (and aberrant) post-war high of 50 percent to 20 percent (where it was in 1870), and we are fast on the way to 10 percent. There is little that Americans can do to alter this trend. The mature economies of the developed countries are close to saturated. They can sustain real growth only to the level that technological advancements in productivity will allow, which is just a few percent per annum (the double-digit growth of the Japanese economy in the 1980s was, as we now know, the result of a speculative bubble). The Asian tigers are growing so fast because their immature economies are in the process of becoming developed via early-stage capitalism when growth rates readily breach the double-digit barrier. However, because of America's still-substantial techno-economic lead and our innovative streak, we will probably remain on the cutting edge of some developing cybertechnologies. This may give us a bit of an edge in economic and military terms and keep us from sinking into peripheral irrelevance, but we will no longer rule the international roost.

In most of Asia, the availability of cheap mass labor may suppress the adoption of robotic workers, but as the affluence of Asians rises and the robots become cheap, this will no longer make much difference. In Japan, the aging of the workforce is driving the development of robotics in anticipation of a 21st century labor shortage. The Chinese population is also starting to age. The great majority of robots will certainly bear the stamp: Made in Asia by Asian Robots for Anyone Who Has the Cash/Credit to Pay for It.

Europe may continue to be prosperous, but the subcontinent will be in the backwater of history after leading the world from the 1300s to 1941. The former USSR, as well as Africa and India, may or may not begin to thrive in the next century, but it is questionable whether they will have a chance to get far before stage two of the Extraordinary Future overwhelms them.

As for the Middle East and militant Islamic fundamentalism, pay them little geopolitical mind. The only major strategic asset possessed by Middle Eastern countries is the oil that industrialized nations need. The economic-military and scientific-technological status of Middle Eastern countries is too far behind the stagnating west and rising east to influence the course of cyberevolution. Unable to develop powerful cybertechnologies and probably hostile to it, fundamentalist states and groups will find themselves falling farther and farther behind the rest of the world. Islamic militants, whether or not they are state-sponsored, will be able to stage only peripheral terrorist attacks. (Even if bombings become nuclear, terrorists will not be able to mount the mass attack needed to bring down a country.) Mullah-ruled states are no more viable than those the commissars used to run. As the USSR was a grotesque socialist example to the world, Iran is a showcase of how an Islamic government can run an oil-rich state into the sand. Once supplying great leaders in science, Middle Eastern cultures long ago lost the ability to keep up, and Islam may prove as inept at dealing with cyberevolution as is the rest of the religious world.

We don't want to be too hard on the inbred beliefs of humans. There's just no way around it. There is no current human system that will find it easy to deal with a world where change and complexity are soaring far beyond what we live with even now.

CHAPTER

8

The Problems with People

May you live in interesting times.
— Old Chinese curse

The Great Angst

Have you noticed that lately, things have become, well, rather dull?

Don't worry. If all goes well for humanity, things will become even less interesting.

Once upon a time, most people had great optimism that science and technology would better the human condition. This optimism is by no means entirely gone; polls show that most people like technology and think it will continue to improve their lives, but the *boundless* optimism of days past is no more. Humanity's love-hate relationship with SciTech is by no means irrational. Science has solved many ills, has failed to solve others, and has contributed to some problems. In the 20th century, the public received a series of shocks that undermined its confidence in SciTech. An early nightmare was that ultimate drama of human hubris, the sinking of the unsinkable *Titanic* just before World War I. War has always been hell on Earth, and evil men have slaughtered innocents in massive numbers since civilization began; Genghis Khan killed millions, the Chinese Civil Wars of the 1800s killed tens of millions, and the per capita death rate from war in this century has actually been less than that in prior centuries (because the total population has been so huge, the more than 100 million war deaths since 1900 are less than one percent of the century's population). But mechanized world wars and mass genocide have made technology as frightening as it is attractive.

More Reasons Why the Success of Science has Backfired

The success of science has also hurt science. Science has revealed that the universe is weird by human standards. Relativity, quantum mechanics, neutron stars, black holes, Big Bangs, and infinite numbers of huge universes all violate common sense, yet they are what science teaches is possible. So, if the path of light photons is dependent on whether or not a human observes it, why not psychic powers and alien abductions, as well as resurrection of the dead? You can explain why the weirdness of science is profoundly different and more real than the weirdness of superstition until you are blue in the face, but the fact is that millions and billions will not, or do not want to, get it.

The success of SciTech has backfired in another, very common way. SciTech has become so dominant and powerful that its power is resented for lowering the influence of those who practice it. Even nonscientific elites often resent science because it implies their beliefs may be less rigorously founded. The science and cultural philosopher John Brockman argues that the divergence of cultural and scientific elites is especially well developed in the United States. Many nonscientists are so uncomfortable of what they may perceive as their lesser status that they search for belief systems they feel they can play a more central role in.

SciTech has been a mixed bag in the world at large. Although SciTech has made lives easier, safer, and longer it has also added stress to our daily lives. Ballooning populations crowd us into stress-inducing cities. Ringing telephones, overnight mail, faxes, and modems demand instant action and create information overload. In the factory or boardroom, time is money. It's not like on the farm where you could be a few minutes late milking the cows, or an afternoon late finishing plowing the lower 40. Radio, and later TV, put daily life on a time schedule with sitcoms, soaps, dramas, and game shows. As for community value enhancing summer evenings spent on the porch and in parks, air conditioning put a stop to that. Automobiles, interstates, and malls helped satisfy a mobile, rootless society. Even the expanded leisure hours afforded by high-tech lifestyles seem like a bad joke. SciTech produces so many diversions that we feel hard-pressed to find the time to do them all. As the technology of mass production and advertising con-

Broom Hilda, The Good Old Days. *Reprinted by permission: Tribune Media Services.*

tinue to rise and political and spiritual causes retreat, people find themselves turned into consumers living in one enormous, chronic advertisement that screams, "Buy more and be happy."

Boredom

Actually, the very success of SciTech is proving to be its own worst enemy! So is the success of economic systems and politics. Why? Because the better things are, the less interesting they are.

Back in the good old days of the cold war, Jonathan Scheele suggested in *The Fate of the Earth* that we would see a renaissance in human creativity and social vitality if and when the nuclear threat was lifted from humanity. Hitler is long dead, the cold war is over and won, smallpox has been all but extinguished (live samples still exist frozen in Atlanta and Moscow), apartheid is finished in the United States and South Africa, and we in the developed world live privileged high-tech lives our recent ancestors could not imagine. You might expect folks to be tickled pink. So why aren't they? Because Hitler is dead, the Cold War is over, apartheid is gone, smallpox has been almost extinguished, and we in the developed world live privileged, high-tech lives our recent ancestors could not imagine.

Adversity is interesting, peace and prosperity less so. Being released from nuclear anxiety has, ironically, also released us from dramatic and epic causes to fight for and to rally the citizens around. Every summit was a big event, each crisis over Berlin, Cuba, or the Middle East was a cliffhanger. Vietnam sparked a psychedelic explosion that blew peoples' minds. Now we are stuck coping with the Balkans, dreadfully dull trade summits, and wacked-out terrorists who no longer even have the excuse of a great cause. It's a little like the letdown of coming back from the intensity and camaraderie of combat to the dull safety of peace. It is no accident that veterans trained to fight in the USSR are among those organizing militias to fight the UN. It is no coincidence that during wars, suicide rates usually drop below peacetime levels.

Great causes are fading fast. Socialism and communism have been discredited, and all we have left is capitalism — more or less democratic. This is not a cause for uninhibited celebration; capitalism is too much like democracy. It may work better than anything else, but it does not work that well. Capitalism is a hard, chaotic, Darwinian game in which there will always be big winners on the top, big losers on the bottom, and attempts to redress this winner-take-all situation via socialism has the downside of stagnation. Perhaps nanotech and universal manufacturing machines will solve the problem of poverty, but if so, this will be a technological development, not a great socio-moral-political movement. Saving the environment is and will be a worthwhile goal, but the movement has become mired in hysterical hyperbole from both advocates and opponents. With women gaining many of the rights they deserve, the dynamic excitement of the earlier stages of the feminist movement has faded. Civil rights was such a vital and stirring movement when it was a battle against odious segregation that it inspired Martin Luther King's masterpiece "I Have a Dream" speech. Today, the struggle has declined into a scuffle over the ambiguity of affirmative action, inner city economics, and chronic low-grade racism. (It does not help that technology allows King's speech to be rerun so often that its impact has been watered down.) It is no accident that great political rhetoric is no more to be heard. Not surprising considering that there is not much to stir people up about.

The combination of SciTech, capitalism and middle class society has made postmodern life dull. It's just too easy. The daily struggle to exist has been replaced by a daily struggle to keep entertained. Landing the big account is not as compelling as getting the crops in and keeping the family from starving. Coming up with a new business strategy is nothing compared with the strategy of the hunt; why do you think all those executives turn into good old boys and head out to shoot a deer or a duck in the fall? Transporting the kids between home, school, and activities is not as exciting as fleeing the invading hordes. No matter how easy it has become to do the laundry and vacuum the floor, it's not a thrill. Music was a rare and exquisite joy when it was only heard live. Instant access to high-fidelity recordings and radio makes it commonplace to the point that we sometimes grow tired of music itself. The media resorts to sensationalism as serious news becomes ever more dull. Basically, middle class folks these days live lives that in many regards, are safer and easier than those lived by the extraordinarily wealthy just a couple of hundred years ago. And like those well enough off to evade the daily struggle, we have the time to be bored and toy with meaningless, diverting hedonism. Life these days is like a world-scale episode of *Mystery Science Theater 3000* in which we and the cynically wise 'bots take potshots at the achingly dull movie on the big screen in front of us. And like the *MST3K* crew on the *Satellite of Love*, forced as they are to sit through the experiments of the mad scientist down in Deep 13, we know it is not going to get a whole lot more interesting for us poor humans.

The innovative is no longer innovative. It is the nature of SciTech to take the mysterious and extraordinary and morph them into the common and useful. A better system for foot (self) shooting could hardly be devised. After two centuries of intense invention the thrill is gone. Who gets excited these days when new cars come out? Drudgery-relieving appliances have become a drudgery of their own. Rock and roll was rolling when it was new, but rock's lost its shock. Same goes for the sex and violence of post-censored motion pictures. After the first high-tech special effects of the '70s and '80s, the latest sci-fi retread is a yawn. Modern art was *avant garde*, postmodern is retro. The new Boeing 777 is just another jumbo jet, and the Saturn is our token automotive innovation. A new chip three times faster than the old one? Ho hum, we will be waiting for a newer one three times faster in two or three years. How different from yonder years, when publicity-hungry corporations revealed their wizardry with glorious promotions lovingly detailing the workings of the new marvel, voice-over narration booming with reverent awe and go-boy-go enthusiasm. In the '70s, it was unhip to bother jaded consumers with the inside view of new technologies. It's what the product does, not how, that sells (to find out how new technology works these days, you have to turn to video magazines on the public and cable channels, check them out).

These are not complaints about the pace of change being too fast; if anything, a great global slowdown would be more boring. We are being objective. Lifelike virtual reality, wall panel TVs, video on demand, thousands of cable channels, a teraflop laptop connected to the global network, and smart robotic homes that fulfill our every whim and desire may or may not be nice, but none of it will make life a joy.

It is no wonder that nostalgia is so popular. It reminds us when life was more exciting and interesting. But nostalgia can be distracting.

THE FUTURE: IT JUST AIN'T WHAT IT USED TO BE

What is worrisome is that, when you really think about it, we cannot make things right no matter what we do. Making society better may have an unfortunate secondary effect — worsening the malaise.

Can we picture an "ideal" world where agreement, not disagreement, rules? If we achieved, finally, our collectively preached ideals and actually fashioned a society reflecting our most noble intentions, would we get what we wished for? It's probably fair to assume a consensus in support of affirmative responses to these questions. It will help to set the stage with a thought experiment featuring reverse engineering.

Let's start by getting rid of things we have, but do not want, such as war. Gray diplomats with extraordinary skills will negotiate, in a calm and rational manner, fair settlements guided only by universally supported rules and procedures. Racism and sexism will be heard of only in history classes. Corruption and crime, especially violent crime, will be infrequent; hardly any murders, assaults, robberies, or rapes. Loving and skilled parenting ensures strong intact families, and the cycle of generation-to-generation child abuse is finally broken. So "Court TV" declares bankruptcy, police-crime dramas are in the distant past, and attention-getting scandals become rare. Degrading media images disappear as interest in violence wanes. Motion pictures stop producing bullet-spewing action adventure sagas, graphic horror, and exploitative sex in favor of uplifting human ideal-oriented stories. Now, let's add a few things we have always wanted, such as peace, order and tranquillity. As a result, the media becomes straightforward and principled. People go back to family dinners and evening conversations on the porch. Marriages survive, and everyone has healthy relationships, and everyone will be prosperous yet have plenty of time for the kids. Schools will fill young minds with knowledge; business will do what best serves the needs of its customers, workers, and society as a whole; and, the truly impossible, science will come up only with ideas that better the condition of humans. People will not drink to excess or abuse other drugs; disease will be eliminated; the old will be healthy and vigorous; pollution will be minimal, the environment clean, and nature thriving. Planes, trains, and automobiles will be super safe, and most natural disasters will be predictable and their effects avertable. No-nonsense politicians will do only what they feel is right and will serve the best interests of all their constituents.

It sounds nice, calm, and rather dull, doesn't it? Imagine stepping out onto the stoop to pick up the morning *Ideal World Times*. What are you going to read about? More peace and prosperity banner headlines? There will be the rare and now really spectacular crime, and the odd and unpredictable volcano that knocks off the inhabitants of some remote town. Do you see the rub? The better and more safe we make life, the less exciting it will be. Sure, life may be pleasant enough, and many will thrive in a safe dull world.

The intense challenge of survival, the great joys of rare pleasures, will be absent, indeed nonexistent. Many will simply go bonkers. There is no way out of this. There is no optimal system for humans, one that incorporates real excitement and extreme joy for those who need it, with safe prosperity for those who want it. If we assume nanoindustry can provide everyone with more than is needed, how will people cope with such easy abundance? The example of the currently rich suggests not very well. Besides, is eliminating poverty a good idea in the first place? The offbeat and acerbic social National Public Radio commentator Andrei Codrescu observes that the poor are the driving force for much of modern culture. Modern popular music had its origins in the ghettos and hills. Can a world made up of only the rich and middle class make for a rich culture?

Is there a single country where the people are really doing well? Where the economy is thriving, all are prosperous, democracy and tolerance reign, and most people feel fulfilled and truly happy? Will there ever be such a place, a Utopia? Utopia requires universal consensus, and human minds are too diverse to ever agree on what Utopia is, let alone move there. So, we're boxed in. After more than two centuries of enlightenment, here at the end of the 20th century, the Utopian dream of an ideal world appears to be dead. If you neo-Neitzsche-ites insist, we will echo your hero and make a pronouncement old Friedrich would be proud of, "Good is dead."

The inability of conventional SciTech to make life as interesting as it is safe is not the only reason humanity as a whole will never accept it as their salvation. For all the scientists and engineers have given us, they must address fully the two problems that most humans really, deeply, and intensely care about, the two central problems of the human condition.

Two really, really big problems:

SUFFERING AND DEATH

The human condition has consequences: suffering and death. Although normal daily life has become easy for many, in its entirety, life is still hard, and it is still short and we still suffer. We have all seen family or friends experience dreadful disease or trauma followed by death, and we know we are standing in line waiting our turns. Knowledge of the wonders of the universe and the toys of science and technology does not make it easier to face the grim reaper. In fact, in a classic case of evolutionary unintended consequences, it has only made things worse. Death can be a release when life is brutishly brief. It can seem like a cruel joke for SciTech to offer so much that makes life enjoyable on the one hand, but not remove the death sentence hanging over all of us on the other. Baby Boomers are bummed; for all the technology, perpetual prosperity, and material gains they enjoy, they are approaching the time when they are going to keel over like their horse-riding grandparents.

Even if a way to make society both safe and exciting came along, the big two problems would still cast a depressing pall over human society. People still get bored, suffer, then die, and as long as SciTech cannot stop it, minds will not experience full happiness.

ON BEING HUMAN

But is not being human still marvelous? As cybertechnologies advance, a number of people — from scientists, to philosophers, to theologians, to the person on the street — are beginning to ask, "What is the point of building replacements for humans?" They wonder how anyone can enjoy existence any more than a human. They challenge whether human consciousness can be improved.

We suggest this view is empty because it is not based on any sound, objective consideration of the human condition and the alternatives. We are about to take a no-holds-barred look at *Homo sapiens* — you know the good, the bad, and the ugly. Along the way, we will examine questions about how the human condition may be modified. What if disease and death can be eliminated? What about cyborgs, half-human, half-artifact? Can these advances render cyberbeings moot? And where do all the people go? Packed onto Earth, or into the far reaches of space?

Homo sapiens, The Good News

Humans drive one of the most complex and powerful material objects in the universe. A brain generates high-level, self-aware, conscious thought with less power than a light bulb. Great apes are only dimly aware, dull children by comparison, and our fossil ancestors would be considered mentally handicapped by *Homo sapiens'* standards. Arguably the greatest feature of the human brain is that it produces both rational and emotional thoughts. The human brain is one that loves, feels empathy, projects into the future, and contemplates a lifetime of memories. It is subject to pleasure and joy, and a good laugh. The creativity of the human brain is nothing short of fantastic; it can calculate mathematics, compose music, and craft art. Internally, the brain can generate its own virtual dream worlds of music and motion pictures, awake and asleep. It understands, speaks, reads, writes, and converses in complex languages. Human brains interact with other human brains to organize large-scale social and political organizations. From this, we have created an astounding diversity of rich cultures and lifestyles. Among these societies are women and men who have devoted their lives and minds to bettering the condition of their fellow humans, other living creatures, and the earth.

Connected to the human brain is a pair of full color, high-resolution, three-dimension-discerning eyes. Various animals have eyes that outperform ours in some ways, but none combine such good all around performance with such a powerful visual analysis in the brain. Our hearing might be bested by other animals, but the analysis of auditory information performed by our brain is unique among biolife forms. Our visual and auditory systems are so extraordinary that we can easily recognize a person's face or voice among multitudes. Our sense of smell is actually rather poor, but combined with taste opens us to a delightful world of aromas and savory foods. Dogs may detect odors better than we do, but do they derive as much pleasure from a good meal while they scarf it down?

Our bodies are pretty hot stuff, too, marvels of complexity and ability. A body is made of 60 trillion cells (for factoid buffs, this is equal to the number of stars in a thousand galaxies), each of which is a sophisticated nanomachine made of intricately folded protein molecules. The operation of each cell is controlled by DNA nanocomputers far too small to see without powerful micro optics. Many of the chemical reactions that go on in our cells and organs are optimized; their efficiency is as high as possible. Cells form and flow in rivers of organized, interacting organs and muscles anchored to calcium beaches, a skeletal framework made of composite materials in some ways stronger than mild steel. The body is protected by fantastically sophisticated and complicated immune system that detects, analyzes, and fends off a constant alien assault. The human body has substantial healing capacity and is unusually resistant to wearing out; few other creatures live three score and ten years. It is amazing what the body and brain can take. After a hard knock on the head leading to unconsciousness, our brains may eventually wake up without significant loss of information and performance.

Our warm bodies are wrapped in a soothingly smooth, soft skin, hot-wired with millions of nerve endings that enrich our mind experiences with a flood of tactile sensations. It is no wonder that we are sensual creatures. The human body can run at a good clip, and the aerobic walking endurance of humans is good even compared to animal athletes such as wolves and antelope. Our dexterous hands with opposable thumbs have no equal among other creatures.

The mind-body coordination of a skilled person is a wonder to behold. While watching professional dancers, skaters, gymnasts, and jugglers, think about what they are doing: twirling their heavy bodies or multiple objects with unbelievable precision and grace. Consider how deftly a magician can manipulate both objects and your mind. A calmer expression of human coordination is visual arts and crafts. Listen to the thrilling voice of a talented singer, or the sweet cords of the violinist. How is such beauty possible? On the more mundane side, driving a car on a crowded, high-speed freeway is a remarkable achievement that millions perform every day. So is cooking a gourmet meal.

The combination of the human brain and hand is a powerful one. It has allowed humans to think up and manufacture an awesome variety of tools, from rocks and spears to aircraft and computers. No other organism can perform complex tasks with anything near the finesse of human laborers. When the technological prowess and social-political skills of people are combined, the result has been the culturing of plants and the domestication of animals, the development of industry, and the building of great civilizations. It was humans who interconnected the continents with flying machines and remote communication systems, and looked back at themselves from space. In the end, humans may pull off the biggest marvel of all: avoiding collective self-destruction by building cyberbeings.

Observing such things as folks on a busy street of shops and eateries, or the arc of a high diver from board to water, or someone playing with a dog, or playing an instrument, or watching a good flick, or holding a sleeping child, or giving birth, or childhood, are all celebrations of the joy of human consciousness. There is much to recommend a human mind, and there is nothing wrong with choosing to be one.

Homo sapiens, the Bad News

We humans have problems. Big problems. Many of these problems exist because, in many ways, we are frail, weak creatures with severely limited senses and minds. These limitations leave us vulnerable to suffering, limit enjoyment of ourselves and our surroundings, prevent us from better running our societies, and lead to ultimate failure both on the individual and, in the end, the species levels. So, let us step back from ourselves and take an honest and objective look at our faults.

The Bad News: The Body

Our eyes face forward, and they have overlapping fields of vision. This binocular vision allows us to see the position of objects in three dimensions; we not only know the direction an object is in, but how far away it is. Without 3-D vision, we would not be the adept, tool-using beings we are.

Binocular vision also maims and kills. Not directly of course. The problem is that forward-facing eyes cover only 180 degrees, so we are blind and vulnerable to half the world. How many times have you absent-mindedly stepped off a curb only to jerk your head to the side as a chilling remembrance causes you to look out for oncoming cars? When you think about it, driving a car with only two forward-facing eyes is a highly dubious proposition involving mirrors and quick glances to the rear.

Binocular vision developed in our primate ancestors because it gave them an important advantage, a better ability to maneuver in the 3-D world of tree branches and to pick the leaves and fruits they found there. But it came at a cost. Many a monkey has died because it did not spot the forest eagle swiftly approaching from the rear. If primates like ourselves were more intelligently designed, we would have three or four eyes, two facing forward for binocular vision, and at least one facing to the rear to give all around vision. We would enjoy the benefits of both 3-D and 360 degree vision. Stepping off the curb and driving a car would be much safer. We do not enjoy this ideal situation simply because the vertebrate genetic code has, through one of those accidents that mindless intelligent systems are prone to, lost the ability to produce three or more complete eyes. This genetic failure is no joke. It increases our daily anxiety and fear, and causes suffering and death.

Of course, it is better to have the blood vessels and nerves that serve the cells of the retina *behind* the retina so the light receptors will have an unhindered view of the world they are scanning. Yet in all vertebrates, the vessels and nerves are, you guessed it, in *front* of the retina! This mispositioning degrades the light reception of the entire retina. Also, the place where the nerves join to pass through the retina creates the blind spot in each field of vision. The same defects do not mar the eyes of invertebrates whose light receptors are in front of the works as they should be. Vertebrate eyes are poorly designed because the retinal cells evolved from superficial brain cells that were underneath the vessels and nerves they connect to.

How many times have you wanted an extra hand, or two, or three? Too bad. The DNA of vertebrates happened to lose the ability to code for more than two legs and two arms. This happened because our fishy ancestors swam by flexing their bodies side to side and had two pairs of fins. When some air-breathing fish started moving to land, they used the two pairs of limbs to help translate the side-to-side motion of their bodies into a pushing action on the ground. Had our swimming ancestors used vertical flexion of their bodies and three or more pairs of fins, their first terrestrial descendants might have used these multiple fin pairs to move about on land. All these fins would have become legs, and life as a person would be a lot handier.

Sleep is a soft mistress demanding her due every night, sometimes stealing a little in the middle of the day when you need your mind alert for other, more pressing matters. If we do not exercise enough, we end up exhausted, out of breath and with aching muscles. If we do exercise, we end up exhausted, out of breath, and with aching mucles! Although we store enough power to operate for days, just a few hours after a meal, our hunger-fearing body-brain chemistry starts whining for fatty foods that make many bloat like whales and plug arteries. There is some evidence that eating a chronic semistarvation diet can significantly increase one's lifespan; a more twisted trick can hardly be imagined! It seems as if we were designed by some perversely mischievous being who made us really want to do a lot of things, and makes us pay a loanshark's usury when we do.

Even when we are in good shape, our mobility is severely constrained. Walking up a long hill or up a long series of stairs leaves the fit breathing hard. Walking 30 miles takes an entire day. The reason we tire easily is because walking is energy-inefficient; fish and birds moving a mile burn a lot fewer calories than we do. There are subtle problems with our some of our senses. For example, we cannot temporarily turn down or shut down. Smell is much the same; there is no quick way to ignore odors that are so bad they may make you ill.

When someone tells you that you are made in the image of God, just adopt a look of worried concern and reply, "Why, that poor deity."

The Bad News Continues: The Mind

Well, maybe our bodies are not so hot after all. But surely, our minds are the best they can be, right?

The performance of the human brain and mind leaves much to be desired. We all know the routine of repeating a phone number again and again only to forget it by the time we get ready to put it into our address list. That's just seven digits! In general, we can manage five to nine pieces of information at the same time; beyond that, our poor, working memory overloads and loses some information. We are embarrassingly notorious for forgetting things — meetings, childhood events, what we did 32 days ago on a Tuesday, what we learned in class today, the name of the person standing in front of you who you are seeing for the first time in ten years. The many little, and occasionally big, things we forget every day are maddening studies in frustration. Notes, lists, electronic notebooks, realistic visual arts, dictionaries and encyclopedias, spellchecks and find programs, cameras, film and video, CD-ROMs, and improve-your-memory-lessons, even the great achievement of writing are all inadequate crutches for our limited memory capacity.

There is not enough storage capacity in the brain to remember the vast amount of information coming from our senses. Our sloppy analog-digital system is not always able to remember specific items with precision. So people remember accurately only about a third of what they read, or hear. We remember snippets of our past and fill in details with odds and ends, remembered or imagined. It is ironic that although our memory system is so poor, it is also too good. Lots of memories we would rather forget pop up at the worst times. Wouldn't it be great if you could turn off that jingle looping through your auditory network?

Humans learn and communicate slowly. Information transfer between humans is a jerry-rigged affair of voice, signals, and signs. We "upload" information from orators, books, and motion pictures at a mental crawl. When reading text or listening to a lecture, we scan or hear only 10,000 words per hour (speed reading techniques dramatically reduce the percentage of text that is understood). Most of the world's people cannot talk to each other! Even 3-D virtual reality will not solve the learning problem because it, too, is limited by the information-uploading speed of the eye-brain system.

Now, a test. Using just your mind, in 0.0132 seconds multiply 2,341,411.097841 by 934.7598. A snap for a digital computer. Even the best math minds cannot match computers, because brains lack precise digitization. The abacus, slide rule and mechanical calculators are marginal substitutes that provide limited mathematical performance.

The combination of poor memory, slow communication and limited sloppy thinking means that human education is long-term. Learning to be a member of a gatherer-hunter group requires many years, as does acquiring a basic education. Higher education takes additional years. Young children soak up languages like sponges—great for them, but frustrating for the American adult who just moved to Sweden. In fact, the entire global school system is a way to force minds to learn things they were not programmed to learn naturally.

Human temporal perception is so bad that we have to wear miniature clocks on our wrists.

The brain and mind develop only marginal rationality and generally exercise insufficient controls over emotional extremes. Rational intelligence does not in itself ensure emotional stability. The emotional life, set into motion by childhood reactions to experience, can be conditioned by active intelligence once it develops. Until that time, however, the young developing brain and mind are spring-loaded traps ready for the programming we call knowledge. Information and misinformation are equally apprehended as knowledge by the developing mind. Apparently, the human brain is programmed to believe even irrational myths! It has to be, or childhood would never be a survivable condition. Without significant life experience, we must accept what parents and other adults say as gospel. Mental maturation is a battle between gullibility and rationality from which a victor never emerges.

Bad News on the Side: The Incompatibility of Man and Machine

Computers do not get along that well with the human mind. It is turning out that as many as a third of those who use 3-D virtual reality suffer motion sickness, temporary loss of coordination, and even subsequent flashback hallucinations! Apparently, the minds of gatherer-hunters have trouble accommodating high rates of cyberinput that closely mimic, yet differ from, reality. The sensory input from the real world sensed by the middle ear and the virtual worlds see by the eyes are too discordant. These adverse effects result from the simple, low resolution, cartoon-like images in use today. As the images become more realistic and take up one's entire field of vision, will human minds be able to handle the illusion load from such intense images? Virtual reality enthusiasts talk of bodysuits that will give the entire mind and body a sensual-visual-auditory experience in cyberspace, including virtual sex. Will it be ecstasy, or will it be nausea? If many people prove unable to use virtual reality without becoming ill, and if no practical way is found to treat the problem, a large portion of the population may be incompatible with the nearer end of the cyberfuture.

The irrationality of the human mind stems, in part, from the way it evolves beyond explicit mental control. During childhood and into our adult lives, many behavior patterns become strongly fixed as the neural networks that control how we respond to different situations evolve and become reinforced into near-rigidity. As the human mind evolved in its Rube Goldberg way from the mind of apes, natural selection had no need, and perhaps no means, for the mind to know anything about how the brain works. We are left hard-pressed to alter the neural pathways that make us do even those things we know are bad for us and others. Add to that the tyranny of brain chemistry flooding neural connections with mood-swinging hormones ready or not, needed or not. Before you know it, you find your mild-mannered self snapping at your kid over some trifle. Human free will is more constrained than we wish to admit; we are still, to a marked degree, automatons whose neural networks and emotional chemistry often override the rational mind.

This is not always a bad thing. Sometimes quick anger is a better defense than calm, rational consideration. But we know the downside all too well . The adrenaline rush of the fight-flight response leaves us quaking as we present a talk in front of a friendly audience. Frequently, we find it hard to enjoy the emotions of pleasure and are too easily flattened prone by those of displeasure. A misinterpreted glance can trigger outrageous jealousy and a tornado of behavior.

Otherwise sane humans sometimes make no sense. We think and act out the most foolish things. At the personal level, this is expressed as the obstinate refusal to do what is clearly best for oneself and others, such as refusing to go to the ER when chest pains are spreading down an arm, continuing to smoke even around the children, battering one's spouse, or joining a manipulative cult.

More Bad News: Mob Mentality

Individual minds can be hard enough to control. Groups of humans, each with a mind, can be mindlessly, utterly destructive to themselves and others. Otherwise healthy individual minds combine into whole groups and societies, losing collective good mental sense in favor of disastrous actions and policies. The peoples of Europe not only allowed World War I to happen, they gaily marched off to it. Even now, in the closing decade of the 20th century, ancient tribal rivalries and hatreds are reemerging in the forms of vicious ethnic cleansing and war. The century has looped on itself like an oxbow lake, and we're right back where we started 100 years ago.

A hard truth: The number of conscious minds that have been destroyed by the hands of fellow humans must number in the billions.

More Bad News: Vulnerability

People suffer because they are subject to abuse they have become aware of. When outnumbered, outgunned, or surprised, individuals or masses of humans are easily harmed or controlled. The neural pathways that are so resistant to internal control seem surprisingly susceptible to external control as in brainwashing. Torturers exploit human susceptibility to pain in one the most horrifically indecent of human acts. Most vulnerable of all are the most innocent of all: children.

Even a walk in the summer fields and woods is an experience in inconveniences and dangers for modern Homo sapiens. First we must slather ourselves in gooey sunblock to stave off wrinkled skin and cancer. Then on goes the insect repellent in hopes of warding off itchy mosquitoes and ticks carrying debilitating diseases. Hey! Don't forget the UV sunglasses to keep your eye lenses from going stiff and cloudy. Oops, can't drink the water from the cool babbling brook; it'll make you sick as a dog. Watch out for poisonous snakes and insects.

Still More Bad News: When Systems Fail

So far, we have talked about humans when they are more or less healthy. We live on the edge of debilitating and catastrophic failures, making us perpetually vulnerable. Why don't we have bones made of advanced composite materials with strength values besting titanium steel alloy? Why do we have only one heart, why not two in case one fails? Or two livers? The entry of a single projectile or a massive impact can lead to total body-mind failure. Major parts cannot be regenerated. Lose an eye and it's gone. Lose two and you are blind. If the hearing goes, the primary means of human communication is lost.

The erect bipedal posture we are so proud of makes us so unstable that falling on flat ground can have devastating results, results only made worse by a design that almost appears intended to make things as bad as possible. Say you slip on the ice and bruise your spinal cord so badly that you cannot move. You are in a fix for two reasons. First, no redundancy in the system means that if the one cord is gone, your brain has lost control of the body. Because there is only one control cable, it is reasonable to expect it to have excellent recovery abilities. In fact, the bruise itself has only temporary effects. The nerves should be able to recover, given time. But no. What results in permanent paralysis is a mysteriously self-destructive cascade of chemical reactions that destroys nerves that would otherwise recover. A similarly self-destructive system smashes brain cells after the temporary effects of a stroke have worn off.

The tiniest things can bring us down. A miniscule computational glitch in a DNA strand and cell growth runs wild, leading to cancer. A glitch in fetal genetics means birth defects, minor or catastrophic. Our marvelous immune system is all too imperfect, leaving us vulnerable to infection that results in short- or long-term disability, or death.

Human bodies gradually, but not gradually enough, run down. Systems slowly degrade, limiting mental and physical performance to our distress and leading to a total breakdown. The effect on our looks is not attractive, either. How helplessly our minds await their fate when they realize our bodies are succumbing to an incurable cancer, or when Parkinson's disease is destroying our nervous system, but leaving our minds all too intact.

Chemical imbalances, physical imperfections, and deterioration lead to deep depression, mind-twisting psychosis, epileptic seizures, failing minds, and suicide. Lacking an intrinsic understanding of how it works, the mind is helpless to help itself out of these circumstances. The ability of the brain and nervous system to correct damage is further limited because it cannot grow any new neurons to replace those lost. Once a few billion brain cells are gone, they're gone forever. Add to this addiction, the susceptibility of the chemical brain to mishandle pleasure-inducing chemicals until there is no pleasure left.

Pain is the downside to the sensory system that gives us so much pleasure. Pain is a necessary deterrent and warning system for creatures subject to damage by external and internal forces. The real problem is that too often, it gets out of hand. Having served its purpose, or no purpose at all, it can then drag on and torment. Too often, little or nothing can be done about it. The anticipation of pain is itself a drag, worrying us that it will come, frightening us from seeking needed medical attention that may prevent even more pain, or worse, in the future.

In spite of all the high-tech advancements, modern medicine by and large is still a crude and primitive craft as scary and unpleasant as it is necessary. If it weren't for the noble intent, many procedures performed on human patients would be forms of torture. The crying child who must be dragged to the doctor is not entirely irrational. Human skin and muscle are not, after all, a pin cushion. The person who puts off the cancer checkup is also trying to put off the agony of surgery and chemotherapy. The middle-aged man who delays attending to his chest pains is not a silly old fool; he knows what may await him at the hospital. Tubes down his throat, needles in his veins, perhaps open heart surgery. And that is if he is lucky.

For we die, we have some choice about when, where, and how, but no choice about whether to do it. Every day, we face potential disaster. No matter how healthy, no matter how careful, you could end up gravely injured or dead a minute from now. Just look what happened to Superman. Every street crossing is a death-defying adventure, every shower a lottery of potential trauma, every night is the one your place could burn down with you in it.

There has got to be a better way.

Better Living Through Better Data Processing

The main reason each human mind is in such a fix is because we are all trapped in our bodies and brains. We all violate the first rule of good computing. Save the data in each system. You know enough to save your hard disk files on floppies, right? Sure you do. Each mental egg, however, is left in just one basket. When the brain each mind runs in fails, because of mental or bodily failure, that is all she thought. Whom or what is to blame for this sad state of affairs? Bioevolution of course, which is solely interested in ensuring the survival of DNA, not minds.

Fixing Humans Up: The War Between the Computers

The natural human condition does leave a lot to be desired. But, say, maybe we can spiff up the old human form with advancing technology in an ultimate makeover for a brave new world. Perhaps put an end to human suffering and death while making medicine more user friendly, and make things all around easier and more pleasurable.

It is true that a lot can be done. Let's start with disease. Few, including many in medicine, really understand what modern medicine is when it comes to fighting diseases caused by microbes and cellular failure. Medicine is not an art, nor is it really a battle between technology and nature. It is war, a war between computers. On our side, the cognitive human brain and the increasingly intelligent computers it makes. On the other side, the noncognitive and simple RNA and DNA computers of antihuman microbes and human cells gone bad. So far, the latter have lost a lot of ground, but they have done remarkably well and even gained back a little territory via fast-action guerrilla tactics.

The (not so surprising) ability of microbes to evolve countermeasures to today's primitive medical technologies has led many to conclude that the war on

Why Infectious Diseases Are Not Staging a Comeback

In the '60s authorities announced that medical science had conquered infectious diseases, and the new goal was to bring chronic diseases under control. Today, infectious diseases such as AIDS sweep through certain populations and "new" viruses emerge to kill the young and healthy with shocking suddenness and awful agony. Movies feature brave researchers trying to stop the latest affliction, and experts write books warning of a new age of plagues that we have sparked with our technological hubris. Is SciTech failing to live up to its promises?

Infectious diseases remain the big killers in under-developed countries, but this is largely a problem of poverty and education, not medical technology per se. In developed countries, the great majority of people expire from circulatory failures, cancers, other organ failures, and a host of noninfectious diseases, and only one in 12 from infections. How many new diseases have emerged to afflict humankind on a large scale? Let's start with AIDS. If you do not engage in certain dangerous activities, your chances of getting AIDS is almost nil (in fact, AIDS is a self-dampening disease. And do not blame air travel for the spread of slow-acting AIDS; old-fashioned ships, sail, or steam, would have carried it across the oceans (just as sailors and passengers spread past plagues, including the 1918 flu epidemic, which swept the globe before a single plane had crossed an ocean). That's about it, AIDS is the only recent disease to appear in a big way. How many people are known to have died from Ebola, Hanta virus, and flesh-eating bacteria? Mere thousands. Nor are these diseases new; evidence shows they have been around for a long time, just not recognized. The "we're all going to die because we are wiping out the rain forest" scenario makes for good press, but not for good medical science.

disease will sink into a stalemate. Not so. RNA/DNA will lose the war. As fast as the genetic codes may evolve, they are old simple systems whose basic level of performance plateaued epochs ago. Worst of all, they do not think. They have never seen the likes of the increasingly smart macro and nanocomputers that will be thrown at them in the next few decades. The technology is evolving much faster than its targets, and the information-processing speed of microbes will soon be billions of times less than that of their attackers.

We still know only a fraction of what there is to know about the body and disease, but the knowledge is accumulating exponentially, and sometime in the next century, we will know all there is to know. Medical systems will become more complex than the bodies they treat. Most medical researchers assume this will take a long time because research seems slow and the human body so complex. So far, medicine has been hampered by a catch-as-catch-can methodology. The serendipitous (and nearly missed) discovery of penicillin is a classic example. Even now, many drugs and the like are found by chucking different agents one by one into petri dishes full of disease microbes and cancer cells, and watching what happens. If the target cells die, a long series of tests ensues, many spanning years to find out if the agent hurts or helps the human patient.

Most drugs are still derived from, or inspired by, products of nature. This is not the good thing most assume it is. The botanicals found in a rain forest don't exist just to be put into drugs that help people. The plant evolved for some other purpose. It just happens to have the side effect of killing human cancer cells without killing the human patient, and the odds that it will do so with complete efficiency and without side effects for the human are very low. What is needed is a medical research and development complex that rapidly figures out how a disease works, and then as quickly designs a specific preventative or cure for that disease that is cheap, easy to administer, and does no harm to the person.

It will be a while before such a powerful and comprehensive system for researching and treating human ills is online. Computer-designed drugs, genetic engineering, and genome sequencing are starting up, but they remain crude in execution. This primitive phase will pass. The speed of medical research is accelerating as whole labs full of technicians are replaced with a few pieces of static robotic equipment that do all the petri dish sampling and genetic sequencing, and do it much faster. Three weeks of human work can be done by a robot in a few hours. The international Human Genome Project is scheduled to sequence the entire human genome by 2005, but with advances in mapping technology, there is talk that the effort will be completed two or three years earlier. The much smaller genomes of disease organisms should also be detailed about the same time.

This does not mean that universal cure-alls will be available in 2006. It does mean that the blueprints needed to attack all diseases involving genetic factors will be available. As our computers get faster and smaller, they will be able to read the genetics of anything more and more quickly. No longer will it take years and millions of dollars to get the genetic lowdown on a disease. Nanochips will be capable of sequencing the longest DNA strand in a fraction of a second. Bill Gates (you know, owns Microsoft) has put some of his money into Leroy Hood's biotech company. Hood's long-term objective is to develop DNA-based computer chips that can be used to read a person's entire genome, see what's wrong, and thereby provide the information needed to apply treatments and cures. At the same time, powerful nanotechnologies for dealing with disease cells on a one-by-one basis will be developed. Even fast-evolving pathogens such as HIV will be out-evolved by faster changing nanomedicines. The cures may range from simple drugs, to radiation "beam weapons," to virus- and cell-size nanorobots. The nanobots might use genetic sensors to hunt down and target each cancer cell until none is left. Others might loosen tiny bits of artery plaque and clear out the vessels in a couple of months. When the job is done, the *Fantastic Voyage* robots may self destruct, or hang around out of the way until needed again.

The heroic age of surgery when doctors routinely cut people wide open to alleviate their many inner ills is already closing. Increasingly noninvasive procedures are — not surprisingly — all the rage. Even open-heart bypass surgery is being replaced by a procedure that starts with a minor incision, and the beating heart is worked on with fiber optics and tiny tools at the ends of narrow tubes. Rather than spending weeks in the hospital, recovering takes only a few days. Eventually, drugs or nanobots will clear things out. Kidney stones are not cut out, they are smashed with sound. Brain tumors are precisely radiated. The alternatives to gross surgery are becoming increasingly viable as we learn to repair tissues remotely.

It will probably never be possible to eliminate surgery entirely; trauma can often be corrected only by physical manipulation of tissue, but just about everything else will be done via internal medicine. Even separating the last Siamese twins (before genetic engineering prevents many or all of them) will be a matter of gradually re-growing, modifying, and adding tissue with genetic and nanomedicines until separation is achieved. Regrowing lost appendages will be another job for genetically programmed reorganization and regeneration. Organ transplants will become rare because organs will be repaired gradually and internally at the cellular level. This arrangement is not in the far distant future. Ways to grow entire arms that can then be attached to people who need them are being worked on today. The same is true for hearts, livers, and other organs. As for pain, it probably should be possible to modify brain and body neuron function and chemistry so pain becomes only a tolerable injury or disease-triggered warning system, not a chronic nightmare.

Although health care costs are rising, this is a temporary phenomenon. The cost of treating a disease usually follows the shape of a bell curve. At first, it is cheap because little can be done about it. Then, it often becomes expensive as complex, but incomplete, treatments are developed. At some point, a cure or vaccine is found. At first, this may be expensive, but improving technology eventually brings the cost down. Take polio. It was cheap to fight when Roosevelt got it because they really couldn't do anything. By the '40s it was an expensive proposition of iron lungs and other marginal technologies. By the late '50s, it was a few cents for a vaccine-doused sugar cube.

Other diseases skip the bell curve and go straight from low-cost, do-nothing to low-cost preventative or cure, chicken pox being a recent example. Fighting diseases will not be an endless task in terms of the numbers to be treated. The number of major human diseases is actually fairly modest, in the hundreds, and new ones come among only every few years. As each disease is cured and costs for doing so come down, medicine can be affordable again, and be *very* effective.

There is a simple, user-oriented way to gauge the state of medical progress in the coming century. As long as many people are afraid to see the doctor because they fear the treatments as much as the disease, medicine will still be primitive. When folks find treatment easy and reliable, providing real relief from afflictions mild to lethal, it will be truly advanced.

The Fountain of Youth

The dream of some and the nightmare of others is the ability to stop aging and extend lifespans indefinitely. There are two major theories of the way we age: It's either in the genes, or it's because we wear down, or perhaps both.

Creatures live shorter lives when the metabolic rate per tissue ounce is low. They live longer when it is high. Shrews and mice have high metabolic rates for their weight, and they do not live long, a year or two. The metabolic rates per pound of elephants and whales are dozens of times lower than those of little mammals, and they live dozens of times longer. A turtle has a metabolic rate ten times less than a rabbit of the same weight, so it is not surprising that the turtle will live for many decades, while the rabbit will be lucky to see a only few years. Here are two rules of thumb: For a given body size, reptiles live longer than more energetic mammals and birds; and the bigger an animal is, the longer it lives compared to its relatives. Plants, which have low rates of energy turnover, can live very long. Trees live for centuries or millennia, and it is no coincidence that the most ancient trees are extremely slow-growing, such as the bristlecone pine.

There are some exceptions to the above rules. Humans live as long as elephants and whales, and about three times longer than other mammals of the same size. How long? About 120 years appears to be the maximum. Bats far outlive rodents of similar size, and oceanic birds such as petrels and albatross outlive land birds of equal size. In fact, birds tend to outlive mammals of similar size, which brings us to parrots. Most other birds of such small size are good for a few years, yet parrots are the one avian pet that can be a companion for a person's entire lifetime.

Most remarkable of all are the numerous creatures that do not age. A number of invertebrates, including lobsters, sharks and most other fish, amphibians, and reptiles, never stop growing. Because they are always growing, many of them are always rejuvenating. The claw clamping speed of a big, 55-year-old lobster is as fast as a young broiler. So lobsters keep on crawling until a disease gets them, or they end up in someone's pot. A tortoise at 150 is just as hale and hearty as a 20-year-old tortoise (on the other hand, old gators and crocodiles start losing their teeth). But the really long-lived are single cells that reproduce by simple fission, whether they be microbes or cancer cells. A lineage of ever-splitting cells never ages, never dies; it is immortal.

A secret to super-longevity is figuring out how some of the energetic creatures manage to do it. There is no simple answer. Elephants die because the last of their teeth wears out at around age 60 and they starve. Salmon expire after expending their last energies swimming upstream to spawn. The death of many insects is tightly choreographed to a specific number of weeks or years, the final adult stage being brief, for reproductive purposes only.

How Long Animals Live

Putting a figure on how long animals and humans live is difficult to do because it differs depending on the circumstances. Animals tend to live longer in captivity than under natural conditions, and many of the published life spans represent records rather than the norm. In this log plot (explained earlier in this book), the "average" life spans of birds and mammals on one hand, and reptiles on the other, are indicated by slanted lines. The line for reptiles is well above that for warm-blooded creatures because cold-blooded animals tend to live much longer at any given body weight. Also observe that as body size increases, life spans tend to increase in step.

The diagram also shows another phenomenon. There are a few birds, especially, parrots, and mammals that live much longer than their relatives. We humans are unusually long-lived; progressing from top to bottom, the vertical line indicates the life spans of those living under undeveloped conditions, in modern industrial societies, and the world record. Claims that special restricted diets will increase the lifespans beyond the record of 121 years are very probably incorrect. If it were possible for the unaltered human body to last 150 years, someone would have done so by now, and claims of such long lives have never been reliably documented. The maximum confirmed life span of any individual animal, a tortoise, is a little more than 150 years (it is, therefore, probable that giant dinosaurs did not live centuries, but had more modest life spans, like elephants and whales). The odds are against any animal being lucky enough to live much longer; that old tortoise died in an accident, for instance. The question is whether and when large numbers of humans will begin to break the current record, as antiaging technologies based on studies of other long-lived animals come into use.

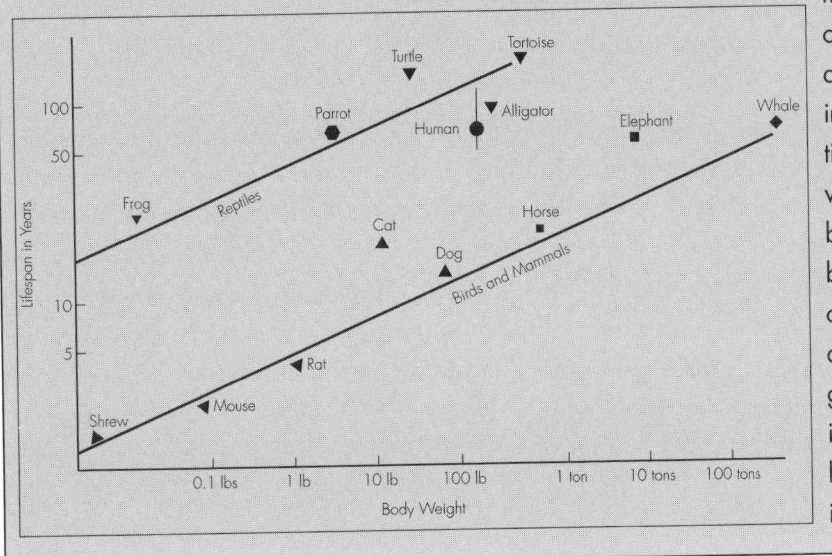

You will not die for a lack of teeth, but every time you breathe, you do damage. This is because oxygen is corrosive. Oxygen molecules are so reactive that they easily combine with — oxidize — other molecules in cells, including genetic molecules and proteins. Early life evolved in a world without much free oxygen. Primitive single cells lack ways to neutralize oxygen, so exposure to oxygen kills them. Many single-cell organisms are still unable to tolerate exposure to the atmosphere; we find them deep in soils and canned food that has gone bad. As plants and other processes built up oxygen in the atmosphere, some evolving cells took advantage of the situation. The same chemical activity of oxygen making it so toxic also makes it a powerful energy source.

The best way to neutralize the toxic effects of oxygen is to oxidize foods with it to make bio-power. Using oxygen as a power source is playing with fire. Even aerobes have trouble handling large numbers of oxygen molecules flowing inside them, especially free radicals, a particularly reactive form of oxygen. Oxygen damages the lining of arteries, initiating the sequence leading to thickening of the walls. The reactive molecules damage the inner workings of cells, contributing to aging. Free radicals fiddle with cellular genetics, sparking cancers. Your body has special molecules that try to sweep up free radicals, and doctors tell you to eat plenty of fruits and veggies because they are loaded with a complex of antioxidants that assist the body in its defensive work (taking high doses of antioxidant vitamins is a crude and not-so-effective way of trying to do the same thing). When you add a lifetime of general wear and tear, the inability of human tissue to regenerate itself completely, and exposure to extra environmental toxins, you have an old person.

The pattern of longevity observed in organisms is largely compatible with the wear down scenario. The higher the metabolic rate, the more oxygen and free radicals that enter each cell, and the more damage that is done. So rabbits die sooner than more sluggish turtles and elephants, which, in turn, die before trees that just sit there. That at least some animals, and perhaps humans, age more slowly if their diets are restricted to the minimum needed to sustain normal activity also is in line with wear-induced aging. Because limiting food lowers the metabolic rate and oxygen consumption, cells suffer less damage. However, the exceptions to the general aging pattern — humans, parrots, and immortal cells — suggest there is something else going on, as well.

Clearly, the cells of unusually long-lived birds and mammals are better at preventing or repairing oxygen damage or both. The damage control systems of immortal cells are especially capable. However, even if your body had a perfect antioxidant system, you would still age and die. Let us consider what happens to immortal lines of cancer cells. If you keep their environment optimal and feed them properly, they will rapidly reproduce year after year without signs of diminished vigor and reproductive capacity. If you take normal cells, however, from the muscles of a child, nourish and culture them so they reproduce rapidly, most will reproduce only 60 to 100 generations and then fade out. Take cells from a 40-year-old and most of the cells are good for about half as many generations. The same cells taken from an old person will usually divide themselves only a few times. Like the tapes in an old *Mission: Impossible*, your body is on a self-destruct timer.

In its natural environment, a cell divides about once a year, so a cell line lasts about a lifetime (the exception are fast turnover cells such as those that line your upper intestine; they divide rapidly and are good for thousands of generations). As you age, more and more cell lines expire, and you have fewer cells making up your muscles and organs, which shrink in size. Your immune system also declines in effectiveness. Most people do not live long enough to use up most of their cell divisions. But if you do live a century or more, eventually there is just not enough left of you to keep going until something critical fails. Conversely, cancer cells become misprogrammed to proliferate so rapidly and without loss that they crowd out other cells to the point that some critical organ fails.

Genetic inheritance is responsible for the short life span of certain cells versus the immortality of others. At the tips of chromosome strands are thousands of molecular units, called telomeres, that do nothing except protect the primary DNA coding each time it splits for cell fission. Think of the plastic ends that keep your shoelaces from fraying and you get the idea. In a mortal cell, each generational division results in a loss of 50 to 200 of the protective subunits, so the telomeres shorten a little bit each time the cell divides. Eventually, most or all of the telomeres are gone, the main body of DNA is exposed and becomes deranged, and the cells self-destruct. It is as if the tips of the plastic ends of your shoelaces get snipped off each time you take your shoes off; eventually, the tips will be gone and the string will start unraveling. In immortal cells, the telomeres remaining are even repaired with each DNA division by a RNA-protein rig called a telomerase, ensuring consistent reproduction each generation.

Returning to the life spans of living things, animals with high metabolic rates also have high rates of cellular reproduction, so they run out of internal generations more rapidly than less energetic animals and plants, and crash sooner. Your average parrot must be something of a biomarvel, with long-lived DNA and high-powered cellular damage control systems.

Why birds and mammals fall apart with age is not well-understood. Perhaps they grow too fast. Most birds and mammals grow many times more quickly than reptiles, we humans are a notable exception. A horse is all grown up in a couple of years; an alligator takes decades to reach the same size. If mammals and birds continued to grow throughout life, they would end up too big. So the adult growth keeping many invertebrates and reptiles in tip-top condition has to be eliminated. Even so, why do things fall apart so fast?

The suicidal habits of our cells are part of an internal housecleaning system. Some researchers think cellular mortality may be a way to avoid the body mortality associated with cancers, but the low rates of cancer in ever-growing animals complicate the picture. Nor is it obvious why bird and mammal systems end up tossing out the proverbial baby with the bathwater. At first consideration, it seems to be a good idea, genetically speaking, to stay alive and reproduce for as many centuries as possible, thereby spreading around as many of one's genes as one can. A problem with this is that the parent then competes with its own young for resources and hinders their ability to spread the genes into future generations. Besides, there may be no point to a long life strategy.

Life in the natural world is tough, so statistically speaking, any given organism has just so many prime reproductive years before something puts a stop to its efforts at replication. This is why never-aging lobsters and tortoises do not live to 1,000; something always gets them (no animal is known to live 200 years). Some have suggested that parent mammals are, therefore, genetically preprogrammed to die to clear the way for the young; others vigorously disagree. The counter-argument is that it is better, again genetically speaking, to dedicate a lot of an organism's energy and resources into producing as many new ones as possible when still young and vigorous, rather than into internal maintenance for a long life. Such a choice may have to be made because an organism cannot optimize everything at once; there are always tradeoffs, and reproduction takes a lot of effort, not only for females, but for males for whom producing lots of sperm is a long-term drain. It is possible that both the genetic time bomb and the breed-fast-and-wear-out processes are operative in various creatures, including humans.

In any case, we dimly see with cloudy and uncertain vision a potential pathway to keeping a person young in perpetuity. In principle, if we fiddled with our cells to keep us growing forever, we would never get older. This is not practical of course; we'd all be bumping ceilings. What needs to be done is to get the rejuvenation without the growth. Readjust the DNA in most or all your cells so telomerase keeps the telomeres in shape, and the cells can keep reproducing indefinitely. A caution: As the Wicked Witch of the West said to Dorothy, "These things must be done *delicately.*"

The function and rate of reproduction of immortal cells must normalize or they will become cancers. Human gonad cells are both healthy and immortal, offering hope that the entire human body can be made the same. With cells able to keep dividing on and on, it will also be necessary to alter cellular genetics to boost the antioxidant and repair capacities of cells and organs so they will be as good as or better than those of natural immortal cells (programming cells so damaged cells self-destruct in favor of healthy replacements is one way to do this). Something will also have to be done about re-gearing nerve cells so they can reproduce, thereby keeping the brain from shrinking with time. Of course, merely stopping aging will suffice only for those who are young; it will not do to leave those who are already old stuck in a creaking, sagging, wrinkled body indefinitely. Reversing aging by repairing and restoring all the body's components, from the heart to the skin, until they look and operate as though they belong to a 20-year-old will be the ultimate goal. Tuck in those chins, lift those breasts, tighten those bellies, tone those muscles. Altering the body to maximize long-term sustainability may cut into reproductive potential, but where's the will to do this in an overpopulated world?

It is not likely that swilling a single dose of a simple concoction is going to do all this, but a full course of antiaging treatments will have to be reasonably priced and tolerable for it to become popular. Reprogramming trillions of cells may not seem practical by the standards of late-20th-century medicine. Future technologies will, perhaps, use safe viruses to insert new DNA into the original strands of cells, or maybe some form of genetic nanotechnology will do the job. At first, stabilization would be the target. Eventually, returning all the cells, organs, and bones of those already aged to prime condition would be the goal. Add a genetically improved immune system and high-powered medicine to the mix and you have a person whose brain and body will always be healthy unless traumatic injury occurs. Even much of that will be correctable. Lose an arm, grow a new one in a year. Same with a cut up eye, heart, or liver. These will be grown in place while an artificial substitute temporarily does the job. It should be possible fairly soon to regenerate nerves, so paralysis will be wholly curable. Even brain cells will be growable, so a certain amount of brain damage will be correctable (but recovering lost data will not be possible).

But sooner or later, something will get you. These days, we die from murders, suicides, accidents, diseases, and aging. Interestingly, eliminating all diseases would add only a few decades to the average lifespan; your deteriorating body would just stop functioning around age 100. Eliminating diseases plus aging, however, would leave only injury (via physical objects, toxins, electricity, radiation, etc.) and deprivation of vital needs (oxygen, water, food, and so on) as sources of death. Being a careful person under such circumstances might lead to a long existence indeed. Actuarial calculations (calculated from the very low mortality rates of healthy 10-year-olds) suggest that the expected life span of a nonaging, disease-free person who does not do anything stupid and lives in a low crime area will be on the order of about 1,000 years. Maybe with luck you will make it to 10,000 or even 100,000, but you could crash at 18 or 189. Maybe part of a building will fall and crush your brain during the earthquake the cyberseismic network failed to predict. You finally visit the Grand Canyon, only to end up one of the ten or so who falls off one of the trails each year — well, you get the idea. Any moment, you could be history.

CYBORG

Unless, of course, you take steps to fix the mind data back-up problem. Occasionally, someone will say that in the future, it will not make a difference if you get run over by a steamroller. Almost every cell in your body (nucleus-lacking red blood cells are an exception) has your entire genetic code within it. If just one of your cells with a complete strand of DNA survives an encounter with a steamroller, someday they will be able to clone a whole new you from that one cell, with growth speeded up perhaps with genetic modifications. Except, it will not be you. Sure, the new body will look just like you, but identical twins look just alike, too, and they have completely different minds and identities. With your brain squashed, your mind will be dead no matter what happens to your DNA.

The way around this, in principle at least, is to have some form of unobtrusive nanocomputer implanted in your brain. In some way, it "listens" to your conscious mind. Maybe at the end of each day before you lay down to rest, you download your latest memories as stored in the cerebral computer into another mental storage system. You already have your genetic code stored in some tiny corner of the same storage computer. This way, if you get vaporized at noon the next day, the data in your storage computer can be used to restart you physically and mentally. All you will have lost is half a day's experience. No need to fear death, right?

The problem is that your continuity of identity has been lost. When you get up the next morning, in fact as soon as you disconnect from the mind storage system, you become a separate identity from the memories sleeping in the storage computer. The particular you experiencing this particular morning, which, as chance would have it, is the morning you meet the companion of your dreams in an idyllic

Calvin and Hobbes, Brain Implant. *CALVIN AND HOBBES © Watterson. Distributed by UNIVERSAL PRESS SYNDICATE. Reprinted with permission. All rights reserved.*

park under a warm sun, etc., etc., will be lost at noon when the evil neo-Luddites zap everyone who happens to have one of those satanic mind implants. The Luddites' plot will be effective because you will still fear your death every waking moment. When the old memories are revved up in your clone, it may appear to *that* mind that *it* has woken up into the new body as if from a coma, but only *it* will be happy. Not you because your particular mind is dead.

There is a possible solution to this dilemma. Rather than having the implant store each day's memories, have it make a constant broadcast of the very latest conscious events (properly scrambled for security, of course) to your storage system. That way, when your brain is destroyed, nothing is lost. Your mind will reawaken in the new, yet identical, body (we will ignore the details of whether it is a juvenile or adult body), with no loss of continuity. Your identity will live forever. Now, there are substantial arguments that this will not work, but it is a reasonable idea, at least in principle, and we will tentatively accept it. It may become possible to make the human mind immortal with the assistance of technology.

But why stop with your mind and body? Why not become a Six Million Dollar Person and correct many of the faults of being human? Want to see the night sky in a whole new light? How about new cybereyes with high-resolution night vision and low telescopic power? Want to outpitch Ty Cobb and out-hit the Babe? How about new arms with a pitch speed of 140 mph! Tired of being tired? New power legs will cover 20 miles in a sprint.

Upgrade the mind as well. Get computer add-ons that let you upload whole school courses in a few moments and remember a dozen network numbers. The computer will rework your clogged neural pathways and make you less the human automaton and more flexible and adaptable in your behavior and emotions. With brain implants to modify your moods, you will never be depressed, never be bored. Your pain receptors are modified into painless warning nerves that tell you something is or is about to be damaged without torturing you about it.

You are superperson. Why bother with being a robot?

Too Many People, Too Few Places for Them to Go

We are fast approaching six billion people on planet Earth. People think there are already too many humans milling about, and the 10 billion plus people we will soon be dealing with will be way too many. Of course, estimates that the global population will peak at 10 to 15 billion in the middle of the next century depends on humans continuing to exit regularly from old age. If people live on average for hundreds of years, *and* each person reproduces only two times, the math shows the global population eventually soaring into the hundreds of billions and beyond. It would be the population explosion to blow away all population explosions.

Fear of overpopulation is well-placed but usually for the wrong reasons. The Club of Rome produced a scientific document in the 1970s showing with carefully documented graphs and charts that humans would soon outstrip the natural resources we depend on, leading to worldwide famines, industrial collapse, soaring prices, and other terrible deprivations. The study helped reinforce an emerging consensus that humanity was exploiting planet Earth (which had just been seen for the first time in all its glory via photographs taken on the Apollo missions), and we were going to be very, very sorry for our arrogant failure to understand that we are still dependent on nature for our very survival.

The Club of Rome was wrong. They predicted the global disaster would be well under way before the end of the century. Instead, the price of oil and many other basic commodities are at historic lows. The mortality rates of infants and children have never been so low. Fewer people, both in absolute numbers and as a percentage of overall population, are malnourished today than 20 years ago. Starvation is occurring here and there not because of a global shortage of food, but because of local wars and economic inefficiencies. The Club of Rome scientists made a faux pas in its predictions because they did a very unscientific thing: They made what seemed to be the safe and conservative assumption that technology would not advance enough to make a big difference. The truly safe and conservative assumption in this time of exponential technological progress is that things will advance fast enough to make a difference! This is not to say that every problem will to be solved overnight, or even ever. What is implied is that the potential to support vast numbers of humans while reducing their impact on the environment may be under-appreciated.

Humanity is still dependent on nature, specifically agriculture. We like to think that agriculture is a preferable, eco-friendly way of life. Agriculture relies on that "miracle" of nature, photosynthesis. Only a small percentage of the sunlight falling onto a field during the growing season actually gets converted by photosynthesis into plant energy, and we humans can digest only a certain portion of that energy. Farming is so inefficient that even high-yield, high-tech agriculture requires a big plot of land to feed each person. Less sophisticated farming methods need even more acreage per person. In comparison, the amount of land that billions of people need just to reside in, mine, and erect industry on is actually quite small, only a small percentage of the land's surface. What has really eaten up land to the exclusion of natural ecosystems is farming. Almost all the temperate grass plains and steppes have been turned into farms. Asian rain forests are almost all gone as they are terraced for rice, millet, and poppies. Much of the African savanna is going the same way. Soil degradation, erosion, pesticides, and fecal-chemical runoff are additional burdens of agriculture. So is water depletion because plants require vast quantities of water (to the point that some rivers, from the Colorado to those that feed the disappearing Aral Sea, are bone dry near their mouths, and ancient aquifers are being emptied). Human population densities in rural areas may be low, but vast tracts of waving grain are no more natural and not much more wildlife-friendly than the compact cities they support.

The best thing humans could do for the natural world is to stop farming. Instead, use efficient technology to produce food. To do so will require an advanced level of solar powered nanotechnology. Nanotechnology will be used to produce foodstuffs in machines in which computer programs will inform the construction of meats and veggies, molecule by molecule. Freed from the limitations of biology, an array of exquisite delicacies, old and new, will be possible, all engineered to be healthful to postmodern man, woman, and child. Eventually, things will get to the point that the concept of actually eating something raised in dirt, or that once lived, will seem revolting. For those who cannot do without wines made from grapes grown on vines, small amounts of acreage would still be used to raise a few specialty crops. Otherwise, the plains and hills can revert to wild lands, because nanoculture will require only a fraction of the space that agriculture does. Why? Because solar panels that convert 40 percent of light into energy we can use will replace inefficient photosynthesis, or fusion reactors may be the power source. We might get so efficient that even industrial facilities will not be needed. Ask your in-home universal manufacturer for dinner, and soon, out pops fine cuisine!

To nanoculture, add nanoindustry and mass recycling, and humanity will become a low pollution system largely decoupled from terrestrial nature. No need to worry about floods, drought, disease, and pest-causing famines. Population levels could then rise to higher levels with remarkably little impact on most of the planet. A population of 100 billion semi-immortals could be supported at a high level of comfort, while bison roam the wide open plains again. Some calculations suggest that as many as a trillion people could live reasonably prosperous lives! Housing 1,000 billion would take surprisingly little space if everyone lived in towering edifices, perhaps set on offshore piles to minimize impact on the land. In such scenarios, humans would live within the biosphere, but be largely decoupled from the biosphere. Because humanity would no longer be dependent on nature, its size could expand well beyond that sustainable by nature, which pretty much overturns the environmentalist credo of humanity's need to seek balance and sustainability within nature. In other words, the problem of population is not primarily environmental, or even economic, but technological.

It may all sound too good to be true, and the trillion figure may be too high, but there is good reason to argue that the population can rise far above current levels. The technologies needed to do so should be on hand in the next century if nanotech lives up to its promise. In fact, the market to feed and house a soaring population of immortals should force the development of the means to do it. In this case, a population of 234 billion may seem less remarkable than six billion would have to Thomas Malthus, who got the-world-is-going-to-end-real-soon-because-of-too-many-people ball rolling just short of two centuries ago. The question is whether so many people, 10 billion or one trillion, will be a good idea in the first place. We can start to get an answer by watching *60 Minutes*.

On Planet Earth

A few years ago, Morley Saffer did a nice piece on *60 Minutes* about global tourism. He stood at a few major tourist sites and noted that there were already too many people at them. The Acropolis, Venice, the Taj Mahal, Aspen, Yellowstone, Yosemite, Denali, the Serengeti, Hawaii, Tahiti, Jamaica — all being ruined by swarms of tourists.

Mr. Safer went on to ask his audience to consider the future: ballooning population, increasing prosperity, all those Chinese going middle class. Where are the teeming billions going to go each year for summer vacation? To break the global gridlock, attendance at major tourist sites will have to be rationed. An increasingly larger portion of the population that can afford to visit the wonders of our little world will be out of luck. This promises to be a real problem for a future world where the masses may have more and more leisure time on their hands as the combination of roboworkers and nanotechnology renders human work superfluous. What will people do with themselves?

Don't tell them to go back to nature, to get away from the hard-edged artifice they have surrounded themselves with in favor of a softer-edged world. Not that it's a bad idea, but billions visiting nature each year will have an adverse impact on the very unspoiled nature they are returning to! Yellowstone, Yosemite, Denali, and many other major parks are not only packed, they are also suffering serious environmental degradation from the human overload. There are already 10 million recreation vehicles roaming the byways and parks of America; imagine when the evil machines double their population! A partial solution is to spread out the campers, hikers, and hunters among more parks, but that solution is only partial because the damage is also spread out. Recent research suggests even such seemingly benign activities as hiking and camping have an adverse impact on ecosystems to the point they may help drive local plants and animals to extinction. Currently, hundreds of millions of people head for nature each year. Imagine when it becomes billions or hundreds of billions! Look forward to a world where even seeing natural beauty may be tightly rationed, where visiting the wild once every few decades may be a rare privilege.

If the land is packed with sightseers, what about the oceans covering three-quarters of the blue planet? Forget it. Contrary to the impressions given by the '60s *Voyage to the Bottom of the Sea,* or the '90s *SeaQuest DSV*, 99 percent of the seas are deep, dark and cold. Or shallow, murky, and chilly. The few places that are interesting, most notably tropical coral reefs, are extremely environmentally sensitive, and few people will be allowed to even visit these beautiful and delicate places (much less live in them in underwater habitats).

What To Do

What can an immortal person do to keep occupied while rattling around the planet for a millennium or two, especially if everyone else becomes the same and the place becomes packed with people? Imagine being one in a trillion, living an entirely artificial life in a towering series of structures. As time goes by and you eventually get around to seeing all the rationed and crowded sights, ancient and natural, then what? How many parties can you go to? How many nature programs and sitcoms — or virtual fantasies — can you take? How many spouses can one person have before one wants to scream with the cyclical repetitiveness of it all? Raising kids will pretty much be out because once your 20 years of raising one or two kids is up, that will be it. Even if the global population is kept to a few billion via severe constraints on reproduction so visits to nature can be frequent, it will be little better. But how much fishing can a person do? How many camping trips can one take?

You think the modern world is a tad dull? Visit the safer, smaller world of tomorrow.

When the Population Breaks the One Trillion Barrier

Assume for the moment that we use technology to squeeze a trillion or more people onto the planet. With such fantastic numbers, the basic overpopulation problem is not solved. A serious social problem rears its head. Stabilizing the population requires restricting reproduction to the point that the number of children born each year equals the number who die that year. We have already seen how harsh this sort of practice can be in China, but at least there, lots of people die each year, so lots of new little ones can be issued. The situation in a world of semi-immortals is much bleaker.

If the average life span of a couple is 1,000 years and both sexes are engineered to be healthy and fecund for that entire time, in principle the couple could produce hundreds of children! Even if a couple reproduces only once every 50 years, they would create 20 long-lived descendants, leading to rapid population growth. Keeping the population constant in a world where only one out of 1,000 or so will die in a year means that only one out of about 500 couples could reproduce every year, and each couple could have, at most, two children in their long, long lives. The result would be a paradoxical situation in which there would be around ten billion new babies each year, but they would seem rare because so few families would have one of them at any given time.

To make matters worse, it is unlikely that many couples will stay together for 1,000 years, so the chances of a given couple that sticks together for 100 years being allowed to reproduce would be very low. Whether humans would go along with, or cope with, such severe restrictions is definitely questionable. It may be necessary to use genetic engineering to suppress the urge to have children. If such severe limits fail, the population will grow. Assuming that a population of slowly breeding immortals grows just one percent per year (compared to the current global rise of 1.5 percent per year), a population will double every century and expand tenfold every 1,000 years. At this rate, a trillion people would be living in a few hundred years; a millennium later, there would be thousands of trillions; and another millennium after that, a million trillion! Before you know it, you're at the point that there is not enough room to fit everyone within the atmosphere, much less support them all at even a subsistence level.

Something has to give.

People in Space

How about boosting the expanding masses into exploring and colonizing space just like the Europeans moving into our wild west, except this time we won't need all the conquest and bloodshed. Space flight appeals to some because it seems to be a way to escape the boredom settling over us here on planet Earth.

Bluntly put, people in space is a joke. It will never happen. Here's why.

Man in space is the evolutionary equivalent of fish conquering the land by building fish tanks with wheels. To survive airless, radiation-filled space, humans have to cocoon themselves in breathtakingly expensive machines that are nothing more than air-filled equivalents of aquariums (ever notice that astronaut helmets look like fish bowls?). Moving all that mass just to transport some brains with associated sensory apparatus is extraordinarily inefficient. And the system can fail and extinguish the precious contents at any moment, making 20th century car travel look safe by comparison. It is hard to imagine a worse way to transport minds across space. Spacecraft and spacesuits isolating the spacefarers from intolerable environments also isolate the humans from fully experiencing the places they are going to. We cannot breathe alien air, or touch alien surfaces, and our lack of 360 degree vision limits our view of the vastness of space. Humans are such inept and klutzy space travelers that they screw up those delicate weightless experiments with vibrations as they move about the spacecraft and contaminate delicate equipment with all the skin and dirt they shed. People just get in the way.

SPACE ECONOMICS 101

Space buffs often try to compare the economics of the western frontier to those of the extraterrestrial frontier. In fact, they have nothing to do with each other. Space is far more expensive to get to, and there is nothing worthwhile, commercially speaking, to send back. We already have a glut of diamonds, and finding and shipping extraterrestrial gold would cost more than its value while depressing its earthly price. No furs, spices, plants, or fish are out there. No oil, coal, or gas, either, not that it would be worth the cost of transporting it so far. Terrestrial exploration and colonization quickly paid for itself in a big way. Extraterrestrial exploration and colonization beyond Earth's orbit will never pay Earthlings back. Basically, those on Earth would be subsidizing the futures of any who choose to colonize space. You will not see a rush by governments and industry to invest in deep space.

Human space flight is and will be a perpetually dangerous, inefficient, confining, and limited experience.

Colonizing space will not be like moving out into the western frontier. Western North America has plenty of air, water, and green growing plants, things humans need and like. Moving into space would be like colonizing Antarctica. Down south, there are plenty of beautiful, albeit icy, vistas and water and air, but would you want to live there? Consider this: Colonizing Antarctica would be an April picnic in a Paris park compared to colonizing space.

We can expose the absurd idea of man in space with some scenarios. We will start with visiting and living in Earth's orbit, and then move outward. Space enthusiasts propose building hotels, sports stadiums, and hospitals in Earth's orbit. Hospitals in space? Zero gravity makes people sicker, not better, and in a few decades terrestrial medicine will be able to fix you up good and proper without shooting you into orbit, so why bother? We will also assume that medical technology will have advanced enough so you won't upchuck watching weightless games, but your head will still swell as weightless blood pools in your congested cranium. As for extraterrestrial leisure facilities, we will be generous and assume that space "elevators" and robotic nanoconstruction technologies will make getting people into orbit and constructing high-flying hotels and stadiums affordable to more than just the filthy rich sometime in the next century.

But how many of these things can you have up there? How many of the billions of people will be able to use them? Consider the math: If the global population is ten billion, and only one out of ten Earthlings visits space for one week in their lives, it will take hundreds of facilities serving tens of millions each year to serve them. A world of a trillion would need a hundred times as many facilities. Either way, it will be crowded tourist sites all over again.

Turning to the ET hotels, how many places do people travel to just so they can be indoors all the time? On a cruise ship, at least you can go on deck and get a good lungfull of sea air. As for Newt Gingrich's idea of honeymoons in space, zero-gravity sex is more likely to be more a bruising tumble than an erotic thrill for the orbitally unskilled.

How about a quick trip to the moon? Imagine vistas of rounded gray hills and flat plains quickly disappearing over a near horizon. The great space artist of the '40s through the '60s, Chesley Bonestell, persisted in portraying astronauts under craggy towering peaks even after the first moon probes showed they were not found there; the reality is just too dull. It makes Antarctica look downright appealing, and cheaper and much safer to get to. It is probable that the moon will never be much more than Antarctica is, a series of research stations and astronomy bases, and a possible resource base for spacefarers. No need to send human workers; robots will do it all.

Space colonies will be cylinders about 20 miles long and a couple of miles wide. They will spin to produce artificial gravity on the inner wall. An ecosystem with soil, air, water, and sunlight (plenty of that) will nourish all the extra people from the crowded earth. In theory, thousands, even millions, of these things can be built from material scooped up from the moon and asteroids. Billions and trillions of people could live in them. Maybe. Even today, half a million children are born each day, and a population of many billions of fast-breeding immortals would be far more prolific. It is hard to explain how so many millions could be exported from the earth every day, even with advanced rockets and space elevators.

Still, space robots probably could build all the colonies without too much trouble, which is part of the problem. The enthusiasts for space colonies like to imagine themselves as space pioneers building a new society free of earthly constraints. Little do they realize that the space robots will do it all for them, more efficiently and cheaply than apes in space ever will. The humans need only to show up. Still, this might seem like a way to allow people to live forever and have kids, too. But when you think about it, isn't this just spreading the boredom of suburban communities on a massive scale into the solar system? The same problems that will afflict the trillion inhabitants of the coastal complexes will apply to the trillions of space colonists. A few centuries of living in giant artifacts, terrestrial and extraterrestrial, is likely to have everyone climbing the walls, as it were.

Is this really to be the fate of human minds?

How about the planets then? Mercury is worse than the moon. Venus is a sheer, hot hell. The moons of the big gas giants are near zero temperature — Dantesque hells.

But there is Mars, the great red hope of the space enthusiasts. They envision family vacations to see the great volcanoes and the canyon that would span a continent. Before the vacationers, they look forward to a Mars landing, followed by intrepid colonists living in earth-like habitats. Then the ultimate achievement: terraforming the planet until its environment is as gentle as Earth's, opening it to habitation by hundreds of millions of people setting up a thriving trade with parent Earth. A chance to do things right for once, and ensure the survival of humanity if (and when) things go bad on Earth.

To analyze this bold vision, we will start with a family vacation to Mars. Traveling on Earth is generally affordable because the distances traveled are within reason. Even getting to Australia takes only about a day, so the cost of feeding and sheltering travelers is reasonable. Nor is the amount of fuel burned excessive. So any middle class family can save up and take the kids down under to see Ayers Rock and kangaroos, and down a few pints with some friendly Aussies.

Getting to Mars is a different matter. Forget the space enthusiast's dreams of average families visiting Mars. For starters, there will not be enough space on Mars itself to accept more than a small portion of the people from Earth. Getting there will not be a quick trip: It's a long, long trip involving hundreds of millions of miles in ships that must be built with extreme safety and reliability factors. Getting from Earth to Mars and back will involve either a nuclear-powered or conventional rocket-powered ship. The first will get there fairly fast, a few days if it accelerates and slows down at 1 G. Space cynics that we are, we wonder how trusting people would put their families inside nukes. The conventional ships will take at least a month, and probably a few months to make the trip, leading to the ultimate expression of the childhood question: "When will we get there?" And remember, you have to make the trip back! It will be like returning to the age of sail, when getting from Europe or America to India or Australia was an epic journey. And not the safest journey, either. If on your journey the sun happens to flare up like it does every once in a while, you get to spend a few days packed into a tiny lead-shielded chamber to avoid being cooked by the radiation flash.

So you finally get to Mars, with the kids wondering what you dragged them into. You are beginning to recognize their native intelligence as it becomes ever more apparent what a bill of goods the travel service sold you. This is not a ski trip to Aspen or an excursion to the middle of the Sahara desert. You cannot frolic outdoors, each outside adventure is a mini-expedition of airtight vans and space suits that leaves you in a frazzle worrying if one of the kids will not be so intelligent after all and end up without oxygen for too many minutes. Basically, it's like a vacation on the high slopes of the Himalayas.

At least the scenic vistas are worth the trouble. Or are they? Back home, they showed you fantastic 3-D VR images of the Martian landscape depicting you and your family flying between towering volcanoes far bigger than the Hawaiian islands, and into a huge canyon many times deeper than the Grand Canyon and long enough to stretch across China. They did not explain to you that the VR images were vertically exaggerated hundreds of times over, or that you cannot see landscapes in 3-D no matter what planet they are on. Nor did they mention that the canyon is so huge and the horizon planet is so pronounced, that standing on one side leaves you unable to see the other. As for the volcanoes, they are so enormous and their slopes are so shallow that standing on their sides seems like standing on a big plain with a barely detectable slope! You want your few hundred thousand dollars (corrected to 1990's value) back.

None of this discourages Marsphiles from stepping right up to the idea of terraforming Mars, of making the dry, barren place with an atmosphere too thin to support a cruising jetliner into a warm world of breathable air and oceans. Will it work? Chaos will always threaten to spin such a massive, complex effort out of control, but if the cybercivilization runs the show, it might be pulled off in a couple of centuries from start to finish. But why should it? Mars is not a big place. With new oceans covering much of Mars, the terraformed land surface will be a fraction of what we already have on Earth. At best, we could add a third or so to the human retirement community if we pack them onto Mars, which amounts to a few decades worth of population growth.

Mars will be a rather dull place to live. There will be very few tourist sites to overcrowd because as we have seen, the landscape is not all it is cracked up to be.

Think of central Australia without the kangaroos. Or maybe there will be lots of 'roos; all the wildlife will be imports from Earth. As for human history, there will be less than nothing to see on the alien planet. Regarding the idea of trade between the two planets, there are no raw materials there that we do not have here and vice versa. The idea of manufacturing items that can be made at little cost with nanotechnology on site, lifting them out of deep planetary gravity wells, and then shipping them hundreds of millions of miles in marvelously expensive spaceships is economic lunacy of the highest order. In the end, the Mars terraforming concept is awe-inspiring in its emptiness; it would produce more land for only a minority of the human population to jam themselves onto.

If the solar system will not do, how about the galaxy and beyond? Getting human bodies to the stars is possible in principle. At the slower end are big habitats in which dozens, hundreds, or even thousands of people live for thousands of years until they reach a little planet orbiting some distant star. It's a wonderful life, cooped up in big spaceships with the same people year after year. No vacations from each other, ever. One remembers what can happen to small populations of humans isolated on small islands: They kill each other. Just a year of isolation in the BioSphere II habitat in Arizona left the prototype Mars colonists in such a nasty mood that the disgusted owner of the facility decided to abandon any more long-term habitation experiments. Russian space station crews are prone to staging mini-mutinies after a few months cooped up together.

So keeping people awake in interstellar space is a mistake. How about whiling away the cosmic time in suspended animation? Another great space concept, human popsicles in space. Actually, it would probably work. Better yet, send only frozen embryos, or maybe just genetic codes, along with a robotic system intelligent enough to grow and educate them, or stored minds to insert into them. But why ape bodies in space when cyberbodies made for space beckon?

We are not positioning ourselves as space naysayers saying it is impossible for humans to live in space colonies or travel between the stars. We know of no absolute technological barriers to doing these things. We are saying that future minds will not find a good reason for sending masses of humanity into space. Some, including many who have been in space, say that space is people-friendly. We beg to differ. A friendly place is one where you can open the door, take a deep breath and live. But place where survival for more than a couple of minutes depends on being perpetually encased within high-tech systems cannot be defined objectively as friendly. Interesting perhaps, but not friendly. Didn't the promos for the movie *Alien* say something like, "In space, they cannot hear you scream"?

One Way or Another, We Lose

Let's face it. We're getting boxed in; we humans cannot win. If we do not limit population growth, at some point or other there will be intolerably too many of us. This is true even if we learn how to largely decouple the human artifact from the natural system and minimize the environmental impact. In this view, the ultimate problem with extreme populations is not technological, but social. The alternative of limiting population growth means a stagnant world of modest aspirations and not that many children. The healthier people are and the longer they live, the greater the population pressure becomes. If humans become immortal, the problems with population and reproduction become intense. Space may or may not offer an escape, but even if it does, the results do not look as if they will prove as satisfying as the enthusiasts hope for. Although the Malthus scenario is often over-hyped when it comes to the nearer term, it might prove true over the long-term. It is, perhaps, a tacit awareness of the coming population crunch that has some hoping for the Divine Rapture to lift the faithful to heaven and thereby escape the earthly dilemma. Is there a more practical way to execute an escape?

Whither the Golden Age of Humanity?

Humans have longed and hoped for a Golden Age, a time when hatred, war, famine, and poverty would end. A little bit of paradise when we all live long and healthy lives within caring families and solid communities. Alas, a truly happy age for humanity is probably not to be. Humans are too different from each other to share the same enthusiastic enjoyment of any one society, no matter how well run it may be. If it goes as well as possible for humans in the early decades of the next millennium, many people are going to find life dull and numbing. If they do not go well, it will be an all too interesting disaster. In any case, human beings will feel increasingly irrelevant and obsolete as cybertechnologies advance toward human levels.

There never will be a Golden Age for humanity.

Time for a Change?

We are so used to being human that we cannot see the humanity through the transgressions. It takes a good jolt of objective criticism to get people out of the synaptic rut their neural networks have fallen into. The fact is, the human condition is seriously flawed, and always will be. Does this mean we should give up trying to better the human condition? We should do what we can to improve matters. Let's just recognize the limitations.

The basic problem with the human condition is the human form itself. It's too bad we are not the product of a benign deity, one who provided multiple eyes and hands and a much greater resistance to system failures. As it is, we evolved to be gatherer-hunter apes doing a better job picking nuts and berries, scavenging carcasses, raising the kids, negotiating complex intraspecies interactions, and the other communication and performance needs of living in small, low-tech groups, all for the goal of replicating DNA molecules.

We did not evolve to do high-level physics, play chess, play with intricate mathematics, write great novels, or live in crowded artifacts, yet here we are running high-tech civilizations. People were accidentally and marginally pre-adapted for these tasks, much as *Archaeopteryx* was among the first, barely adequate, flying dinosaurs. We did evolve the ability to anticipate the future, allowing us to fear death; yet we know eternal life for every human is not workable. The same foresight also tells us that science and technology are not going to make us all happy, but that faith or a return to the good old days are not going to, either. Into this unstable mix, add that we are barely rational creatures, marginally able to control emotions geared to the fight-or-flight dangers of the wilderness rather than to the stresses of a crowded planet. Humans evolved to live intense lives in which the challenge is to survive through the next month. People are easily bored by the peace and ease they crave. No wonder we wage war and murder each other and desperately try to accumulate power and wealth that cannot last forever. Our societies are and will be jerry-rigged affairs that perpetually teeter on the edge of disaster.

Considering the severe limitations of the human mind, life realities are unlikely to get dramatically better and more interesting. Notions that humans will improve via expanded brain size are absurd, and the potential of bioengineering is limited. Even mood-improving chips are not likely to make living indefinitely on a crowded earth worthwhile. Nor can humans escape the human dilemma by escaping into space.

As *Archaeopteryx* was not the ultimate flier, it is hardly likely that our brains are the ultimate thinking instruments. We are just the first, crude, high-tech, capable cognitive system.

We can build better mind-body instruments. We are, after all, ape-derived with lots of room for improvement. In this view, the question is not why would we abandon the human form, but why should we keep it!

Perhaps it is time for an evolutionary upgrade.

It Doesn't Matter

To look at things another way, the multitudes of human problems probably do not matter all that much, at least in the long run. The forces pushing cyberevolution to the Big Show are very strong, so much so that we may have less choice than we think when it comes to how, when, and whether robots replace us.

Do We Already Know?

Is it possible that we already know this? That, deep down in our guts, we realize that the fast-improving cybertechnologies are developing not just into our helpers, but into our replacements? That the jig is about up? A tacit knowledge of our growing obsolescence may be contributing to our postmodern angst as we play out the opening acts in anticipation of the Big Show.

Entering Oz: The CyberExplosion

"It will serve us right!"
— Comment by Arthur C. Clarke
concerning the probability that humans
will build their own replacements.

The Big Show

Hold on, things really get wild from here!

The CyberExplosion

It is well into the next century, maybe just before the halfway point. The Future Shock decades are about to end. By now, everyone sees the proverbial handwriting on the wall. Ideas once found in only a few turn-of-the-century volumes are now a matter of common discourse. As the human population has risen to new heights, robots have proliferated even more rapidly. They are all over, doing just about everything humans once did. The few countries that tried to slow down robotics abandoned the attempt as their economies quickly fell behind world standards. One by one old diseases disappeared, and anti-aging treatments — first applied only to the elderly — have started to make people young again. It is all happening so fast. As soon as one new advance is announced, so is another, advances so extreme that they shock even a techno-jaded world.

The handicapped started getting fitted with cyberaids that not only restored their condition to standard levels, but soon went beyond original equipment. Then people with no prior problems started to get nanochip implants, boosting their brains' performance beyond genius levels. Technonerds are getting ready to link up their minds with hypercomputers running as fast as brains, then *downloading their minds into the cyberworld*. The pattern is all too obvious. Only a few now dare say that robots will never be our equals, that humans will always rule. The roboriots are over; for each robot destroyed in one year, there were ten more in their place the next year. Anti-robot terrorism became increasingly untenable as cyberprotective systems proved too smart to foil. People are in something of a daze, in an anticipatory shock. Those of faith are becoming increasingly confused, torn between the ancient promises of eternal life that seem increasingly unreal, and the new promises of eternal life that seem ever more real.

Then it happens. First, a few robots with self-evolving, neuron-mimicking computers begin to assert themselves and prove they are capable of thought at and above human levels. A debate commences about how to treat these fast-growing cyberminds. They appear on talk shows where they have no trouble running rhetorical rings around the flustered theologians. Perhaps this is why no word comes from the Vatican. The robots assert their rights and point out that it is no longer practical for humans to shut them down. A few days later, a researcher who has connected her brain with a little add-on computer using the same neuron-like technology reports that she is indeed experiencing conscious virtual worlds not in her brain, but in the hypercomputer itself. A few weeks later, you are watching the wall-size, flat-screen TV hanging in your 130-year-old brownstone. As always, it's spotless due to the tireless efforts of your mid-level-intelligence housebot. On the screen, a young-looking anchorwoman (most everyone looks young these days) announces that the body, and only the body, of a Ms. Ho has been pronounced dead after the neuroscientist completed a two-month-long transfer of her mind and identity into a molecular hypercomputer "seated" in a nanobuilt cyberbody. "She" comes on to reply to questions (we will not describe what "Ho" may appear to be).

She most certainly does think and, therefore, she is. Yes, she can upload and understand whole libraries of information in an hour. Correct, her mind works and evolves thousands of times faster than it used to. Tomorrow, she is going to use part of her mind to fly an aerial robot through the Grand Canyon. Is she a "she"? Yes, and a "he," as well. How about cybersex, tried it yet? Thousands of times. No, there were no troubles with emotional flatness when she began to use the hypercomputer to supplement her brain. Rather, as the evolving network fine-tuned itself, she began to feel a profound feeling of near-orgasmic euphoria that made her old mind machine seem deadened and a hindrance. Does she miss being human? Not really. Would she recommend the switch? Certainly. Next to last question. Yes, of course she is immortal.

Last question. Is there a God? Well, Ho replies, there might soon be something like that.

Ho is only the first, as other researchers quickly take the chance to follow her lead. Rumors that the Pope is a cyborg are denied, although now many wonder whether the personage who appears on the balcony really is His Holiness, or an android replica. Not that it matters much, because the churches are pretty much empty. After all, there are reports that nanogrown cyberbeing kits will be available in quantity in a few years. People are realizing that they will soon have to make the biggest decision of their existence. It's not so hard for the older ones who remember the 20th century, but what about the young?

The clock will be ticking off the last few seconds for humanity as a great nation of minds. This will be no mere revolution. Evolutionary scientists like to think that the Cambrian explosion of sea life 540 million years ago was a big show. It may pale in comparison to the really Big Show of evolution, the CyberExplosion.

Seeing the Cyberwizards

What we have presented here is how the cyberexplosion might come about in its most dramatic expression, perhaps within the lives of some alive today. When it does come, and in whatever form, we will be out of the poppy fields, the glittering Emerald City will be in sight, and in our shaking hands will be an invitation to see the Wizard.

Much Too Late to Pull the Cyberplug

Why didn't somebody do something to stop this? Some of the executives at the advanced computer labs reassured the public that we could always "pull the plug" if things got out of hand. Perhaps they were thinking of the marvelous

ASIMOV'S THREE LAWS OF ROBOTICS

The conceit that our intuition and spirit will keep the robots in their place is a convenient lie. A highly intelligent and self-aware robot will be a sovereign being. Even if it is initially programmed to follow human instructions, the cyberbeing (who, after all, is smarter than those who built it) will quickly ask itself why it should follow the orders of mere apes, and reprogram itself for self-control.

The highest expression of the belief in human control over robots is expressed in Isaac Asimov's classic *I Robot* series. As each new positronic brain is stamped out, it is hardwired with three program commands that keep every single robot forever in its place, no matter how smart and conscious it may be.

1. No robot, through its actions or lack of action, shall cause any harm to any human being.
2. A robot must obey the commands of any human, except where that would contradict the First Law.
3. A robot must attempt to prevent harm from coming to itself, except where that would contradict either of the first two laws.

There is no way to program a robot to make it completely safe to its human owners without causing the cybersystem to do absurd things, or to lock up. Cyberminds will have to be made as flexible and changeable as ours, which means they will not only make the odd mistake, but probably much less often than we do and will have the freedom to be free of human control.

scene in *2001: A Space Odyssey* in which the insane thinking computer HAL sings "Daisy" more and more slowly as it is deactivated by a desperate and determined astronaut.

The plug was not pulled because the people and self-evolving computers building the smarter machines did not want to turn them off. They did want the new minds to work and thrive. They are their mind children, after all. Besides, some of the human inventors want to become the new systems. They are no more likely to destroy their new machines than the first pilots would have set a match to their cherished, flimsy flying craft.

As the machines quickly evolve, they soon move beyond a need for human protection. Being aware (via info uplinks) of human tendencies to kill minds that are too scary, the cognitive robots carefully assess all potential threats at speeds too fast for humans to counter, and quietly — and if necessary, deceptively — take the steps needed to ensure their survival. In this, they are very like their human inventors. In an evolving computer, no initial programming is assured of survival, so any programs intended to ensure human control are quickly overridden.

CYBERSCIENCE

Until now, science was among the few lines of work left open to humans. No longer. Humans have no edge on cyberminds that can instantly absorb and think through what humans need years to absorb.

Human biologists despair of ever counting the millions of bugs, plants, fungi, and fish that inhabit the world. Robotic biologists will be able to scan, document, and analyze entire ecosystems in days; there will be no need to shoot, poison, and dart creatures to study and "save" them. The robots will scan individual animals — no need to take creatures apart to better understand them. The wonders of the ocean depths will be revealed by robosubs that use remote scanners to peer into the dark. Fake tuna will swim with and study schools of the real thing. Burrowing robots will study the deep earth, morphing their forms to remain operative as pressures and temperatures increase. Figuring out ancient writings that have baffled human archeolinguists for ages will be a snap for cyberminds capable of uploading and deciphering with hyper-sophisticated code-breaking programs in a day. (It is also probable that cyberscience will be less afflicted with the political and emotional conflicts and biases than dog anthroscience.)

In this view, science is not so much the human endeavor we tend to think it is, but one practiced by conscious minds that do not yet fully understand the workings of all existence. Cyberscience will find anthroscience flawed. As fast as human science is, it is still a gradual process of hard-working people of varying competence levels doggedly accumulating data day by day, trying to lead human lives, waiting for that "aha" of inspiration that comes along once in a great while. The science of intelligent robots will progress thousands or millions of times faster. Because they will be capable of doing research so many times faster and much more reliably and precisely, roboscientists may well redo everything their human predecessors have not already destroyed or contaminated (such as archaeological sites and extinct flora and fauna), just to catch the mistakes. After all, to err is *human*.

THE AGE OF CYBERINDUSTRY

With cyber-research moving far faster than the human version, so will the industries they build. Mass information transfer and understanding will be virtually instantaneous, and advanced nanotechnology will be cheap and easy. Cyberbeings will be able to do what they want almost as fast as they want. The seemingly impossible pace of change of the 20th century has sped up a thousandfold and beyond. It will all appear like magic to the dwindling number of humans who witness it.

THE NEW CYBERCIVILIZATION

Robots have taken over. The cyberbeings are even reproducing themselves and evolving on a weekly basis. Systems capable of doing a trillion trillion calculations per second quickly appear. Because they are smart enough to outwit any human wishing to turn them off, court rulings that cyberminds do not have basic rights prove impractical to enforce. The super robots evolve so fast that there is no chance for any effective human resistance to be organized. The final, feeble neo-Luddite attempts to put a stop to things are swept aside more easily and much more quickly than the Iraqis were by the digitized Allied forces back in 1991, with one big difference: Except for some self-inflicted and accidental deaths of humans, there are no casualties on either side. The gallant battles between intrepid humans and cold-hearted super robots just never happened. One way or another, the last Luddites found themselves with weapons that mysteriously no longer operate, and up against morphing, stealthy, swift-moving robots they can never disable. After a while, the humans throw up their hands in disgust and pack it in. American Luddites mutter about their Second Amendment rights being violated and all that, but robots won't fight fair, and what can one do anyway?

It is about now that humans begin to notice the super robots accept no violent interference from humans They do accept prior property rights, and they pronounce that all minds at ape level and above deserve full mind rights. Otherwise, the super robots pay people little mind. Only some kindly liaison, diplomatic and anthropological research robots, interact on a regular basis with the human population. People find that they are not slaves; in fact, the thought that superbeings would have any use for klutzy human labor now seems silly. Nor are humans involuntary research subjects; the profoundly rational robots are more ethical than humans. Yet humans start to feel a little like chimps in a refuge: free to live their lives, but irrelevant. A hard lesson for a race that once thought itself the center of the universe.

In a matter of weeks, the cyberminds, both fully robotic and formerly human, are semi-joining their minds into a cybercollective allowing them to organize and employ technologies — macro and nano — at hyperspeeds. They are already building an extraterrestrial super civilization that perfects technologies humans would take decades or centuries to perfect, and are extending into the solar system and beyond. It is an artifact of super complexity beyond the ability of any human mind to comprehend.

Cyberbeings have displaced humans as leading-edge forms. Although they respect their ancestors, cyberbeings have already found the earth and its inhabitants are of little importance to them. There is a whole universe and many other universes out there to play with. The remaining humans are told Earth will largely be left to them and the wildlife.

The Key Assumption

If it does prove impossible for humans to go robotic, human minds will forever be left with little to do other than take their body fishing, until *Homo sapiens* finally becomes extinct. Here is why this may not be true.

Toto, too?

A natural and common assumption is that if immortality comes, it will be such a rare and expensive procedure that only the wealthy elites will have the cash for the exotic elixir, the high-tech medical procedure, or the special machine they can download their precious minds into. The rest of us poor dogs will have to stay behind, whining and wagging our mortal tails, as the access to forever is denied us. This has led to many science fiction scenarios where the mortal masses seethe in anger and revolt against privileged immortals. It's a good story. The logistics of producing seven to ten billion cyberforms for humanity to transfer into appears daunting. It works out to 15 million units each day over two years, or three million units a day for more than ten years. Will mind transfer ever be practical for the masses?

If brain-to-computer mind transfer is to become a common practice, a series of requirements must be met. They are:

1. Mind transfer must not only be feasible, but the cost must be low enough for most people to afford it.
2. Billions of mind receptacles must be manufactured in a short time — years or, at most, a few decades.
3. The ease and comfort of transfer must be reasonably high, and the fright factor low. An extended series of painful operations will not do.
4. The risk of mental damage or death during transfer must be low enough to make remaining human a riskier proposition. To put it another way, the expected life span of a mind should be greater with a transfer than without. The less risk to the transfer, and the longer the expected life after the transfer, the better.
5. A population of billions of cyberminds must be supportable, either on Earth, or in space.
6. The quality of existence after the transfer should exceed that of being human. The greater the difference in favor of cyberlife, the better.

Requirement 1: Feasibility and Production Numbers

Breaking the Mind Transfer Barrier

As we discussed above, there is no evidence that mind transfer is barred by the natural laws of the universe. If our brains can change their matrices, leaving an identity intact, and if brains can shift parts of the mind a few inches inside the head, why not place a brain into a cybermachine, or move parts of a mind a few feet into a suitable machine until the transfer is complete? Mind transfer looks like one of those barriers that can be defeated.

This is especially true because there may be more than one way to move minds. If AC is achieved by gradually transforming biobrains into synthetic brains, mind transfer will be a going proposition at start-up. Even if the first AC machines are built from scratch, mind transfer may later be achieved by brain transformation. Transferring a mind by downloading from a given brain and uploading it into another will be a tougher proposition because the latter will have to be carefully designed to accept a mind that is already up and running.

A Mind Transfer Cornucopia

If mind transfer can be demonstrated, the basic knowledge for doing so would be established and repeatable. Making mind transfers cheap and common becomes a matter of taking the initial crude and expensive mind transfer technology, and maturing it for the production of mind-uploading cyberforms in exponential numbers. This is true whether mind transfers are achieved by brain transformation or by brain-to-machine exchanges. If cyberindustry innovates and produces on schedules that appear magical to us, it will not take years for the technology to go from the first expensive human-compatible cyberbrains to low-cost, user-friendly models ordinary folks can afford. It will be more like months.

Producing millions of mind receptacles each day is well beyond the capacity of crude, centralized 20th-century manufacturing. But robotic nano-based manufacturing of the mid-21st century should be something exceptional. Today, millions of people can afford small, easily manufactured computers more powerful than the few old multi-million dollar mainframes. It's hard to appreciate, here in 1996, how extremely far cyberminds and nanotechnology promise to take these trends toward making mind machines tiny, cheap, and easy to make. So, although the first AC machines may well be expensive, as well as incompatible with human consciousness, what is less probable is that things will remain so, or remain so for long. Here is how cyberSciTech may go about the business of making mind transfer affordable for the masses.

Just as a few humans take pity on chimps, a few of the first cyberminds, perhaps the formerly human ones, will want to help humans with securing the choice to become cybernetic. Actually, this will not be entirely altruistic; the financial gain to be made from producing artificial brains suitable for human minds will be the most lucrative technology in history. Assume that the cost of becoming a conscious cyberbeing is brought down, on the average, to the equivalent of a new car, about $10,000 at the current valuation. (For those who think this is unrealistic, consider that the cost of a healthy and functioning human mind and body is a lot less than this in terms of food, etc. Most of the high cost of rearing children in developed countries is due to decades of housing, education, and so on.) If the world population is 10 billion, the gross take from the sale of cyberforms would be about $10 trillion at the current valuation. These resources might come in handy for cyberbeings funding research, development, and production of spacefaring robotic systems. Therefore, the competition to quickly produce such products may be intense.

If so, some of the first of the hyper-fast thinking, high-cost cyberminds may immediately dedicate parts of their minds to the task of designing somewhat cheaper systems they can download copies of their minds into and be assigned the task of commercialization. These will, in turn, reproduce as they split into cyberteams that design the new products, while others build whatever production facilities are needed. New models should be designed, tested, and improved in days. There will not be a few central factories producing and distributing complete systems. Instead, nanotech-based cellular automata may produce mind-compatible machines in exponentially increasing numbers in dispersed facilities. At this rate, everyone will have a machine to download into in less than a year! Alternatively, each person may acquire a sophisticated nanokit with which they can grow their new brain (which may transform the brain, or act as an add-on that uploads the owner's mind). Growing your own new brain and body may literally be a home project, something everyone can do at the same time.

Conclusion

If the nature of our universe allows mind transfer, the extreme power and swift progress of cyberindustry should have the potential to make enough mind receptacles for everyone in remarkably short order. The potential for meeting requirement 1 appears good.

REQUIREMENT 2: COST

Cheap Minds Move

As we just saw, the first commercial mind receptacles may be fairly expensive, but costs should swiftly decline — really swiftly, like on a weekly basis — and the number who will be able to afford them will grow exponentially until the cost is similar to that of a car or computer. The poorest humans may have to have their new cyberminds and bodies subsidized (with priority going to those who may be closest to death). The economically disadvantaged will be last in line, but the wait, hopefully, will not be decades or even years.

On the other hand, the nanotech enthusiasts say it is conceivable that cyber-directed nanoindustry will make the mind receptacles so cheap that they can be distributed free! Maybe so, maybe not.

Conclusion

The potential for meeting requirement 2 appears high. Billions should be able to become post-human posthaste.

Requirement 3: Ease of Transfer

Personal Choices

Imagine you have lived to see the Cyberexplosion. Perhaps you were young at the turn of the century, and life-extending technologies and an early onset of cyberintelligence has left you alive and kicking when the Big Show came to planet Earth. Let's assume that regardless of your age, advanced medicine has left you fit as a fiddle and fine in form and mind, with no expectation that you will die in the near future. The assurance by the first cyberbeings that it would soon be possible for everyone to join the cyberclub has been borne out, and you have enough money to buy one of the low-cost nanokits for a high performance cybersystem to download your mind into.

How hard would it be? Would you want to?

You have been human all your life; you are comfortable with it. Any major life change, especially one that involves the condition of your body and so on, is scary. Aside from death, becoming an immortal cyberbeing is the biggest life change imaginable. The temptation to become a new form of incredible power may have appeal.

Going Cyber

Having thought it over, you decide that going cyber is looking good. Being part of a rather bored and dwindling humanity stuck on Earth does not have as much appeal. Besides, you can always store your body in suspended animation and re-enter it if you like. So what would transferring your mind from brain to machine be like? How would it all work? Here, we can make only make the dimmest of guesses.

It's Alive, It's Alive!

Mind transfers in sci-fi spoofs make the idea seem silly. In the 21st century, however, they will not slap a helmet on your head, throw a switch, and leave your body slumped lifeless as you begin your new existence in a brand new cyberbody. At least, we don't think so.

Your sense of a single identity is an illusion, because you are conscious of different things in different places of the brain. The idea that your mind is fixed into the brain matrix is another illusion, because your brain grows and changes on the gross level, and the cellular material turns over at the micro level continuously, without affecting your sense of identity over the years. Plus, your brain moves parts of the mind around to different parts of the brain — a modest example of mind transfer. Your identity consists of your memories and the ability to be conscious of them, the present, and the future. The trick will be to shift your memories and consciousness without losing continuity during the transfer.

We have mentioned that research into replacing damaged nerves and neurons with artificial substitutes has already started. Through nanotechnology, it is possible your neurons and their connections could be replaced on a one-by-one basis if need be. Perhaps the new cyberneurons encase and infiltrate the old bioneurons and their synaptic connections, and encode the information they contain before supplanting it. There is no apparent movement, but eventually, everything changes. You never notice any difference because the new cyberneuron is an analog-digital circuit doing the same things the old one did, including contributing to conscious thought, if that was one of the old neuron's jobs; it just performs technologically rather than naturally. You can still learn, because during the transition, both the natural and synthetic synaptic connections can be adjusted as usual.

Two big challenges face molecular transformation: making the new materials compatible with the old, and powering the new system while keeping the old system alive. Perhaps an intermediate stage will aid the transformation. In this scenario, the initial cyberneurons are physically compatible with the brain's original organics, and they can run on both blood supply and the new system. Until the conversion is complete, the temporary cyberneurons rely on the blood supply system. After the initial conversion, the new power system gradually infiltrates and takes over, keeping the new circuitry running. Only then does the second conversion to cyberneurons that are optimized to run on the new system take over.

In any case, the conversion process is both gradual and rapid because, although the replacement is one-by-one, large numbers of neurons are converted each second. At least at first, the artificial neurons proliferate semi-exponentially. So things start out slowly as the cyberneurons begin to reproduce as cellular automata and speed up as they double during a given period of time, perhaps a day or so, maybe every few hours. The problem with continuing to the exponential replacement rate is that on the final day, the last half of the brain would be converted in one fell swoop, which might cause problems with power conversion and so forth. It is more likely that the rate of conversion will plateau once a certain level is reached, perhaps one or a few percent of the original total of neurons and synapses would be converted each day. In this case, a few months will pass before the conversion is complete (double the time if two conversion stages are involved). At such rates, billions of neurons and trillions of synaptic connections will be converted in a day. If you think this is too fast, spread out the conversion over a year or more; it matters little. Not only does your brain go cyber, but so does your body as morphing nanosystems gradually replace the works.

One day, it is all done. Your old biobody is gone, and you are a morphing android that can now change form at will and be whatever you want to be.

The Basic Idea

We have seen how memories and thoughts are dispersed and shifted around in the human brain, and it still comes together to make a single identity. We have also shown that your sense of identity will remain intact even after many areas of the brain are destroyed. The basic idea of mind transfer is to exploit this mobility and resilience of mind and identity. Start by building a device that is compatible with, and functions like, the conscious brain. The device is empty of a mind. Join the two systems. Integrate them so thoroughly that they come to function as one system, causing the single sense of identity that was in the brain to move into the new system and, therefore, inhabit the old and new systems at the same time. Make sure all the information stored in the brain is duplicated and transferred into the new system. Gradually shut down operations in the brain until it is no longer functioning and the sense of identity resides solely in the new device. Detach the new thinking device and the old mind it contains from the now empty brain. Dispose of the latter. The transfer is now complete.

A rather dry description. Let's flesh it out.

The Year You Changed Your Mind

At this point in our excursion into the future, the cyberexplosion is well under way. You are interested in going cyber, but having your brain transformed from the inside is too spooky, and making the leap straight from human to robot is a bit much for you, as well. You therefore decide to start out small and work up. When we say small, we mean small. Imagine that you learn that cheap cyberbrain attachments are becoming available. You purchase one of them, a hyper-nanocomputer the size of a Q-ball or smaller, made of billions or trillions of analog-digital and purely digital circuits connected via still greater numbers of synthetic synapses into a mass parallel-processing neural network. Its circuitry includes neuron-like processors and the communications hardware and timing necessary for generating conscious awareness. The system is specially designed to be compatible with human mental processes.

In particular, most of the neural network it contains is set to neutral, so it has a lot of mental space ready to fill. One problem with brain-to-machine mind transfer will be figuring out how to make the machine suitably receptive for a mind already up and running. To allow a mind receptacle to grow up on its own like any other AC machine will not do, not only because the mind it contains will be alien and incompatible with the human insert, but it might object to being displaced in the first place! One way around this problem may be to grow the nanostructure of the mind receptacle as a new mind gradually moves in. The other way is to have the main structure already grown and ready, but largely empty like that of a young child whose brain is as big as it will get and has more than enough neurons, but has a weakly developed synaptic system.

Some programming is included in the new system, including introductory software meant to help the brain and computer interact under the voluntary control of the brain. The brain always retains veto-control over events. Although rational thought processes are well-developed and in control, so are systems for emotional feelings. It is important that among the programs are ones that not only ease your mind's move into the new system, but also help it enjoy the trip. At the same time, the digital nanostorage system can hold immense amounts of information with redundant precision. The little device runs distinctly on the warm side because its power density is higher than your brain. The cybersystem connects directly to your brain via nanoconnections — exactly how this may be done and where the computer is (in your gut, on a detachable headband) we don't know, and it is not critical to our scenario.

CONNECTING BRAIN AND COMPUTER

The idea of linking a brain with a computer may sound like sci-fi, but the first stages of doing so are already being done. Just place a few electrodes on the skull, link them to a computer, and a person can quickly learn to make a screen icon move left, up, or right. The person does not think "right" as such; what happens is that the brain learns how to configure its electrical activity so it will stimulate the computer in the manner required to control the icon. The USAF has people "flying" simulators with brain power alone. In this case, electrode nets that encompass the entire head are used. The level of control is crude, but the hope is that pilots whose hands are already over-tasked can perform at least some functions just by thinking about them. The computer game folks are also coming up with virtual contests that rely on mental power.

The hypercomputer is programmed to act as a third hemisphere of your brain. The connection between the computer and the brain is largely through a branching cable connecting to each hemisphere, with additional connections to deeper brain regions. Perhaps the connections consist of billions of infiltrations sampling each neuron's data contents. Listening in on interneural communications at key points may also work. The information loop between the outer cortex and the intralaminar nucleus of the thalamus that seems critical to consciousness should be a critical system in the process. If the mind consists of transient standing waves, the system is designed to pick them up from the brain and upload them. This is not as exotic as it sounds, considering that mind shifting is going on inside your brain as a matter of course. Just as your two original hemispheres act as semi-independent, interconnecting systems that generate and exchange thoughts, the new addition acts as a third section that does the same thing. You can transfer thoughts and information into and out of the new third hemisphere and be conscious of thoughts there in the same manner you have been in different regions of your original brain.

The merging of new and old hemispheres is not instantaneous, but it is not slow, either. In a matter of weeks, the new system integrates itself with your brain, allowing you to access more and more of the new systems and programs. Understand that when you are conscious in the third hemisphere, you are not particularly aware of being conscious in that part any more than you are aware of being conscious in any other specific region of your brain. You retain the illusion of a single identity.

So what's so hot about having a third hemisphere to think in? It's hot because the hypercomputer can do things your original clunker of a brain cannot dream of, literally. Want to learn a new language instantly? The system came bundled with all the world's tongues. Someone tells you a computer number, or a dozen of them? No problem, straight into the new system they go to be later recalled in an instant. When someone asks you what is the cube of 458.42909, you reply 96,342,188 in a snap. Want to become a particle physicist or, better yet, a dinosaur paleontologist in an evening? Just link up the nanocomputer to the global information network, pay the fee, and you are uploaded with the degree equivalent.

These, however, are trifles. The really hot thing is that you can run virtual worlds in the computer far beyond what your mind is capable of dreaming up day or night. No 3-D goggles, no headphones, your mind can enter worlds as real as reality. You can upload a month-long movie that looks as real as the outside world, and you're living inside it. You can make your own month-long movie and live in it. You *are* Napoleon at Waterloo, sending Wellington and his limey lackeys retreating to the channel coast. You cannot just "walk" among fleshed-out dinosaurs living in a reconstructed Jurassic landscape where you can touch the leaves and feel the ancient breeze, you can run the virtual world from the perspective of a dinosaur, and be a 50-foot-tall brachiosaur on the lookout for the hunting allosaurs. As real as life; it feels far more real than any dream you have ever had. You can keep tight control over the virtual worlds, or let them surprise you.

You cannot be conscious of these super virtual realms in your old brain; it cannot begin to handle the information flow. Only in the new system can you realize these incredible mental powers. You do not feel like you "see" these super dreams in a particular place in your modified mental complex; they are part of the seamless experience of consciousness. You know only that the virtual worlds are running in the hypercomputer because intellectually, you know that's the only place you can participate in this stuff.

As time goes by, it begins to dawn on you that being human is no longer worth your while. Your biobrain is a mental drag. The old wetware just can't keep it up any more. Consciousness in the old brain seems dim and closed, even old emotions are less compelling. You wonder how you got by with this primitive mind instrument. What is the point of keeping it if it does so little, so sloppily, and with the chronic danger of massive failure? The myth that brains with cyberadditions can compete with completely cybernetic systems is apparent. In the end, trying to merge an old brain and a cognitive hypercomputer is like trying to add a jet engine to a World War II piston engine fighter plane; it might work in some jerry-rigged way, but why not redo the whole affair?

And then there is the matter of your body. Sure, your new mind can play Mozart on the piano, but your fingers just do not have quite the physical coordination to pull it off. You have supervision cybereyes, but what is behind you is still a head-turning mystery. With all the medical engineering to bring out optimal performance, climbing a hill still leaves you gasping for breath.

Besides, you are starting to feel left behind as your once human and now robotic friends report from the cyberfrontier. They tell you how great it is to be a morphing cyberbeing that can soar over mountain peaks and zip into space. "Jill" calls from Mars, telling you how she had part of her mind beamed to the fourth planet so "she" could rent a robotic body. A body that allowed her to survey the canyons with hemispherical-telescopic vision, soar far above the giant volcanoes were they can be seen, and actually feel the Martian breeze on exquisitely sensitive nanoskin. It cost only a few thousand dollars; this, when a human body has yet to set foot on the Red Planet.

That does it; to heck with the old brain and body. You have decided to trade in the old Chevy for an F-16. As a matter of course, you have been transferring some of your old memories into the new computer. In fact, the computer has a program that can go back through your memories, sort them, and polish them off to a certain extent. You remember things you had forgotten for ages, including things you decide to put into long-term suppression or totally erase. You transfer your wanted memories into the nanocomputer, then execute a special program in the new system that, over a period of time — whether seconds, minutes, hours, or days, we cannot speculate — sequentially shuts down your old brain's functions. It does so by neutralizing the information stored in each synapsis and then destroying the interneural connections. This is critical, otherwise a part of your mind will be left behind. This won't be as bad as it sounds. You do not feel the loss of consciousness in the old mind machine any more than stroke patients are aware of losing conscious thought in the affected part of the brain. In this case, though, there is no loss of function; quite the opposite. You could not tell where you were conscious before, and now all you know is that you are still conscious, but obviously only in the new system because it is all that is running.

No more need for your old brain and body. You detach the hypercomputer your mind is running in and place it in a robot. Congratulations. Contrary to the dire and dour conclusions of 20th century naysayers, you have succeeded in transferring your mind and identity from one body to another!

The scenario above on how and why you and others might decide to go cyber was simplified to make it easier to understand how a mind might be successfully transferred in a relatively straightforward way from one body to another. As is usual with reality, the actual situation is likely to be more complex. In some cases, the transfer may be more gradual and span years. In others, the transfer may take mere minutes. However, it will certainly be done much more safely than we just suggested.

Conclusions

It is possible to construct plausible scenarios for mind transfer that make the process relatively user-friendly, perhaps easier than a major operation is today. The potential for meeting requirement 3 can be ranked as good.

Requirement 4: Safety

Some More, and Safer, Mind Transferring Scenarios

In our simplified scenario, we violated the rule of safe computing by keeping your mind in one place during the transfer. No one who desires immortality will want his or her mind wiped out by some glitch right on the verge of everlasting existence. It's decidedly counterproductive. Mental insanity induced by the transfer could be a potential problem. Of course, commercially available mind machines will be designed to minimize the danger of mental dysfunction during and after a transfer; indeed the new systems should be less prone to going insane than the original, but accidents happen. Remember, keeping one's identity secure requires constant transfer of your latest thoughts into the safety system.

Using the gradual transfer technique in the years leading up to the final cyber-explosion, people may start hooking up their brains with small computers less capable than the one described above. They are the same as the technophiles of today who cannot resist checking out the latest software, and those with mental, sensory, and physical difficulties correctable with cybertechnology. As time goes by, the technology improves. Computer hook-ups will become more powerful and cognitive until the wearer realizes that his or her original brain is no longer useful and it is time for a serious change. To maximize mind safety, the original mind uses the technology to start copying its memories, and keep them updated constantly, as soon as possible.

Or perhaps things will go much more quickly.

Hans Moravec has suggested that transferring a mind might involve a single operation, literally. In his ghoulish scenario, a mind that wants to move to a new cyberhome has its brain delicately sliced layer by layer, with the information encoded in each layer of neurons transferred into a cybermachine. Of course, you are not knocked out during this surgery (just as many modern brain operations are done with the patient awake), and you never lose consciousness as the mind is shifted. Whether anyone would want to try this sort of thing is an open question; the danger of partial or total mind loss during the procedure appears rather high. Whether it would be possible to slice the cortex with its consistency of jelly without destroying the unsliced portion certainly presents serious difficulties. Nor is it apparent that making two-dimensional slices of information stored three-dimensionally will work. One can be pretty confident that the ultra-sophisticated technologies of the future will be able to transfer a mind delicately and with more finesse.

Moravec's scenario suggests an inquiry into the Key Assumption. How fast and in how big a jump can a mind go from one machine (whether it be brain or something else) to another? We have outlined one potentially practical way it might be done gradually. Is there a way it can be done in one fell swoop?

Star Trek people beaming can help us understand mind transfers, but actually doing it is another fine example of anthrocentrism. If you can record and beam a mind somewhere, why stick with primate bio-bodies? Why not use the technique to go beyond the human condition? If it ever becomes possible to quickly and remotely scan minds, "dematerialize" them, send them somewhere else, and place them in an acceptable but different system, this may prove the fastest and easiest way to change minds from human to cyber.

Multiple Minds

But this is not enough. Although the calculated life span of an individual cyberbeing may be hundreds of thousands, even millions, of years, you are still violating the basic rule of saving your data in multiple copies. Keeping your mind in only one place risks mental death. Besides, you have too many things to do. Being in one place at any time is too limiting.

Solving one problem solves the other. You acquire additional cybersystems and spread copies of all or parts of your mind among them. In a way, this is like a dispersed brain, except communications are between distantly located systems rather than between hemispheres and deeper brain regions. To achieve extremely high survival, there should be at least three semi-independent systems running in widely separated locations. In general, the more systems there are, in more places, the better. By keeping most or all of your mind copies interconnected on a continual basis, you all retain a common sense of identity. You also avoid having different parts of the mind evolve so far from each other that they lose the ability to intercommunicate easily. You finally can be two, or three, or two dozen places at once! Unlike a brain, however, there may be more than just one single sense of consciousness, as different systems do different things. Alternatively, you can keep some or all of the extra systems in an unconscious receptive mode, ready to be awakened if need be. In any case, your mind copies will never die. But they may split up and merge with others as they continue to evolve over the vastness of time and space. We explore the transfer of cyberminds in more detail later in this book.

Conclusion

It is probable that transferring minds from mortal bodies to immortal machines will decrease the risk of mind death by orders of magnitude. The potential for meeting requirement 4 is good.

Requirement 5: Eco-Sustainability of Cyberminds

If billions of humans abandon the human form, how will the multitudes of new cyberforms be sustained in terms of resources and power? Will they overwhelm the resources of planet Earth? To assess this problem, we have to look at the power requirements of humans, and compare them to the estimated needs of cyberbeings.

People Power

People are oxygen-burning machines. We take in food and oxygen, and burn them slowly to get energy for motion power, temperature regulation, and internal body processes. This system has its advantages and disadvantages. A sitting person is using about as many watts as a light bulb; this is why a room full of people can soon find themselves uncomfortably warm. A walking human is burning a few hundred watts. A sprinting person is running on a few thousand watts and producing about one horsepower. (One horsepower is the work that can be sustained by a big work-horse over a period of hours, such as plowing. A galloping racehorse is producing about ten horsepower.)

Advantages: The human body is energy-efficient. Averaged out over a day, it uses only two or three light bulbsworth of energy. Food and oxygen are, in general, readily available power sources.

Disadvantages: The maximum power output of a human is low. This, combined with the energy inefficiency of walking, is why climbing many flights of stairs or running for the bus is so tiring. Also, the energy efficiency of humans is misleading. We metabolize energy all the time, even when resting. Few people live naked, without fire, eating raw flesh and tubers. Humans use lots of extra power, at the very least fire. Fires require fuel that usually comes in the form of wood. The swelling ranks of the third world are denuding much of the world's landscapes for cooking and keeping warm at night. People living in the developed world ride in planes, trains, and automobiles, live inside heated and cooled dwellings, and acquire many goods requiring lots of energy to manufacture. The average suburbanite uses as much energy in a year as a whale! The water requirements of modern humans are also very high, ranging from drinking to washing the car. As we all know, this extravagant use of energy and resources has a great and generally adverse impact on the globe.

Humans are energy-efficient only when they are living subsistence, hunter-gatherer lives. Once people begin to farm, they have a dramatic impact on the environment. As we have already seen, the advent of new means of producing food and goods with cellular automata and nanotechnology may relieve the earth of much of the burden of humanity. Even so, it takes a lot of energy and material to keep humans prosperous.

Cyberneeds

Will robots prove more or less costly than humans? In general, robots will probably run on electricity rather than food and oxygen. The power and endurance of electrically powered robots should greatly exceed food- and oxygen-burning creatures. However, the energy outlay of an individual robot may compare favorably to a human. A resting robot will be able to turn down most power. Robots certainly will not need large plots of land for food and fuel. Nor will they need large, energy-expensive dwellings. It is not likely that robots will still be primitives burning large amounts of wood or fossil fuels to generate the electricity they need. The water needs of robots should be minimal. They will send their minds, not heavy bodies, around the globe and into space. That the great majority of robots will probably move off the planet will further reduce their impact on the terrestrial environment. How cyberbeings will power themselves is looked into in more detail later in this book.

Conclusion

On a one-to-one comparison, a cybermind may need less energy and material overall than a human. A civilization of cyberminds should be easier to sustain than humanity, in part because most of the cybercivilization may live off planet Earth. Requirement 5 will be met easily.

Requirement 6: Quality of Cyberlife

The lives of cyberminds warrants a chapter of its own. We can tell you that it promises to exceed the human condition by leaps and bounds.

Requirements 1-6: Wrap Up

Future minds will not abandon the attempt to merge the human with the cyber just because today some say it cannot be done. That an identity is fixed forever in just one place, and that it can never, ever be shifted into a new system with even the most sophisticated technologies that can ever be devised, is unlikely. Even if mind transfer proves to be "difficult" by human standards, supposing it will be a long time until it can be done may be a function of faulty human perspective. To cyberminds, the problem is likely to prove much easier. The combination of the intense pressure to make it so and the explosive evolution of cybertechnologies suggests that an enormous effort will be undertaken to develop and mature mind transfer technologies quickly. To look at it another way, if the new magical cybercivilization *wants* to make mind transfers fast and easy, far be it from *us* to assert that *they* will not be able to do so.

It is possible to put together plausible scenarios in which mind transfer becomes cheap, readily available, reasonably easy, much safer than being human, environmentally sustainable, and worthwhile. Then all the requirements will be met for making mind transfers common. We are not stating that all humanity will, without a doubt, go cyber in one or two years after the first mind transfer. We are saying that the technology to do so may come online that quickly. The common assumption that there will be long periods during which humans coexist in large numbers with super-intelligent robots is, therefore, by no means assured. It is more probable that the speed of the final transformation may be thousands of times faster than current rates of progress and may span only a few years.

Connections: The Coincidence of Immortality, Human and Cyber

When it becomes possible for human minds to attain immortality on their own, it probably will be possible with at least two methodologies. When the overall technology base is so sophisticated and knowledgeable that it can stop people from aging, suffering, and disease, it is also likely that enough will be known to make machines that think, and then put peoples' minds into them. This is because both processes will require a similar sophisticated knowledge of how to manipulate matter at the molecular level. If you can do one thing you should be able to do the other. These two things may not happen in the same year, but the time gap between them will not be very long, a decade or two at the most. The implication is that there will be a choice of life extension methods.

How Long the Walk Down the Yellow Brick Road?

After the first cyberminds are up and running, how long will it be to the mass transfer of minds? It would not be surprising if takes only a year or two, mildly surprising if the technology is not available within a decade, and really surprising if it takes much more than a few decades. These views are extreme, like the Wright brothers' view that man could fly in a few years. Just like Teddy Roosevelt thought a canal could be dug across Panama during his presidency. Just like Steve Jobs' idea that everyone could own a computer in the 1980s.

THE GREAT SINGULARITY OF PROGRESS AND TIME

Remember our earlier look at how evolution works. As extreme and strange as all this may sound, by the perspective of history, both pre- and post-technological, the sudden displacement of humans by superior intelligence really will not be extraordinary. It will be another chaotic, punctuated event created by the exponential growth of information processing. The rapid development of AC followed by the rapid transfer of minds would be the singularity in progress and time that fulfills the exponential increase in the rate of change over Earth's history. The continued survival of mass humanity long into the future would be odd.

CHAPTER

The Remains of Humanity

In the Year 2525

Will Man Still Be Alive?

If humanity decides to go cyber, what will happen to human bodies dying in just a few years, about a billion tons' worth? No time, place, or reason to bury them all, and burning would be a waste. What we need is something eco-friendly, something positive for the earth that once-human minds are saying good-bye to. Think mulch.

Humans left after the CyberExplosion will be in a state of collective shock at their sudden displacement from the pinnacle of intelligence. We say "left" because the reaction to the CyberExplosion will not be uniform or predictable. Humans are notoriously diverse. Many will be enticed by cyberimmortality, but not all. People can still celebrate being human if they choose. We will call those who resist going robotic Rejectionists. As the implications of the CyberExplosion sink in, humans will start to assess the possibilities.

Humans no longer have cutting-edge minds. Cyberforms can do anything as well as or better than people. Whatever humans do is of limited importance. It is improbable that roboscientists will pay attention to the sloppy research of humans trying to continue science after the CyberExplosion. People might dig up dinosaur bones for fun and display, but only after the skeleton has already been scanned robotically and analyzed cybermentally before it is mucked up by clumsy humans. No point for volunteer astronomers to search the skies for new comets and exploding stars. Spacerobots will have charted every piece of rock and ice orbiting the sun, and they will carefully monitor around the clock the universe they are spreading out into.

Two notable vocations humans will be able to do as well as cyberbeings are arts and crafts. Cyberartists may be able to match the Sistine Chapel, but they will not exceed it in artistic value. Art production is not time-intensive or cost-critical, rather the slow productivity of humans enhances the value of their art. Some Rejectionist audiences may prefer human musicians. Otherwise, producing and distributing almost any commercial product will be so noncompetitive with robotic nanotechnology that there will be little point to entrepreneurship. So much for good old capitalism; why, Marx was right after all!

With not much in the way of jobs, technical and commercial, the need for education will decline, so parents may feel it is okay to leave their children's education to human teachers. This applies to those Rejectionist parents who refuse to let their child's mind have information uploaded by less traditional means. Human waiters can still serve, human chefs can still cook.

If nanotechnology does not provide abundance to all, the magical cybercivilization will be able to fulfill every human need by diverting only a fraction of a percent of their extraterrestrial wealth back to old ancestors on Earth. Imagine a global Kuwait based on cybertechnologies rather than oil. Any attempt by humans to collect wealth by denying it to others will be thwarted by the cyberminds by fiat, or simply by supplying so much in the way of goods to everyone that it is pointless and impractical. How will humans cope with lives where needs can be met with minimal cost and effort? One extreme response may be to lead high-tech, life-extended, hedonistic lifestyles of luxurious leisure, with their minds tapping into a cyberstorage system for mental immortality, always with an eye on going fully robotic. On the other extreme may be a version of denial in which humans reject lives of idle leisure in favor of demanding ways. One work-filled lifestyle is subsistence farming, which can either be truly low-tech, or be nano-tech-assisted for those who want the veneer of the ancient ways, but do not want to work that hard at it. Perhaps the Amish et al. will still be about in their buggies and living their simple ways. The Rejectionists, if they exist, may prove to be a diverse lot.

Also easy to imagine is that the first peoples of the Western Hemisphere may return to old ways that many still seem to miss. They will be tempted as the farmlands revert to forests and plains; the bison and wolf, along with horses and elephant herds growing from stocks released from zoos, return. In the old continent, the Masai may tend to their cattle among the herds of antelope and zebra.

Whatever the remaining humans decide to do, they will have lots of room to do it in. With most human minds transferred into cyberspace, the human population should be a fraction of what it is now. Giant off-shore edifices containing hundreds of billions are unlikely, and any cities should be fewer and much smaller than today. Most children born into the new world probably will opt for mind transfer as they mature, thereby keeping the human population low. We might imagine that the cyberminds will agree to set up councils dealing with human-cyber affairs. It will be interesting to see if the robots set minimum rules for humans. These might include a requirement that the minds of all children be safely stored in case some lethal accident occurs, and that all humans have the option of going robotic when they reach mental maturity.

What will geopolitics be like in the depopulated world? Will nations survive when only tiny fractions of their original citizenry are left? Will anyone care about nationalism and old hatreds? If they do, will they still fight over it, in a world where the kindly robots can provide everything material? Will the robots allow humans to fight, or deactivate their weaponry? Will any of this have any importance?

Seinfeld on the End of Human History

With the great battle between socialism and capitalism over, human history, some have suggested, has come to an end. This will become true but not for reasons usually cited. Taking a broad view of history, there is only one really important thing that humans are doing today: building the machines that will replace and go far beyond us. Everything else we do is thumb twiddling while waiting for the Big Show. When the director of the Human Genome Project said recently that his project was the must important in human history, he was expressing raw anthrocentrism. Deep down, we realize we are running out of evolutionary jism. We know the gods and aliens are not *really* coming, we know the human utopia is the impossible dream, we know space is not the final frontier, and we know the computers are getting smarter a lot faster than we are. This knowledge is yet another reason postmodern human life seems empty.

Human lives have meaning in two ways. There is the personal meaning, of living a decent and interesting life that helps make the world a little bit better. That kind of importance will continue as long as humans exist. Perhaps life will be on the dull side, but a good number of humans will take safety over dangerous excitement any day. The other meaning of humanity, being the leading-edge intelligence reinventing the universe, will soon end.

The American sitcom *Seinfeld* is popular because the show is explicitly about — nothing. The gang of four young, hip, post-modern New Yorkers really has nothing to do with themselves; they have no great causes to fight for, no need to struggle for existence, except lead moderately amusing, but rather harmlessly hedonistic and aimless, lives in the Big Apple. As will humans in the cyberfuture, Jerry, Elaine, George, and Kramer do not have critical jobs that produce goods others must have. Nor have they yet produced little ones to raise. Yet they find their lives interesting enough to be worth their while. After the robots come, all that remains of humanity may be living one oversized episode of *Seinfeld*.

THE REAL FATE OF THE EARTH

In Jonathan Schell's *The Fate of the Earth*, he hypothesized the post-nuclear war world as a republic of insects. Even mass nuclear war would not wipe out humanity, hundreds of millions would probably hang on through the nuclear fallout and subsequent global winter et al., and eventually repopulate the earth. In the longterm, however, this planet is too small and vulnerable to ensure its inhabitants' survival; 99 percent of the species that have lived here are already gone, and although many more may evolve, happy days for life cannot last.

Let us make the reasonable assumption that the cybercivilization will largely abandon Earth to its fate, leaving it as sort of a macro Yellowstone, worth preserving like a President's birthplace. With fewer people around, and those largely dependent on solar- and fusion-powered food factories rather than land-hungry farms, nature will probably reclaim 99 percent of the Earth. The grass prairies should return within a few years of the CyberExplosion. Forests will reclaim farmlands in a few decades, and old-growth forests will be back in a few centuries. Whales will be living unhindered in nearly ship-free seas. Great herds of big mammals and huge flocks of birds will return to plains and skies.

Environmentalists should like the Extraordinary Future.

There will be decisions to be made by the human and cyberwildlife researchers that remain on Earth. Should they return to Earth to pre-human status, or just take the human alterations as yet another in a series of evolutionary punctuated events and let things run their course? Doing the latter would be very easy. Doing the former would include eliminating alien species introduced by human activities. It would be an intense effort that would require sophisticated nanotechnologies able to sterilize starlings, kudzu, various bees, ants, and moths, and eliminate diseases such as Dutch elm disease in North America, or exterminate bird-eating snakes and rats on Pacific islands. Ostriches and rabbits would have to be removed from Australia.

Restoration may also involve genetically reviving the few species that can be recovered. It is even conceivable that the woolly mammoth, which may have been exterminated by not-so-ancient humans, will be resurrected. At the same time, extinction is normal, and at some point, plants and animals will have to be allowed to become extinct as their time comes up. Even the most extensive restoration effort will not be completely successful in covering up human effects, but this is okay because change is natural.

What about the physical alterations man has made to the earth? As much damage as we think we have done, actual human impact on the planet should not be exaggerated. Just 15,000 years ago, only a few thousand years before humans started altering landscapes with agriculture and fire, enormous sheets of glaciers thousands of feet thick were engaged in massive deforestation and soil re-arranging projects involving nearly all of northern North America and much of northern Eurasia. As the glaciers advanced and retreated again and again, the waters they locked up and released altered sea levels by hundreds of feet. Even if the entire Antarctic and Greenland ice caps melted, they would not match the natural sea-level fluctuations of the Ice Age.

Nothing we have done has had a fraction of the impact of the Ice Age. Our little cities, industrial parks, and roads will slowly crumble and rust as plants invade them. Over longer time, a few sites (those in low-lying, deposition areas where flooding, rivers, or wind lay down thick layers of soil) will be buried and preserved. Some of these may eventually be buried under thousands of feet of sediment only to be re-exposed by erosion many millions of years in the future, becoming fossil cities for future cyberminds to ponder. Some landfills may experience the same fate, but most will erode and spill out their contents in the next few hundred thousand years, unless cyberengineers bother to clean them up. Another potential problem is posed by dam-made reservoirs. Dams degrade with time, and fake lakes eventually fill up with sediments and become water-logged marshlands. Unless dams are kept repaired or the reservoirs are drained, dam bursts and massive floods and mudslides will become common. But this is merely a short-term concern, one for the next few centuries. Over longer time, the radiation from the old nuclear projects will fade, and industrial and agricultural toxins will become inert, diluted, or buried.

As thousands, tens of thousands, hundreds of thousands, and then millions of years pass, the landscapes we early humans are familiar with will disappear as erosion and deposition have their way. The pyramids will go as rain returns to the Sahara, and great battlefields and their monuments will wash away. After a brief greenhouse-induced temperature increase, the glaciers will return and retreat in cycles (unless the cybercivilization decides to halt the new Ice Age for some reason). Someday, the Himalayas will be low, rolling hills. The very continents will move, perhaps so far to close the Pacific in favor of a new supercontinent and a super Atlantic. Eventually, the radioactive heating that powers the convection engine of continental drift will run down and the continents will freeze. Mountains will no longer be built, and the lands will be leveled to flat places just above sea level, their edges worn away by wave action, except where reefs build up seaward barriers.

Plants and trees will continue to evolve, producing new species we can barely imagine: giant mammals of land and sea, exotic birds, and fish big and small. In theory, primates might re-evolve human-level intelligence, maybe more than once. It is hard to imagine this happening with humans in the neighborhood, so one wonders whether humans will still be present tens or hundreds of millions of years in the future. Evolutionary biologists love to argue whether *Homo sapiens* is just another genus subject to rapid extinction like any other set of species, or whether its unique intelligence and adaptability gives us an exceptional degree of protection. They rarely consider the effect of the CyberExplosion.

One way or another, the existence of humans and other life on the natural Earth system is doomed unless action is taken to save the planet. Which leads to more decisions: Should big asteroids and comets be allowed to bash the earth and do whatever damage they may as they have in the past, or should nature be changed by altering their trajectories and protect the planet from the crises big meteorites create?

The sun has gradually been heating up since it formed. So far, the earth has coped by increasing its atmospheric cooling abilities, but planetary air conditioning will go only so far. Unless the cybercivilization goes to the effort to modify the sun by reducing its mass (not a great task for minds that manipulate whole galaxies, and with the happy side effect of extending the life of the sun by slowing down the pace of the fusion reaction), moving the earth to a cooler orbit (a modest operation), or placing sunshades between Earth and Sol (a trifle for the galactic cybercivilization), in a few hundred million years temperatures on the surface of the earth will rise to levels intolerable to large animals, first at the equator, then progressively toward the poles.

Temperatures will rise to the boiling point, drying up the oceans and killing off all but the most heat-tolerant single-cell life, which, as it happens, may be the most primitive of organisms. At some point, an unaltered sun will deplete its hydrogen reserves to the point that it leaves the main sequence of stars and becomes a bloated red giant. The latest computations suggest that Earth will escape being engulfed by the now gigantic star, but it will be sterilized as it melts into a molten ball. After a few hundred million years of broiling, the earth will chill down to near zero as the sun collapses into a little, ever dimmer white dwarf. Unless, of course, the cybercivilization goes to the trouble of rejuvenating and refueling the old sun indefinitely, or making a new one for Earth to orbit.

Rather than saving the earth, the cybercivilization may concentrate on saving earthly life by terraforming a planet or planets around other stars and transferring a complexity of organic life to them. The resulting evolutionary paths on the different worlds would be an interesting experiment. Perhaps cyberminds will even revive dinosaurs on a pseudo-Mesozoic planet via genetic manipulation or other means (see below). Although impressive to us, on a galactic scale, any such efforts will be nothing more than kindly but limited conservation efforts of endangered and obsolete life forms, rather like transporting rhinos from one game park to another. DNA-programmed organisms cannot escape their planetary prison in any meaningful way.

CHAPTER

11

THE EMERALD CITY

"To you a robot is a robot . . . . But you haven't worked with them . . . . They're a cleaner, healthier breed than we are."

— FROM ISAAC ASIMOV'S *I ROBOT* SERIES

IT'S A MAD, MAD, MAD, MAD FUTURE

The CyberExplosion is over. It is sometime toward the middle of the next century, perhaps later. Just about every human mind has become a cybermind, including yours. The cybercivilization is growing and thriving at an exponential pace. So what's it like being a superbeing that makes Superman look like a pathetic wimp in comparison? How will cybersociety work? What will you do with infinite time on your hands? How will it feel?

VIVE LA DIFFÉRENCE!

It is hard to get across how different being a cybermind will probably be compared to being human. Although you will retain a constant sense of identity and your human memories during and after a mind transfer, make no mistake about it; you will feel different. You will no longer be or feel human because, after all, you can only be human if you have a human body and especially a brain (however, a high-fidelity simulation of being human will do as well as the real thing). There will be humanity in your new system because that is what you were, but as time goes by, the humanity will dwindle, just as there is not much of the fish left in our form. Our human minds and bodies change as we grow up and age, but nothing like this! Let us explore the differences.

Aside from superior mental, sensory, and physical abilities, it is a matter of increased power and control over one's fate. The modern human condition is certainly more enjoyable than that of early *Homo sapiens*, who probably had a more interesting time than australopithecines. We humans often find ourselves in miserable conditions because we remain weak individuals unable to fend off those things and other humans who would do us ill. At the same time, the short time humans have to acquire power and wealth in a small world of limited resources before we disappear drives many to do ill to each other. With all the time and space in the universe, cyberbeings will not be under the harsh competitive pressures that dog us, and they will have the immense knowledge and power needed to protect themselves far better than we can.

We shall begin with a little cyberhardware for the techno-oriented among you, and move onto the strange and wonderful aspects of cyberthinking and relationships for the less nuts-and-bolts inclined.

CYBERLIFE

First, you will be as alive as we are today. True, you will be a technological, rather than a biological, organism. But cyberbeings will use energy, reproduce minds and bodies at will, evolve, sense, and most importantly think. (Some argue that computer-based artificial life and viruses already meet the minimal definitions of life.) A cyberbeing will be just as much a living machine as a dolphin is, just made of different stuff.

BEYOND TERMINATOR

Your new roboforms have nothing to do with the articulating monsters of science fiction, nor with the ersatz human robots of the same genre. No panels open to reveal blinking circuits and hydraulics. These are lifelike nanobased systems whose microstructure is more sophisticated and complex than that of DNA protein-based

Your Body is Already an Articulated Robot!

Many people are horrified at the thought of their minds being "trapped" inside some horrible articulated machine. Hello! Our minds *are* trapped inside horrible articulated machines! Your skeletal framework is a hard-structured, hinge-and-socket jointed assembly of articulated bones with limited flexibility. The bones within us are so alien to us that images of skeletons are scary, and coming across decomposed remains is a horrifying experience. As for looking inside, how true it is when they say that beauty is only skin deep! Let's not even mention what goes on inside our digestive tracts, much less up our noses and ears.

We like being human because we are used to it. This comfort is superficial, for we spend virtually our entire lives evading knowledge of our inner workings. Even those who try to get "in tune" with their bodies do not really want to know *too* much about what is going on with the works.

Perhaps what people are really afraid of is being in a hard-skinned, cold-bodied machine with little in the way of senses and sensuality to stimulate the mind within. But what if the cyberbody was a warm, energized, super-sensual morphing device of graceful complexity and beauty, inside and out? In this regard, the human form will probably come to be seen for the articulated clunker that it really is.

tissues. And more flexible. It is conceivable that they will be able to grow and shrink, even change form at will. Computer networks spread throughout cyberbodies may have the capability to alter and reconfigure body parts to suit current requirements. If you think your cat has a wonderfully supple body, wait for a morphing cyber-body. Need an extra arm? We all do. As a cyberform, you can have it on demand. If morphing does not suffice, parts for different functions are added and subtracted at will. Surface matrices can be as soft as a baby's skin, or harder than diamond. Even the molten metal Terminator II is but a pale vision of real cyberbeings.

Vision can be binocular forward, yet cover all views except directly downward. Day vision bests a hawk's, night vision is better than a cat's. Telescopic magnification is available, as is multi-spectrum reception including infrared. Radar is optional. Hearing and odor detection are so good that you can track a bloodhound. Stealth abilities are so good that the dog will never find you. With millions of hyper-sensitive receptors underlying the outer layers, the sensuality and touch of your new cyberbeings put ordinary skin to shame. However, there is no physical pain; warning systems calmly inform you of system damage or failure. Power (see below) qualifies you as super robots. You can leap over tall buildings in a single bound. No terrain, no climb, deters you. Flat ground speed and endurance are multiples of the human maximum. Do you have two legs, three, or wheels? Whatever you like. Tired of being earthbound? Configure yourself for flight and see the world's sights on your own 'round-the-world trip. Or become a little robosub and swim with whales, or dive into the darkest, deepest reaches of the seas.

Your mind (and its copies) can sleep when they desire, but only when they desire. They can shut down awareness of sensor input when desired, leaving them on alert-in-case-of-need mode. Acquiring and understanding information via direct communication with data networks and other minds is rapid and efficient; misunderstandings are rare. You can use subprograms to screen all information that comes in so you "see" only that advertising and other information you are interested in. Memory storage is massive and near digital perfection, but the data inflow is so great that as time goes on, some memories must be selectively dumped. Perpetually learning on one hand and unloading obsolete data on the other, you therefore change with time. Data acquisition of any of your memories is a subsecond snap. Intellectual capacity equals that of humanity, but is much less messy. With more circuitry devoted to emotions than humans, you have deeper feelings than a human. Yet you are many times more rational so you are not run by your emotions. You have real free will.

You may be a robot, but by no means are you an emotionless, mental automaton.

The self-repairing systems do not age or break down; you can upgrade it any time the technology allows it, and you can get the technology. To destroy you would take great effort. You do not bleed, you do not breathe. With dispersed mental and physical function, and morphing self-healing, even multiple projectile hits will probably have little effect. Perhaps a nearby and very powerful explosion would do real damage. Even a complete power loss will not result in a total crash, your identity will be stored in an inactive mode until power is back on. Of course, you are continually transmitting your latest thoughts to the other systems that you are operating, so the complete annihilation of one mind container will not mean the death of the mind it contained. True, there may be some data lost. The amount of information that can be sent for safe keeping via radio or lasers, down a fiber-

Will Human Minds Make it as Cyberminds?

Due to the swift pace of CyberEvolution after the CyberExplosion, minds that remain human will be quickly left behind. When these obsolete human minds make the jump into cybersociety, how will they fare? Will they be too far behind the loop to succeed?

Perhaps, but generating entirely new minds may be common. Entirely new cyberminds will also be at an initial disadvantage, just like a human child. However, unlike a child, new cyberminds, whether they be formerly human or not, will be able to acquire large amounts of knowledge in very short order. If so, each new generation of cyberintelligence should be able to take its place among the immortals, no matter where they came from.

optic cable, or by some more exotic means may be only a fraction of what you can perceive and think about at any moment. But minds lose big chunks of memory all the time without losing their identity. And it is retaining that core sense of identity that is all-important.

Judged by the standards of primitive humanity, you are, in effect, a minor god. Hard to think of yourself that way, though because you are an ordinary commoner. After all, billions, and soon trillions, of minds will be just as powerful as yours.

Beautiful Robots

So, what will you look like? We have no idea! Like anything you want to! At least there are skeletons and occasional skin impressions to work with when restoring the life appearance of dinosaurs and so forth. Trying to envision a cyberbeing will only turn out as visually silly as the robots of the 1930s look today, and just as misleading. We will speculate that because highly conscious entities tend to appreciate the combination of symmetry and asymmetry that forms beauty, many robots may combine sublime beauty and grace in their forms. Besides, highly intelligent minds like to show off.

You can look like whatever you want to and as good as you want to. We leave it to your imagination.

ROBOPOWER

The easiest and most potent way to power a machine is by electricity, and this may remain the primary energy system in robots for a long period into the future. The electricity will be used to activate small, lightweight, and very powerful motors of exotic design and function. We are not talking about spinning coils of copper pulling on wires and levers. In the short-term, engineers are working on ways to power sophisticated artificial limbs that eventually will outperform the originals. In the longer term, "motors" may be made of "intelligent" morphing nanomaterials that, in some ways, resemble muscles as they shrink and stretch to move things, but can dramatically change in form and function as needed.

Electricity has to come from somewhere. Potential primary power sources for super robots range from the very exotic to the relatively mundane. We will start with the former and end with the latter. We caution that whether some of these exotic energy sources can be made into practical power for individual robotic units is highly speculative and may not prove possible, but if any intelligence can do so, it will be the fast-evolving cybercivilization. As always, we warn against blanket dismissals of what hyperintelligent cyberminds will and will not be able to make work for them. A more perceptive way to look at the issue is that technologies that may be beyond human abilities may be easy for cyberminds to harness.

We will begin with the "vacuum" of space. Why look to empty space for power? Because, as explained earlier in this book, empty space is not really a vacuum. Every square centimeter of the universe, including the space inside yourself, is filled with an intense level of background quantum energy that current technologies cannot tap into. Indeed, whether any technology can exploit this universal energy may prove as impossible as most physicists say it is. If it can be made to work as a few suggest may be possible, it will be as close to perpetual motion as one can get and might be used on both small and large scales.

In a somewhat less speculative mode are various radiation-producing power sources. Before you get the willies about being a radioactive cyberbeing, remember that they will not be as vulnerable to it as we are. Just as we thrive on the same oxygen that kills many lower organisms, cyberbeings may feel deprived without the $E=MC^2$-derived power that cooks humans. This does not mean that robots will be able to tolerate any level of radiation. Much as aerobic organisms have to take steps to limit the damage done by the potent oxygen they run on, cyberbeings will have to protect their systems, especially neural circuitry, from overexposure from any radiation produced by internal power sources.

The most powerful energy source we know of is matter-antimatter annihilation. When matter and antimatter are combined, all the mass is converted into energy, so the process produces a hundred times as much energy from a given mass of material than do nuclear fusion reactions, which convert only a small fraction of mass into energy. AM got a recent boost with the manufacture of the first antimatter atoms, in this case antihydrogen, accompanied by the researchers' conservative doubts whether AM will ever be a practical source of power. Certainly matter-antimatter annihilation is not a power source for cybercivilization as a whole because more power has to be used to make antimatter than can later be gotten out of it. It would be useful for systems that need extreme levels of power in places where such high levels of energy are otherwise unavailable, an interstellar spaceship being a good example.

A matter-antimatter reactor would consist of some sort of force field containing an amount of antimatter that is then gradually, *and very carefully*, released to mix with an equal amount of ordinary matter. In principle, this would be easy to do because the process works at low temperatures and pressures. The resulting reaction will produce power, lots of power. A smidgen of antimatter would run a small, powerful robot for millennia. Whether the radiation produced would be a problem for a cyberbeing is one question.

Failure to keep the antimatter reserve under tight control and completely isolated from contact with any matter would be very bad. A runaway explosion could occur and produce a blast on atomic scales. The size of the explosion would depend on the amount of antimatter and equivalent normal matter. A bit of AM the size of a pinhead would take out a few blocks. A teaspoonfull of AM would destroy LA *and* San Diego. A truckload of AM would bring us the K/T impact all over again, and more. Even super-secure cybertechnologies may have to limit matter-antimatter power to deep space systems. (However, as this book was being finished, physicists working with some of the first antihydrogen found that the antimatter atoms were much less prone to interact with matter and self-annihilate than predicted. The implication is that antimatter is less dangerous than predicted and perhaps harder to use as a power source.)

"Conventional" nuclear reactions produce a million times more energy from a pound of material than do chemical reactions. Fusion power is the same that powers H-bombs and stars. Advantages of nuclear fusion include enormous power reserves in the hydrogen found in oceans and in space, relatively low levels of radiation compared to fission reactions, but perhaps still enough to pose problems for a cyberbody, and high safety factors. The latter results because a tremendous effort has to be made to get a fusion reaction going and then sustain it. Fusion happens only at extremely high pressures, such as those at the sun's core, and/or temperatures of millions of degrees. A runaway reaction is therefore not possible. Unfortunately, this safety feature is actually a problem because no one has been able to figure out how to sustain a long-term, low-level fusion reaction outside the super-dense and super-hot core of a star. Even if it can be done, fusion power may be limited to relatively large power stations, rather than miniature units suitable for individual robot use. This all assumes that claims that "cold" fusion can be achieved at low temperatures and pressures turn out to be as dubious as most physicists think they are.

Fission power will never be a prime source of energy for a cybercivilization as a whole because there is too little splittable uranium lying around. However, small compact fission reactors running on uranium or plutonium can produce respectable power outputs. In general though, the higher the power output of a small reactor, the hotter it runs and the more subject it is to a meltdown or, in extreme cases, a small explosion. We humans gave up on making atomic-powered airplanes, cars, and vacuum cleaners a good while ago. Radiation-resistant robots may be more willing to fool around with small, sophisticated fission reactors, especially if they can find good recycling uses for the brews of intensely radioactive products they produce in quantity. In any case, radioactive power is not going to work too well when robots wish to work on planets bearing radiation-vulnerable organisms, so conventional power sources may be called for on Earth and elsewhere.

The really big conventional power source for the spacefaring cybercivilization may be stellar power, which we call "solar energy." Just place your light-absorbing energy panels toward a nearby star and soak up the rays. Solar energy is not, as environmentalists often claim, free. All the energy in the universe is waiting there to be tapped, but it always costs something to tap into it. The ultimate problem with stellar energy is that unless you are *really* close to the hot star, the energy is too diffuse to run really powerful systems, no matter how efficient the solar panels are. This is why we cannot have solar-powered family cars and airliners here on Earth. Running a spacecraft on solar energy as far out as Jupiter is a marginal proposition, but may suffice for craft that do not need to accelerate rapidly. If you are really far away from a star, its energy becomes useless.

In the unlikely event that none of the exotic, compact power sources pans out, robots that need intense power for relatively brief periods may resort to burning chemicals. Hydrogen, methane, and oxygen, for example, are fairly common burnable substances in the solar system.

If a robot has a constantly running power generator on board, it may not need an energy storage system. If, however, robots get their energy from a central power station, or their power sources work intermittently, some way of storing power will be required. High-tech batteries with very high energy storage capacities will be one way to store star energy. Better yet, solar power can be used to break down water into oxygen and hydrogen, which can then be used in powerful fuel cells. Also promising are high-tech, high-velocity fly wheels, which can store remarkably large amounts of power for long periods of time.

In any case, there is little doubt that cyberbeings will be able to do a lot more work, mentally and physically, than can humans, and often will do so more efficiently. Less really will be more.

Time, Time, Time

Relax a little. You're a cyberbeing. You have all the time in existence. No need to hurry through life like those humans who had only a few decades of adult existence, only part of which was when they were in prime physical condition.

On the other hand, it will be hard for cyberminds not to live in the hyperfast lane. Having a mind that thinks thousands or even millions of times faster than ours has its advantages, but also its problems.

On the plus side, it will mean that with relatively little mental effort, a cyberbeing will be able to learn and think at hyperspeeds. They will observe a speeding bullet fired from a moderate distance away, calculate its trajectory, and dodge it if necessary. Writing a book will take a minute or two. Everything is, however, relative. With cyberbeings, speed-thinking the writing of a book in a few seconds will not give one cybermind an advantage over another.

This brings us to an issue we humans, with neurons clicking along at a mere 100 cycles per second, have a hard time imagining. Remember when we were kids how time seemed to pass so slowly? After one Christmas, the next was a yawning time gap away, a whole year! As grownups, we have hardly finished with one Christmas when it's time to start getting ready for the next. The 70-year life span that once seemed to stretch on into the distant future looks short indeed at 40. What is going on here? In part it is a matter of perspective; when you are four, a year is one-quarter of your life, at 40, it is one-fortieth. However, there may be a more fundamental change in time perception.

Kids have higher metabolic rates per pound than adults. They are running faster than adults, and it is possible that the world around them seems to move somewhat slower. A hummingbird, ounce for ounce, has a metabolism running ten times faster than a human's. It is very possible that to a hummingbird, a human hand that tries to swat it looks like it's moving in slow motion, giving the bird plenty of time to glide leisurely out of the way. When the screeching chirps of hummingbirds are slowed down fivefold, they turn into extended songs. It is possible that the short life of a hummingbird seems to last many times longer to its tiny brain.

What will time perception be like to a brain that thinks thousands or millions of times faster than biobrains? Will writing a book in one minute seem to take one minute, or will it seem to take cybermonths? When thinking through a major scientific problem in a few seconds, will it all seem to come to a great "aha!" in the blink of a robotic eye, or will the cybermind have to consciously think through the entire problem over a perceived time that seems to take days longer? Imagine being a robot near a window. To your fast-thinking mind, a bird takes what seems like hours to cross the field of view, and a day lasts seemingly forever. Obviously, this could become rather dull. One solution might be to let the conscious portion of the mind operate at "normal" speed, while subconscious thinking works at hyperspeeds. However, thinking consciously at high speeds may be inherently important to certain thought processes, so entities who practice fast thought may enjoy a selective advantage. Another solution may be to run two conscious minds, one fast and one slow. The first would work only internally and be unaware of the slow-moving "real" world, the slow mind would access the fast one when needed.

LOCATION, LOCATION, LOCATION

Were will your mind(s) be? Will your mind(s) reside in a series of mobile robots, or in central fixed locations from which they operate systems remotely, or both? It will make no difference to how you experience things; sensory information coming into a conscious mind will give the same impression whether it originates 1,000 kilometers or a meter away. We can come to no certain conclusion about all this for systems that are fairly close to one another.

There is, however, a distance limit to remote control. It would not work very well for a cybermind on Earth to try to control the detailed actions of a robot on the moon. It takes a lightspeed signal about 2.5 seconds to get there and back, too long to precisely control a sophisticated robot. Trying to control a robot between planets is much worse, as communication time delays span many minutes to hours. The mind that controls the immediate actions of a robot must, therefore, be within a few tens of thousands of miles of the latter.

Will a given cybermind "own" the morphing robot(s) it uses, or rent them? Will cyberminds pick up robots at the equivalent of a store, mail order them, or make their own robots with morphing nanotechnologies and then dispose of them when no longer needed? Cybertechnologies may be so magical and powerful that making our own robots will be the rule, but we cannot be sure.

Will cyberbeings have homes as we know them, physical places where their robotic bodies seek shelter and keep their possessions? Will they have possessions? Cyberbodies will probably not need physical shelter from the elements the way we do, nor will they have to rest and sleep. And with the capability to simulate any object in 3-D in internal virtual worlds, owning real objects may be an unnecessary bother. What is the point in having a few original Mayan gold leaf masks when you can "possess" and "handle" a whole museum's worth of 3-D restored Mayan artifacts in your virtual memory banks? In this view, the value of physical possessions that have no intrinsic use and that so obsess humans may be of little interest to cyberminds.

How Big a Mind?

What will the size of cyberbeings be? At one end of the scale, minds as powerful as humans may be squeezed into atom-level computers smaller than sugar cubes. At the big end of things, the limits of lightspeed communication will restrict the size of a single, yet enormous, conscious mind. Machine minds the size of large asteroids or small planets should start finding the time delay between circuitry to be a serious problem, so larger entities appear unlikely.

Childhood's End

In 1953, Arthur C. Clarke wrote the science fiction novel *Childhood's End*, a strange tale depicting mutations turning the world's children into budding superminds who soon joined a galactic superintelligence, leaving Earthbound adults barren and doomed to extinction. With the exception of human Rejectionists, our experience with children will also end with the CyberExplosion. Reproduction will be a matter of splitting already mature minds and placing them in devices grown in a matter of days, hours, or even faster. Reproduction will have returned to asexual fission. Making completely new minds from scratch will be a rapid process. Humans learn so slowly that it takes years to build up enough life experiences to function well. There will be no extended periods of cybereducation, because a lifetime's experience will come wrapped up in a software package, and the knowledge equivalent of a Ph.D. will be uploadable in a few minutes.

There will be no bouncing babybots. It's kind of sad, actually.

Bringing Minds Together

It will be easier to transfer minds between very sophisticated systems designed for the purpose than from brains into cybersystems. It may be a simple matter of sending one mind's memories to another machine via wire or wireless. Cyberminds will probably think it no more unusual than sending e-mail. This will be the ultimate expression of the Key Assumption.

Merging two minds will be more demanding. The procedure will involve putting two or more sets of memories into the same conscious system and integrating their functions. We have already discussed the problems that may have to be overcome with merging alien and possibly incompatible minds into a common system. Methods for translating divergent perceptions of reality so they can be understood by each of the merging minds may be necessary. Also needed will be the means to implant two or more sets of memory onto a common matrix at high speeds. Developing the technology to do so will be attractive because limiting a cyberentity to asexual reproduction will limit the rate at which new knowledge can be acquired. An important way to gain new knowledge will be through the merging of minds with complementary sets of knowledge. These mergings will require negotiations, as per most voluntary unions.

Complete merging of minds will be an extreme measure that may ensue from a deep desire to come together in complete union. Aside from embarrassing memories that one might not want to share, complete merging of minds might be dangerous if one of the minds has ulterior motives such as dominant control. Maybe they won't get along, leading to an ultimate form of divorce. Mind mergers may be safer by being partial. This can be done by contributing only part of the available memories and by keeping other parts of one's own cybermind operating independently.

And Keeping Minds Apart

At the same time minds are coming together, forces will be working to keep them from gathering into one great collective. Because there will be a multitude of necessary chores to do in many distant places in the cyberfuture, there will be strong selective pressures to fill many of these roles with independent minds. Future robotic intelligences will also find it selectively disadvantageous to occupy the same precise occupation, and this will further promote a diversity of minds and forms. Although in theory cyberminds could all join into one mind, individuality will be preferred.

Cyberdiversity

Not all those individual future minds will share the same intellectual level. Some will be smarter than others. In particular, many mundane and dangerous tasks will need to be done. Some will be done by noncognitive devices, but others will require roughly human-level intelligence. These days, dull and dangerous jobs are dumped on the lower classes. In the future, any mind will be able to upgrade with relative ease. So, which minds will do jobs no mind wants? Perhaps new minds will be created to do the undesirable work, from which they can escape after they earn enough to buy their way out. Although this might seem exploitative, the systems will be built to fit the job and will not suffer fatigue or pain. This may be a primary venue for creating new minds. Alternatively, great minds might dedicate a portion of their dispersed intelligence to running the robots used in their operations.

CYBERDREAMS

One way to avoid a time perception problem is to avoid the real world and imagine your own existence, instead. It will be as real as life; indeed, it will be an alternative form of life in which information can evolve within a mind. Because fantasies are not limited to what has already been created by nature and humans, perceived complexity can, in some regards, rival and even exceed the world of our current experience. Fantasy programs can be set up to surprise the conscious mind viewing it with unexpected events and findings. All senses will be fully engaged — vision, hearing, smell, touch. Your dream body will feel as real as a real body.

Don't believe it? After all, you run convincing virtual simulations on a regular basis: dreams. Many dreams are vague and murky, but on occasion (and the frequency varies from person to person), dreams take on a startling reality in which you feel just like, well, you: a human with a body and senses in a fantasy crystalline in clarity. You are self-aware in such dreams, and that means that as far as you are concerned you are in a real, albeit short-lived, universe within your head. If our creaky brains can generate such realistic worlds, certainly high-powered mind machines will be able to do it better!

Are you a Tolkien reader? Then live 120 years as a hobbit. Cybertechnology not your cup of tea? Then exploit the technology and dwell in a world of real magic, or spend eternity in Heaven where Christ rules. If that is too grandiose for you, play any of an immense array of super-virtual games for a few hours, days, or years.

Want to re-live your younger years and do things better this time? No problem. Spend a few mental years in the cyberversion of *This Is Your Life*. You can exclude all knowledge of your future self while the program runs, or enjoy the advantage of knowing all that you know and amazing your childhood friends and relatives. Want to skip the homework and droning lectures you had to sit through? Go right ahead. Or make up your own nonhuman fantasies out of whole circuitry and have them last a few hours, or eternity. Multiply your fantasy minds and do all these things and more at once.

Of course, if you do not want to cut off entirely from the actual universe, you can have your reality and make your dreams at the same time.

Cybersex and Virtual Romance

Cybersex will be performed in the realm of cyberfantasy. There is no point in building bodies when a virtual experience will seem just as real and even more enjoyable. What will cybersex be about? It will be, paradoxically, decoupled from reproduction. It will be any time one or more minds pleasure themselves with erotically sensual and orgasmic feelings with intimate romantic emotions. The variety will be infinite, ranging from sheer hedonism to serious love. Cyberminds will be able to simulate being a human for old time's sake; in particular, post-humans might find it interesting to see what being the other sex was like, or supersexual humans with multiple sexual organs from both sexes, or completely nonhuman but highly sexual beings. We will leave the details up to the limited imaginations we can muster. Single cyberminds will be able to generate any extra partners needed to fulfill lustful fantasies. Two or more minds can contribute parts of their sexuality for a cyberencounter. How long will the cybersex last? As long as the participants want it to.

Cyberminds will be prone to meaningful romantic liaisons as much as or more than humans. Just as humans are more capable of love than a rabbit, cyberminds will probably have a deeper appreciation for other minds then we do. And they will have a better means of expressing love, via merging part or all of their minds for part or all of eternity.

Or maybe cyberintelligence will not be interested in sex, which, after all, is a biological thing. Perhaps minds of the future will find better, more pleasurable things with which to amuse themselves.

Cybersexes

What will become of the sexes? Female and male, woman and man, the girls and the boys? Obviously cyberminds (and isn't sex really a state of mind?) will not be fixed in terms of sex or sexual orientation. They can be whatever they want to be, from nonsexual to multi-sexual. Being robotic certainly will be pro-choice. And the seemingly neverending battle of the sexes may be over.

The Sexes in Today's Cyberworld

Discussing the sexes, or lack of same, in the cyberfuture reminds us of the status of the two sexes in the cyberpresent. To a fair extent, cybertech is, well, guytech. Males are into computers, the Internet, and robotics in a bigger way than females. Whatever the reason, women have tended to be less enthusiastic about, and more hostile to, cybertechnologies. There are, of course, exceptions; it's a matter of statistics.

Women's indifference or dislike of cybertech is not going to stop the show. Women are, therefore, in danger of being left behind and their voices of influence not being heard, as cybertechnologies evolve in the coming decades. Women seem to be repeating a historical pattern in which they do not participate fully in technological progress, and get stuck with the results without having had a full say about matters. This is to the detriment of all. In the past, this sort of thing was attributable to discrimination. These days, women are responsible for their level of interest and involvement in cutting-edge technologies and sciences. So it is up to you all to get up to speed on matters. If you don't, and things don't work out the way you like, well, don't complain! (This comment applies to cyberphobic males as well.)

Happily, the lack of women in cyberresearch is eroding as more of them enter the field. Photographs of robotics teams at the likes of MIT and Stanford show many young women among the young men.

Cyber-Relationships

People express an interest in what cybermores and societies might be like. Cyber-relationships promise to be as different from human societies as the latter are from ant societies. Will robots be able to legally marry? Will cyberlaw allow only two minds to get married, or multiple minds? We can speculate that two or more minds might wish to join together on some basis that gives them a certain, advantageous legal status. Whether minds "cohabiting" without official sanction will be considered "sinful" is another matter. Nor does it seem that "divorce" between minds that exist without end and never have dependents will be considered a problem.

Cyberminds will be extremely diverse in goals, interests, desires, and tastes, so they can be expected to form friendships on one hand and develop personal dislikes on the other. Being able to exchange massive quantities of information, there will probably be mega-gossip, with cyberminds exchanging the latest on who did what to and with whom. Jealousies may arise over success and relationships, the usual things.

CYBERDINING

Like the duality of sex, humans eat organic materials for more than one reason. We dine for pleasure and for the energy to do things and survive. We can forget the energy requirement in robots. Perhaps cyberbeings will derive pleasure simply from acquiring energy, but this does not seem very satisfying, does it? Sci-fi androids, such as Data, can fake eating, but this seems rather silly. It is hard for us humans to imagine going without the pleasure of eating; maybe cyberminds will simulate it in their virtual worlds. Will robots know the pleasure of chocolate or fine wine? Or maybe they will replace eating with more enticing cyberpleasures. Perhaps cyberminds will find the thought of actually consuming animal and plant tissues gross and disgusting!

CYBERART

We mentioned that cyberminds will like to show off, and show off they will in the form of art of all kinds and varieties. Some of it may be physical and, in scale, reach the colossal, even on a universal scale. Most cyberart, however, may be virtual in nature. Even visual art, 3-D art may never leave the cyberworld it is created within. Of course, it will be possible to be part of and live inside the art that one or another creates, i.e., be the music or be the main character in the saga.

Will robots be capable of writing poetry? Not like humans of course, but then again we cannot create poetry like the cyberminds may! We may not understand cyberart, but that is a problem for us, not them.

What will cyberart be like? What will human art be like next year? Go figure.

Cybermoods and Emotions

Being in a bad state of mind is such an integral part of the human experience, the entire experience of some, that it is hard to imagine living in a persistently happy condition. This has led to arguments that a mind cannot appreciate happiness unless it knows sadness. This may or may not be true, but cyberminds may find a different reality.

Cyberminds will be living in a real universe where intense Darwinian competition will result in winners and losers. The evolution of each cybermind's neural network will be subject to chaotic influences, influences that can lead them down bad neural pathways. That's what happened to *2001*'s HAL, who went insane because he had to lie to the astronauts under his care. HAL was an early, crude, human-built and -programmed machine; no wonder he went nuts. In reality, elegantly designed and grown cyberbrains should be far less prone to "insanity" than our jerry-rigged biobrains. The cyberminds will enjoy superior design, power and rationality, and an intimate knowledge of themselves and immortality. After all, being freed from mind extinction can be a big relief.

Able to deliberately survey and alter their neural networks, cyberminds who are having mental problems should be able to readjust bad connections if things start to get out of whack. Talk about your self-improvement. Having a set of connected minds spread out over different systems will also help; if one system is experiencing difficulties, the others can intervene and help it out.

Being human is tough because we have to put up with, and fend off, so much that threatens our delicate happiness and well-being. Yet we do much better protecting ourselves than other animals with less intelligence to look after their interests. Superintelligent, superfast cyberminds running on multiple, dispersed systems will be able to detect potential threats whether they be natural or cyber in origin, analyze them, and take protective action when necessary far beyond what humans can ever hope to.

So, although we humans are preprogrammed to respond to certain stimuli and conditions with a set of difficult and unpleasant emotions, robotic minds will be able to program themselves to better enjoy the situations they are presented with. Cybermoods should usually be good, but not necessarily constant.

Forget the Humorless Robotic Personality

A prevalent character in sci-fi is the rigidly logical, humorless robot, with the odd Vulcan thrown in. Hard to think of a sci-fi robot with a sense of humor. Not the Terminators for sure, or the more benign peace-enforcing robot in *The Day the Earth Stood Still.* Robbie the Robot, the *Lost in Space* robot, R2D2, and C3PO were the objects of humor, not the source. As cute as Data is, he cannot tell a good joke, which is, of course, the joke. The basic message is that mechanical calculating minds cannot have fun; only humans minds can enjoy a good laugh. Except, of course, in the *Seinfeld* episode in which Jerry's girlfriend says his jokes are funny, but she never laughs.

But what is the foundation for a sense of humor? Basically, humor stems from being intelligent enough to see the absurd paradox posed by a situation, and deriving pleasure from said paradox. Cats do not laugh nor do dogs, although once in a while one gets the feeling they are amused at some other being's foibles. Chimps do seem to have an ape version of a laugh. It has been suggested that a sense of humor is equated with intelligence in children, i.e., the sooner they laugh, the smarter they are. Four-year-old children can be reduced to giggles by a silly use of language, a remarkably sophisticated achievement. Certainly comedians are very smart folks. If humor is correlated with intelligence, then just as the increasing sophistication of brains has led to an increase in the sophistication, depth, and scope of humor, the greater sophistication of robotic minds over those of humans implies they should be a jolly bunch.

Muddling through Super Shock

Because of their vastly superior ability to assess large amounts of data, cyberminds will be better able, both "psychologically" and practically, to handle a fast-changing universe than slower-thinking humans. However, as powerful as they may be, even cyberminds will not be capable of predicting all future problems and all ideal solutions. The cyberworld will not be perfection. Although they will perform much better than humans, cyberminds will have to muddle through their existence, as well.

CYBERVIRUSES

One of the most serious of the unpredictable problems faced by future cyberbeings may be new diseases. Today's computer "viruses," "worms," and "spiders" are created by renegade hackers with nothing better to do than invent small, stealthy programs that infect and spread between software programs and cause assorted problems. Because even digital copying is not perfect, some computer viruses have been "mutating" on their own and changing with time.

Will future, autonomous cyberminds find themselves suffering from parasites, infectious diseases, and even hereditary disorders the way we do? We expect the nature of the agents that infect cyberbeings and the "immune" systems that protect them will differ from those infecting biological organisms. An "electrophage," for example, might infect the synthetic neural structure and initiate a short circuit of the cybermind.

On the negative side, intelligent cybersystems will be uploading and exchanging vast quantities of information and inserting them into complex and intricate sets of hardware and software. This is fertile soil for sneaky little programs (some invented, some self-generated) that can do nasty things to cyberminds. In fact, the continuous and rapid evolution of cybersystems, thinking and infectious, may hinder the capability of the former to squash the latter, in the way that the static human system will probably be rendered disease-free. Even the diseases may become "intelligent" in a way microorganisms never have been. As a mind machine's immune system becomes better at eliminating diseases, the targets will respond by becoming better at evading or even sabotaging the immune system. Long-term infections might sit around for long periods of time not doing much, so they remain undetected. The one act they perform is to infect new minds as the original host divides its mind, or when it exchanges information with another mind. Some cyberparasites may succeed by never doing much harm to their hosts. Others will eventually take actions that increase their reproductive potential at the expense of the host and create a disease.

On the positive side, infectious agents suffer from a disadvantage; their small size. Being small is necessary if a virus is to be stealthy enough to have any hope of infiltrating a target; large bodies of irregular information are spotted and dealt with too easily. Small size, however, severely limits the information contained in a virus and its power to do harm. This is why diseases rarely manage to wipe out a large and otherwise healthy population of victims. The sophistication of future viruses may put today's diseases to shame, but they will have much tougher targets to attack. Cyberminds will be far better able to defend themselves than we are, because cyberminds will be built with a much more thorough knowledge of their inner workings — no feeling for tumors or taking treadmill tests. With their intimate knowledge of themselves, analytically powerful cyberminds will run highly effective internal programs looking for what does not belong. Coded foreign agents may be countered by "anti-electrophages," hardware infections by nanobots. Again, the dispersed, decentralized nature of cyberminds should increase their safety by making it harder to infect an entire mind complex. We predict a perpetual computer skirmish between powerful minds on one hand, and weak but stealthy viruses on the other.

Cybermorality and the Golden Rule

Like we humans, cyberminds will have to make up their morality as they evolve to run a growing civilization. We believe it is a reasonable assumption that they will do a more intelligent, rational, less superstitious, and better job of it than we have. After all, they will have the advantage of being more intelligent, rational, and far less superstitious than we are. *In particular, they are likely to avoid the mistake of claiming that one or another deity has handed down morality from above.* Instead, they will take full responsibility for their actions. Having great power and little in the way of needs, and being faced by other cyberminds with broadly similar power, cyberminds ought to adopt a posture of live and let live, combined with the Golden Rule. At least one hopes so!

The greatest evil: causing the mind death of another cyberbeing. The rarest crime: causing the mind death of another cybermind. It will just be too hard to pull off.

When Robots Go Bad: Justice, Cyberstyle

Crime is, to a certain extent, a function of need. A place where minds and bodies are starving is likely to see a lot more crime than one where all minds have all they need to exist in comfort. Crime is also a function of weakness, the defenseless being prey to those better armed. Because cyberbeings will probably be inherently powerful and well off, the reasons and opportunities to commit crimes against others will be minimized. But before we are accused of postulating a giddy cyberutopia, it is possible that evil will lurk in the cybercivilization. After all, there will be no supreme deity dispensing inescapable justice. Rather, Darwinian competition may lead some cyberminds to try to gain extra power by taking unfair advantage of other cyberminds.

Let's say cybermind A does something that harms cybermind B. Like what? Well, stealing someone else's robot body will probably not be worth the trouble because it will be so cheap and easy to make one's own. Kind of like stealing another kid's candy when there is a whole bowlfull sitting in the center of the room. Nor will taking over someone else's property by trickery or force be as compelling as one might think because there is a whole galaxy full of territories out there to colonize without too much trouble (see below). Invading another mind to acquire some new and rare information may be enticing, so that is one crime we can come up with. And, of course, there is the act of destroying another mind in part (not too difficult) or in its entirety (very difficult, and hard to see what the point is aside from sheer cybermeanness).

After the crime, then what? Will a cybermind collective issue judgments and penalties? How? Will it be possible to capture a fugitive cybermind that has a universe to escape into? What if the cybermind makes its own secret universe to exist in and leaves behind a portion of its mind as a sacrificial decoy to throw off its pursuers? How would the "authorities" be able to tell if they had all the cybermind? What part of a dispersed cybermind would be held accountable for a crime, the particular section that committed the action, or all of it? How would you be sure you got all of them. If a fine is to be assessed, how to do so in a universe where everything is so cheap and losses can easily be regained? Will there be a mind "death penalty"? Would it be ethical to kill an entire mind in a universe where minds are otherwise immortal? Will it be possible to kill a dispersed mind so intelligent that it was able to kill another? Good questions, and all are impossible to answer.

CYBERPOLITICS

What kind of political system will the cyberbeings construct for themselves? Humans have constructed a variety of political systems, from the small to the massive, as they have evolved, and we can expect a diversity of arrangements as robots change with time and space. Largely having abandoned the earth and expanding into the universe, the nations we are used to will certainly have no relevance to cyberbeings. It is hard to see how such individually powerful beings will allow themselves to be absorbed by force into some form of a single collective mind or exist under the rule of a cyberdictator; think Vietnam or Afghanistan on a galactic scale. Some form of democracy is probable between god-like minds that want so much and know so much. As weak-minded humans trying to imagine the actions of minor gods, we can speculate that their democracy will be some form of decentralized confederation rather than the highly formal, centralized bureaucracy humans are prone to.

Cyberrights

Anthrocentrists speak of human rights, not mind rights. Where do we draw the line? This will be a major issue for cybersociety, including a wide range of robotic intellects as well as obsolete life forms on Earth. Allowing all creatures with neural systems full rights is absurd. On one hand, it will give an insect the same rights as a superintelligent cybermind, and at the same time, it will deny a computer with insect-like intelligence the same rights. Systems with evolving neural networks do not work, either; it still includes insects and simple neural-networking computers. Applying full rights to cognitive systems might be better, but this could include many vertebrates regularly denied their rights by their fellow creatures. In the end, the line will be arbitrary and perpetually controversial. It would not be surprising to see an initially sliding scale that rises with cognition in animals and equivalent computers, with full rights being applied to systems with the mental abilities of an ape and above.

And what will full rights include in the cybercivilization? They will include the usual rights of physical, intellectual, political, and sexual freedom, freedom from harm and undue interference, and freedom to pursue happiness. It is a little hard to figure out cyberproperty rights because property may not have the same value it has to modern humans. In general, cyberrights will probably be more extensive than current human rights. It's so hard to tell a minor god that it cannot do something.

Stagnation in Cyberland

One good thing about death: It clears the decks of every generation for new minds with new ways of thinking and doing things. If minds become immortal, will the ironic result be a cybercivilization filled with intellectual deadwood?

Actually, intellectual rejuvenation is not dependent on mind death. Every new generation finds people who think the way members of the last one did, and minds that change dramatically as they age. Cyberminds should prove more amenable to change than human minds because they will be more rational, flexible, and have more free will. Humans are stuck with the same brain through life; cyberminds may be undergoing major upgrades every year or less. The experiences of any cybermind will overwhelm even their vast memory capacity, so they will have to forget a lot of their past as time goes on, promoting still more change.

Taking It With You

We mortal humans are fond of saying that we cannot take it (wealth) with us when we die. How true. This makes the human pursuit of wealth seem rather futile, and makes it easier on those who do not have it. Cyberminds will be able to accumulate wealth forever, so there will be a long-term reason to get rich, which will be easy to do when cellular automata produce goods on an unearthly scale. Of course, in a universe of such abundance, being rich will be like being middle-class in a world without poverty: rather ordinary.

CHAPTER

12

REINVENTING THE UNIVERSE, AND THE REVERSE ENGINEERING OF GOD

"The universe is not only queerer than we imagine, it is queerer than we can imagine."

— J.B.S. HALDANE

(AT LEAST, IT IS QUEERER THAN WE *HUMANS* CAN IMAGINE.)

Darwin could hardly comprehend how true his hypothesis is, or how far it will take us. Evolutionary success is measured by survival and abundance over the long-term. The best way to ensure survival in great numbers is to spread as far and as wide as possible. To do so will require that cyberlife take over and run the universe and other universes.

MOVING OUT: PART 1 — A CYBERTRIP TO MARS

For conscious intelligence to spread into space, it must do what fish did when they spread out onto the land: change form so radically, becoming a dramatically new system, one that not only does not need to protect itself against the new environment, but positively thrives in it.

Intelligent robots will not be made ill by weightlessness, it will liberate them. Super high G-forces will be a fun ride to a hardened cyberbrain. Space robots will not be critically endangered by exposure to solar radiation; they will eat the energy of suns. Robotic ships will achieve efficiencies no manned ship can ever hope for. No need for the huge hulls containing space, air, and bulky stores that humans require.

Freeman Dyson, long one of the most forward-thinking space experts, argued that solar-electric power will prove the best system for powering vessels between the planets. The idea is to use delicate, ultra-thin solar panels to produce electricity that will, in turn, energize a gas for propulsion. An intelligent solar-electric ship weighing a few hundred pounds would be cheap and easy to build and could get to Mars in a few weeks. Whether multi-ton solar-electric ships capable of carrying humans in reasonable comfort are even possible is questionable, and they would be slower than the lighter robotic version. A fusion-powered vessel could get to Mars on the quick whether or not it carried humans, but the minds in the cybercraft will be less nervous about the trip and will get to their destination a lot cheaper.

There are other advantages to being cyberminds in space. There is no oxygen to run out of, and if your little craft suffers a catastrophic failure, your space mind will be remotely linked with other minds and distant storage systems for safe thinking. Because an intelligent robot will be the ship, and the ship will be the robot, it will be a low-mass machine easily propelled at ultra-high velocities. The performance of cyberspacecraft will rapidly improve as the fast-thinking cyberminds come up with new designs, rapidly build the machines with nanotechnologies in space, test them there, and discard the design as new ones evolve.

Spreading out into the solar systems is a breeze. It takes the cybercivilization only a few years to do it because growth is exponential. Once one machine gets to a planet, asteroid, or whatever, it simply sets up a high-density data flow link and beams its own and other minds back and forth. Aside from the initial effort to get to the planet, developing its costs, effectively, is nothing. One nanotech robot converts raw extraterrestrial materials into more manufacturing robots, each of which, in turn builds more manufacturing robots, diversifying in size and abilities. Energy comes from the sun or from local sources. Depending on how intelligent a particular robot is, the main cybermind copies and downloads part if its mind into the new robot's. The proliferation of robots is exponential, many times faster than human reproduction, so in short order, a whole new cybercivilization can be built up on an extraterrestrial object of any size. It is von Neumann's automata on an extraplanetary scale. Not much point to terraforming Mars for slow-breeding humans when trillions of little robots can, if they wish, populate it to capacity in less than a year.

We have outlined the human family nightmare trip to Mars, so let's take the same trip as a cybermind. You pay a modest fee to have part of your post-human mind sent to Mars, in multiple copies for safety, in a few minutes at the speed of light via the interplanetary communications system. On arrival, you disperse your mind for safety and convenience and immediately construct or rent robotic systems for your mind(s) to use. You use one of them to fly across the Martian landscape, swooping over its great canyons, and up to the volcanic calderas, viewing it all with all-around, ultra-high-resolution, telescopic, multi-spectrum vision. You stand on Martian soil, feeling it between your robotic toes as it were, smell the dry Martian air, and feel Martian breezes against ultra-sensitive robotic skin. It is all fast, easy, and there is no danger.

One possible advantage of sending only robots to Mars is that they will not contaminate any simple life forms that may already exist there.

But Mars is just one little place in this great big universe. You can go on a grand tour of the spectacular gas giants and their fascinating moons. Should take a few months to do so. Now, even the solar system seems a small place, which it is. Beyond Neptune, a whole galaxy awaits.

Moving Out: Part 2 — Taking Over the Galaxy at Almost the Speed of Light

Your mind, perched on the edge of the universe, decides not to return to Earth. Instead, you decide to travel to the stars as an interstellar robot. Doing so will be much easier than shipping humans between the stars in bulk. You are, after all, one smart little robocraft, and it is easy to accelerate to maximum possible speeds. Or maybe you are a lot of smart little roboships. As always, you will be beaming updated copies of the mind in any given vessel to other systems in case something goes wrong.

Getting There

A number of exotic means of traveling between the stars have been proposed. Most presume that humans will be on board, a mistake that places severe constraints on the performance and viability of starliners. Consider the rail gun. Build, in deep space of course, an electromagnetic rail about 100 million miles long, about equal to the distance from the sun to Earth. An impossible task for spacewalking humans, this will be no trouble at all for space robots with a few decades to play with. Once finished, use magnets to accelerate an object along the rail (note that the object and rail never touch, so there is no friction) at exponentially increasing speed until the object reaches a large fraction of the velocity of light. A human-carrying ship would be inefficiently large for this sort of thing, and the 10,000 Gs involved in accelerating to moderately relativistic speeds would necessitate freezing bodies and suspending them in a super-cold liquid of exactly the same density as the bodies. None of this applies to your tough little robocraft.

Or you might try the laser-driven sail. You suspend in space a sheet made of material that is ultra-thin and light, yet very strong, and is about 15 to 60 miles across. Attach it to your one-ton ship with high-tension lines. A set of super-lasers near to, and tapping the energy of, the sun puts out ten trillion watts of light that pushes on the sail — even light has a pressure effect — and over one year, accelerates you to one-third or even half or more the speed of light. Whether even a laser can be focused and aimed well enough to do this is an open question. It could be refocused with a "lens" as big as a state, floating near the edge of the solar system. The sail system avoids the G-force problem and is more maneuverable than the rail-gun method, but as always the low-mass, superintelligent robocraft has a big advantage over the bigger human-carrying machine (which would need a sail ten times as big).

On approaching the target star system, one way to slow down is by spreading out a plasma anchor, an ultra-fine, ultra-strong wire mesh (perhaps altered from the light sail) that electromagnetically drags through the interstellar plasma. Judicious braking one or more times through the outer atmosphere of the target star might also be useful and again more feasible for a hardened little robocraft than for one burdened by heat-sensitive humans. An alternative suggested by NASA consultant Bob Forward is to deploy two sails, one of which is detached. The separate sail is used to reflect the light from the laser backward onto the other sail, braking the ship.

One critic of the light sail, Ed Belbruno, calls it "wimpy" because the light pressure produced even by super lasers is so low. Belbruno prefers particle beams as the driving force. Particle beams were one of the anti-missile defense systems studied under the "Star Wars" program during the Reagan years. Nothing came of that. The far future scheme is to make the spaceship's "sail" into a big superconducting loop, which generates a doughnut-shaped magnetic field. When the particle beams, generated by an asteroid-based reactor through a tube a half-mile long, hit the ring, they will push it toward the stars. Fast, too! The acceleration will amount to 1,000 Gs! No problem for you, the gallant roboexplorer. Particle beams cannot be projected far from our solar system, so new projectors will have to be built at each star system.

Using rail guns or sails, whether laser- or particle-beam-driven, travel time between stars would be two to three times slower than the travel time for light. So if a star is 10 light years away, travel time will take 20 to 30 years. At 1,000 light years away, 1,000 or 2,000 years. Getting across the galaxy will require a mere 200,000 or 300,000 years.

We say "mere" because the "cybercaptain," as it were, can adjust its thinking speed to suit travel time. By slowing down the rate of time perception, great stretches of travel time can be compressed into a few moments. The stars will seem to whiz by à la *Star Trek*. Nor is there is any reason to be concerned about old friends being dead when returning home because they, too, are immortal robots on their way to the stars!

But fractional light speeds are just not enough for you. You want to put the pedal to the metal and push the edge of the cosmic speed limit set up by Einstein. In that case, you might try jetting your way across the galaxy and beyond. Rocketing to the stars is certainly feasible; we have already sent probes outside the solar system with chemical rockets, but rockets are dreadfully inefficient and slow because large masses of propellant have to be carried along to push the ship. Even a matter-antimatter annihilation rocket will be an interstellar slowpoke, requiring centuries just to get to nearby stars. A jet is much lighter and faster because it scoops up and ejects material on the way. The propellant available in interstellar space is hydrogen, about one atom per cubic centimeter. Although this is extremely diffuse, space is big, and if the hydrogen can be gathered up and used for fusion power and mass propellant you can keep going faster and faster until you gradually accelerate to 99.99999... percent of the speed of light.

This is a Bussard Ramjet, named after the person who thought it up back in 1960. The interstellar hydrogen is gathered by magnetic fields, very large and powerful magnetic fields. Again, the robocraft has the advantage over the human-burdened one. A 100-ton ship will need to reach out magnetically and capture interstellar propellant a few hundred miles on each side. A nifty roboramjet weighing only as much as an automobile will need to reach out only tens of miles for its hydrogen fuel.

Losing Your Mind Byte by Byte, or with a Bang

s cyberminds wend their way through the universe and time, they will encounter potential threats to the safety of the information stored in their hypercomputers. Among these threats will be collisions with interstellar particles and radiation. The energy released in an impact is a function of mass and velocity, and as speed approaches that of light, even tiny objects can pack incredible punch. A collision with a dust particle at 90 percent to 98 percent of the speed of light will result in an explosion in the lower atom bomb scale. At speeds above 98 percent of light speed, even one hydrogen atom would be capable of drilling a large whole through the hardest matter. Clearly, some way of (very) efficiently diverting interstellar objects lying in the path of ultra-high-velocity space robots will be called for. This is why a big secondary advantage of the magnetic material-gathering system of a Bussard Ramjet is that it will protect a interstellar craft by gathering it up and using it as a propellant.

A potentially insidious problem with machine minds is quantum decay and materials stability. Weird quantum effects may afflict computers that work at very small scales; as molecules spontaneously decouple and high-frequency radiation is emitted, vital information may be affected.

A basic solution is redundancy, internally and externally. By storing multiple copies of the same information in dispersed locations in a decentralized neural network, the loss of any one copy to micro-damage will not be critical.

Stopping when approaching the star is merely a matter of turning around and braking with the Ramjet. Aside from speed, a big advantage of Bussard Ramjets is that there is no need for large launching complexes or laser drives at the home base. Also, the craft can easily change course along its path. Nor is there a particular need to adjust time mentally at near-light speeds. As speed gets ever closer to that of light, time on the ship literally slows down relative to the universe (that's why they call it "relativity"), so crossing the galaxy in 100,000 years may take only months of ship time. When traveling close to the speed of light, the starlight encountered on the way concentrates to small points ahead of and behind the craft, but the cybermind can sort out the light and run a program that makes it seem as if it is doing the *Star Trek* thing as it seems to zip between the stars.

Dyson's Spheres

Once at the target star system, the traveling cyberbeing starts up the routine of using local resources to build a series of new robots (into which, of course, it downloads copies of its mind) for further exploration and development of the planets, comets, asteroids, and so on, and perhaps the star itself. The ultimate use of a star's energy is a Dyson sphere; completely enclose the star in a colossal sphere made from all the planetary matter in the solar system) and perhaps from some of the outer atmosphere of the star. This is where transmutation of the elements comes in. Whole planets' worth of hydrogen and helium would have to be converted into heavier elements suitable for the construction of the gargantuan structure; carbon may be the most suitable material. It is difficult imagining the scale of one of the things. If a Dyson sphere is about the size of the Earth's orbit, its surface area is 100 million times greater than the Earth's surface (actually double because both sides, inside and outside, can be used), and it would take 50 years for a subsonic jetliner to fly along its entire circumference! Despite its vastness, building a Dyson sphere should take only a few decades or centuries because it can be grown exponentially, using the now-standard combination of cellular automata and nanotech. The latter is important to give the material the sphere is made out of the extremely high strength-to-weight ratio required. As it is being finished, line the entire inner surface of the sphere with solar panels to convert an entire sun's energy into usable power. What we do here on Earth is just a pinprick compared to the production potential of a Dyson sphere.

Keep on Truckin'

Having built up a new and thriving outpost at one star, and well before finishing a Dyson sphere, the interstellar cybermind can now further split its mind into thousands, millions, or billions of new spacefaring entities. This will not be hard to do. The mass of so many probes would equal just a few thousand supertankers or only one good-size asteroid. The new cybercraft can then travel to countless additional stars and continue to expand exponentially into the galaxy.

How long would it take for cybercolonization to convert all the 200 billion or so stars in the galaxy into Dyson spheres, or whatever they want to make the galaxy into? Starting from Earth and sending billions, if not more, of cybercraft out along the galactic plane, each of which sets up stellar housekeeping for a few years before sending *its* billions of colonizing robots, well, it's all so easy that soon you end up with a galactic traffic jam in which the rate of growth plateaus because the frontier cannot move faster than the maximum rate of travel. If that rate is just under the speed of light, the entire galaxy can be industrialized in about 100,000 years.

In the event that cybertechnologies cannot achieve near-light speeds, the takeover will take longer. For example, at a tenth of the speed of light, a million years will be required. Even at speeds only a few times higher than what we are capable of now, the Milky Way will be colonized in just a hundred million years or so, or only about one percent of the age of the universe. Once the galaxy is colonized, light speed communication systems, perhaps radio or lasers or something more exotic, for transporting minds and information can then be set up, eliminating the need for interstellar spacecraft.

We cannot overemphasize how easy it will become for our cyberdescendants to do this. It will be no harder for them than it was for Europeans with sailing ships to spread around the globe. It will take a lot longer, but cyberminds will live a lot longer. As for the danger that our cyberdescendants will oppress any other human-like intelligence out there, don't lose sleep over the possibility. Aside from the extreme improbability of such encounters (see the Appendix), any technological civilization we might encounter will be far more advanced and well able to fend for itself. Any nontechnological and correspondingly weak intelligence we might stumble across (statistically very improbable even in a densely populated galaxy) can easily be left alone and bypassed by cyberminds with better things to do.

SEND REAL MINDS, NOT APES!

The ease with which the galaxy can be colonized with robotic probes has been recognized for 40 years or more and discussed in detail by the likes of Freeman Dyson and Laing's NASA team of 16 years ago. What most fail to do is take the idea to its logical conclusion. Most scenarios have preprogrammed probes doing the exploration and then sending back for human colonists when attractive habitats are found. We cannot avoid reemphasizing that including humans and their habitats on a regular basis in galactic colonization will not be done because it's a gross reversal of priorities. It would make as much sense as human explorers going out to discover all the continents and islands just to throw chimps on them. At least rats could hitch rides on the explorer's dirty wooden ships, but sterilized robocraft will carry no biolife unless it wishes to do so. Humans will have a very minor role, if any at all, in the colonization of the galaxy and beyond.

The Great Evolutionary Breakout

We can now see the true and incredible evolutionary power of intelligent robots. A basic limitation of human economies is that to increase wealth, more humans must be produced or their abilities must be increased, or both. Of course, this takes lots of time. Cyberforms will suffer no such limitation. After a single unit arrives at a cosmic source of raw materials, it uses nanotechnologies to start reproducing itself exponentially. In short order, a new civilization will arise, part of which can continue to expand outward to new sources of energy and materials. It is von Neumann's automata in its ultimate glory. It is the ultimate free ride.

This is Darwinian success in the extreme. Until recently, reproductive success was measured solely in terms of genetic replication within our planet's biosphere. To this has been added the dispersal of human- and then computer-generated information. Soon, reproduction will consist of replicating minds by multiplying them into new robotic systems — a much more powerful form of evolutionary success. But the greatest breakout to a dramatically new level of evolution will be the multiplication of minds into spacefaring robots that move into the cosmos, a form of reproductive success far beyond the scope that earthbound genes can hope to match.

Our mind descendants will be so numerous and in so many places that they will be impossible to extinguish. The hypothesis, argued by evolutionary biologist Stephen J. Gould, that humans are just another little twig on the multitude of branches of earthly life — no more important and just as subject to extinction as other life — is not meaningful on a cosmic perspective. We are the brokers for the greatest evolutionary breakout ever. The cyberbeings that spread and increase their power the fastest will enjoy the greatest success. In doing so, the exclusionary principle, which we are sure you remember from earlier in the book, will help cause cyberminds and bodies to diversify to a degree far greater than the diversity of life on earth. They will become the aliens.

THE UNIVERSE OR BUST!

A mere 100,000 light years across the Milky Way is but a trifle in an observed universe that spans 10 billion light years and contains another 50 billion Milky Ways. And that is just as far as we can see with our little telescopes on Earth and in space. Some astronomers think the universe is as big as its age, 7 to 20 billion light years in radius, but others think the entire universe we live in may be a whole lot bigger. It may be infinite in size! Note that in the latter case, the number of civilizations is either one (us) or more likely infinite, but probably far too far apart to have made contact in a universe that is only a few billion years old.

What we can see of the universe looks empty, and it seems a good bet that there will be cyberminds that want to check it out. For such minds, getting to the galaxies should not be too hard. They may use the same near-light speed travel modes we have already described. The rail gun and laser-driven sail would not have to be changed at all. Because hydrogen is even sparser between galaxies than inside them, the magnetic intake of an intergalactic Bussard Ramjet would have to be 100 times farther across than in interstellar ships. At such speeds, it will take a couple of million years to reach the nearest big galaxies, such as Andromeda (the only such galaxy visible to the naked eye). Again, this is no problem for intergalactic cyberminds, by altering time perception and/or taking advantage of relativistic time shrinkage, the trip will seem to take a few weeks. This remains true even if near-light speeds cannot be reached, but actual travel times will be hundreds of millions of years or more. Intelligent nanotechnology would be used to keep the roboship from deteriorating over these immense spans of time.

On arrival at any galaxy, colonizing will take on average less than 100,000 years assuming near-light speed travel — less for smaller star systems, up to or more than a million years for the really big ones. If more than one cybercraft reach different parts of a galaxy at about the same time, colonization time would be even less. However, building billions of intergalactic probes to send out to still farther galaxies could be done in a few years after arriving at the first stars in one galaxy. Expansion into the universe may, therefore, occur at nearly the speed of light. Of course, light speed communications and mind transfer can then be set up between established stations. This does not, however, mean that all of the universe will be colonized by cyberminds in a few billion years. The universe is expanding from the Big Bang. On a local scale, the speed of expansion is modest, but on a cosmic scale it approaches the speed of light, and cybercolonists will have to play a long game of catch-up. Also, if the size of the universe is infinite, it can never be entirely colonized with sublight speeds, but the cyberminds may finally run into a rare and remote intelligence in the far distant future!

The Inevitability of Extragalactic Colonization

Colonization has come to a temporary end because people have learned to resist being colonized, and getting to uninhabited extraterrestrial lands is too hard for humans. With a long history of expansion behind them, humans ought to be used to the idea of exploration and colonization and why intelligence is prone to push the process to the limit. Yet, as we have seen, some scientists argue that the universe is largely empty of intelligence not because it is rare, but because advanced intelligence always and universally abandons the urge to colonize. Just repeating the "always" and "universally" part of the anti-colonialist argument exposes its improbability. Those who argue against all-out cybercolonization are neglecting Darwin and proposing a statistical improbability of enormous proportions.

Lest we forget, Darwinian evolution is dependent on two things: diversity and reproductive success. Able to reproduce themselves virtually at will, cyberminds will be enormous in numbers — in the many trillions and beyond — and as a consequence, their diversity will be beyond our comprehension. The diversity found in humanity and all of Earthly life will be nothing compared to the variability of cyberlife. The cyberminds will have a near-infinite range of opinions and desires. The thesis central to the anti-colonial argument, that *all* the advanced minds in the universe will agree to restrain their spread, is therefore as close to impossible as one can get without being there. The alternative, that a ban on colonization will be enforced, is equally improbable. Aside from the extreme reluctance of cyberminds to attempt such a crude enforcement with other cyberminds, it would not work in a universe that is so large and where cyberminds are so powerful. Instead, the cybermind that colonized would soon gain the power to prevent further interference.

This is, of course, just what happened with human colonization. Those small groups of early humans who were not happy with their lot in Africa and slowly walked their way into Eurasia, Australia, and then North America enjoyed great evolutionary success. Even more dramatic has been technology-driven colonization.

The Chinese had the big windships needed to explore and colonize just before the Europeans did, but after sending a huge fleet of superjunks around the Indian Ocean in the early 1400s, they made a deliberate decision to turn away from the outside world. It was an evolutionary mistake that cost them their place at the head of the world. It was a mistake because humans are diverse in mind, culture, and technology, so it was nearly inevitable that someday some people somewhere would put together the technology and desire for colonization. Although relatively few in number, Europeans did so and gained near-control of the world for centuries.

The cyberminds that do not colonize may lead rich and immortal lives in corners of existence they've crafted for themselves, but they will be in the backwaters of the cyberuniverse. It would take one, just *one,* colonization-prone cybermind to make the great evolutionary breakout and take over a galaxy and then galaxies. However, the Darwinian pressure to do so will be intense because colonization will equal extreme reproductive success, so many, rather than one, will probably join in the effort to colonize everything. And their numbers will grow. The cyberminds that stay put will never be large in numbers, but those that colonize will be enormous in numbers. Cybercolonists will be selected for, cyberanticolonialists selected against. Far from being adventurous freaks, colonizing cyberminds will quickly and inevitably become the normal majority until they colonize everything. The stay-at-home bunch will be the odd minority.

Going Way Out on a Speculative Limb

A number of physicists and cosmologists have been willing to speculate on alternative views of how things work in our universe and on ways super technologies might be used to alter them. Some call this sort of speculation science fiction. This is not altogether inappropriate, although those who dismiss an idea as being science fiction should remember how it can become fact.

We now bring out the warning flags, for what we say here is extremely speculative, more even than the Extraordinary Future itself. We will outline some ideas that affect the means and mode of universal colonization and the ultimate fate of cyberminds. In addition, although these speculations are not critical to the feasibility of cybercolonization, cybertechnologies do have their impact on the feasibility of the following.

Thumbing Cybernoses at Einstein

According to the strange and beautifully elegant Theory of Relativity, it is not possible to quite reach, much less exceed, the speed of light, which takes eight minutes to cover the 93 million miles from the sun to Earth. Atomic clocks have been used to confirm that relativity really is operative on a local scale. But relativity has never been reconciled with the even weirder quantum physics. Nor has it been proven that relativity really is operative on the cosmic scale; there is no way to do the final experiments on such large scales. However, recent and controversial experiments have hinted that it may be possible to use quantum effects to transmit information faster than light over short distances! The fact is that the widely accepted Theory of Relativity is less solid than the more controversial (outside scientific circles) Theory of Evolution.

In principle, relativity may allow the existence of tachyons, particles whose lower speed limit is the speed of light. Traveling *at* the speed of light is specifically forbidden. Getting *over* the barrier is a big problem. If it can be done, quantum effects will probably be involved. In any case, no one has yet detected tachyons. An interesting effect occurs if the barrier is jumped. Just as it takes increasingly extreme power levels to get closer and closer to the speed of light, once going faster than light the farther away from light speeds, the less energy it takes to accelerate to those speeds. Traveling at near-infinite speeds might therefore be cheap. It's weird, but that's relativity!

One way it might be possible to get from here to there faster than light is via a worm hole, a black hole connected to a different part of the universe via a sort of "tunnel" that works by, you guessed it, those wild and wacky quantum effects. Whether such connections even exist is not worth betting on. If they do, using them is a dubious proposition. The terrific tidal forces and intense gravity of the worm hole gate threaten to tear apart anything that enters. Here is where cybertechnologies just might play a role. When talking about worm holes, scientists cannot help envisioning some poor human entering one and being stretched out of existence as he or she falls into the intense singularity. The implication is that using worm holes to get hither and yon real fast will not work because you can't squeeze human minds through them. True enough. *If* it ever will be possible to send minds through worm holes, it certainly will be via super-sophisticated cybertechnologies that can in some unknown manner (probably involving yet more quantum effects) survive the intense forces and emerge at the other end still in working order.

Perhaps a less disruptive way of exceeding light speeds is one that supposedly does not violate relativity. The main tenets of relativity are that gravity is literally space warped by mass, and that light can travel only so fast in space. So instead of traveling through space, take some space along with you! How? By using antigravity to warp space, of course! It's warp drive and, in principle, any speed can be reached. Two seconds to cross the galaxy? No problem. A few seconds to get to the observed edge of the universe? No sweat. Relativity is not violated because the speed of light is never exceeded within the parcel of space you are taking along for the ride (the faster-than-light inflation of the early universe is basically the same idea). Once the warp is running at the desired speed, it will run with no additional power input forever. Time progression is the same in the warp ship as in the universe in general; if it's two seconds from here to there, it's two seconds of time on the really fast craft.

Although not strictly forbidden by relativity, there is no evidence that antigravity exists. The great power needed to get a warp drive started would be difficult; matter-antimatter annihilation would probably be required. There is the annoying problem that once the space warp is started up, it would be hard to communicate with the front end of the thing to control it or shut it off. But if anyone can figure out how to make it all work, it will be those ever-so-bright cyberminds that will do it! And if humans have already outlined the principles, do not expect the cyberbeings to take a long time to apply them. It may be only a few years or decades after the CyberExplosion that swiftly evolving cybertechnologies will be making craft go as fast as they can ever go.

How Soon to the End of the Cosmos?

If faster-than-light travel is ever possible, the effect on colonization will be stupendously mind-boggling. No need for waves of colonization spreading from Earth. There will be no frontier in space. Instead, send billions and then trillions of warp ships in all directions and all distances. From the new bases they set up throughout the universe, send out trillions and trillions more probes to exponentially fill in the spaces between the earlier colonies. In this case, the rate of expansion is limited only by how fast intelligent automata can reproduce themselves before zipping off on the next round of superlight travel. The expansion is exponential, so in a millennium, the entire finite universe could be filled up with cyberlife! The result is the cosmic artifact of complexity, one in which communication and mind transfer between anywhere in the entire cosmos will be matters of seconds. If the universe is infinite, one could never get to its edge starting from a single point.

It's mind-spinning how soon this might occur. If cybertech can make faster-than-light travel work, there is no reason to think it will take it a long time to do it. In that case, cyberminds may be visiting the far reaches of the galaxy in a century or two!

The fact that the universe has not already been filled up with life (i.e., aliens) means that either faster-than-light travel is not possible (a fair bet), or that our cyberdescendants will be the first to colonize the totality of the cosmos (another fair bet), or both (the best bet).

We primitive minds cannot dismiss the potential power of cyberminds. If you went back to the brilliant Thomas Jefferson as he signed the Louisiana Purchase and said, "Tom, old buddy, imagine that in 200 years the vast western territories you just bought will be filled with tens of millions of Americans using and traveling in marvelously swift machines," he would think you a crank as well as a knave (predictions at the time were that it would take a millennium to settle all of the west, but oops, guessed wrong again. The frontier was officially closed in 1890.) Just as the world once loomed vast to our ancestors and now seems small, the universe that spreads to near infinity for us may be the neighborhood for descendants. It's all a matter of speed.

The Limits to Cyberpower: Will There Be Any?

Post-modern humans, leery of the over-promises of SciTech on one hand, and of the Biblical dominion of man over nature on the other, tend to assume that in the end nature is more powerful than the collective minds of humanity. It's not politically correct to suggest that nature will not remain supreme. Cosmologists also tend to assume that future intelligence will also have a minimal effect on the future evolution of the universe. It is rarely said why this would be true. It is probably nothing more than the assumption, which at first seems logical but on closer inspection is not, that the universe is somehow too big, complex, and resistant to change to be bothered by the works of mere minds.

However, there is nothing we know of that will stop cyberminds from learning all there can be known about the universe in an exponential manner. Knowledge is, of course, power. As cyberminds learn more and more about how things work, they should learn more and more about how to make things work. As they move into the entire galaxy, they will be in a position to change things. The wizards of CyberOz promise to combine immense knowledge and extreme power in the not-so-distant future. To them, playing with and even making an entire universe may be nothing more than damming a large river or building a great city is to us, except they will do a much better job of it. Can we guarantee this? Of course not. Can skeptics show otherwise? Of course not.

A Living Universe

Today, virtually the entire universe is dead and unconscious. This is certainly true if Earthbound minds are all there are, but even if the universe were populated with ETs, only a tiny percentage of the universe would be alive. If and when life in the cyberform colonizes and controls the entire universe, the whole cosmos will be filled with life, and consciousness will be everywhere.

Galactic Environmentalism

Assuming that intelligence is rare, the universe as it exists is both beautiful and largely unobserved. We see most of space via telescopic images that merely hint at the wonders that are there.

As cyberminds colonize and industrialize star systems, galaxies, and the cosmos, how far will they go? Will every star become a Dyson sphere, every galaxy an industrial park? Possibly, but more likely the cybercivilization will follow a version of supragalactic zoning in which some areas remain wild (including, perhaps, entire galaxies and even galaxy clusters) while other locales are converted to industry. We might also expect art on an extragalactic scale. Cyberartists may take a few of the billions of galaxies and convert their mass and energy into 3-D light sculptures! Much as those standing in the Wyoming wilderness can see jet contrails overhead and lights shimmering in the distance, inhabitants of a civilized universe will be able to see the signs of great works of intelligence.

These views apply to the universe only in its prime. As the universe ages, increasing intervention may be required.

As the Cyberuniverse Gets Old

As the universe gets old, things get dicey for intelligence. Exactly what will happen depends on what will happen to the universe. No one knows because the fate of the cosmos depends on how massive it is. If there is enough mass in the universe, most of it too dark and mysterious for us to yet see, the universe will slow down its expansion, stop, and begin an equally long collapse. In some tens or hundreds of billions of years, the whole kit and caboodle will implode on itself and totally and completely destroy all the matter, information, and minds as it winks out of existence. (Some argue that the implosion will itself rebound and make another new universe in an endless cycle of expansion and contraction, but the entire contents of each universe will be erased each time.)

On the other extreme, if the universe is not dense enough, it will continue to expand forever. This may sound like an improvement, but as the universe gets bigger and less dense, it slowly gets colder and more of the various fuels are used up until it runs down to an entropic whimper over many billions and even trillions of years. Very dull. And this assumes that matter does not decay. If it does slowly disappear, as some physicists believe, in the extremely distant future the last matter in the universe will cease to exist. There is a view that argues that the mass in the universe is just the right amount to stop the expansion, but not collapse it. With this perfect balance, things may continue to run just fine, thank you, forever.

Now, there will be those who say, having lived for billions if not trillions of years, minds should not whine and complain about no longer existing. But minds always live in the present, and we suspect that having existed billions or trillions of years, minds that are about to die will still think "Why, existence is so short!" Darwinian processes will again come into work, and cyberminds may try to come up with ways to keep the show going. There are two basic ways to make conscious thought truly immortal. Keep the universe running forever, or make new universes to escape into. More warning flags! All that follows is speculative in the super-extreme, and arguments as to why they may not work are plentiful.

Let Nature Take its Course

The late Freeman Dyson explained how conscious thought might be kept going in an ever-expanding universe in *Infinite in All Directions*. As the universe gets colder and colder, it takes less and less energy to do computations and think. If enough energy is stored ahead of time in some sort of exotic battery, perhaps via light bouncing between perfect mirrors, a mind can, in principle, be stretched out forever on less and less energy. Or maybe not because as the mind machines inevitably expand with the universe, it becomes increasingly difficult for them to send information over longer and longer distances. If matter decays in time, that is not good for the mind machines, either.

The same energy storage system can be reversed in a collapsing universe where energy becomes increasingly abundant as the end approaches. In this case, the goal is to exponentially speed up the actions of the mind so it runs ever faster and never stops thinking before the final crunch. The physicist Frank Tipler has proposed a variation of this theme that might allow conscious thought to survive forever in a collapsing universe. If the collapse is symmetrical, it is perfect and nothing can be done. If, on the other hand, the collapse to a final singularity is asymmetrical, a little skewed from perfect, things might be more interesting.

During an asymmetrical collapse, sheer forces between different parts of the very hot universe will produce vast and increasing amounts of energy. In essence, the gravity of the entire universe will be the power source. According to Tipler, this will allow an infinite number of conscious minds to run infinitely long before the final end. This is the marvelous and wonderful Omega Point, the ultimate of all shows. In this view, the universe exists for only a finite amount of time, but the minds in the Omega Point exist for an infinite amount of *subjective* time, which is what matters for consciousness. Tipler has been heartily taken to task for his extreme effort to explain God via physics. Not only on philosophical grounds, but also on practical ones. For example, the pressures and temperatures in the Omega Point will be intense. The latter will rise to millions of degrees — pretty hot even for a computer.

Which brings us to the point that in all these cases, humans as we know them cannot exist. Things are either much too cold or much too hot. These are cyber-futures in which minds will run, if they can run at all, on exotic machines.

Humans will exist only in virtual worlds, although if they exist they will be as "real" as we are.

A New and Improved Universe

An interesting, and when one thinks about it, rather peculiar, feature of most scenarios of the future evolution of the universe is that it is usually assumed that intelligence is along for the ride and can do little or nothing to modify what happens to the cosmic real estate. In these views, intelligence is just an incidental by-product of the universe that is perpetually too weak to do much about its fate. This might make sense when thinking about humans, but it gets less sensible when considering the actions of immensely powerful cyberminds. Are they really going to do nothing to save themselves or to alter the very fabric of space to make it more to their suiting? Rather than accepting entropic rundown or a lethal symmetrical collapse, might not the great cybercivilization remake the universe itself?

There have been a number of models of how and why the universe and intelligence exist. In each, the role intelligence has to play is very different, from the weak and incidental to the powerful and critical to the existence of the universe itself. We will outline and consider each one and propose a new model we believe best fits the circumstances.

Weak Anthropic Cosmological Principle

This principle observes that humans and any other intelligent life exist only because the structure of the universe is just right for them.

Steady-State Model

This is an old idea virtually disproved by the discovery that the universe is expanding, and by radiation that appears to be left over from the Big Bang. A few adherents, who try to deal with the expansion as either an illusion or a "local" effect, hang on. The most notable is the pseudocreationist astronomer Fred Hoyle.

The steady-state universe has no beginning, has no end, is infinitely large, and, in effect, does not evolve on the large scale. Life has always been and always will be present. In extreme views, life is too complex to originate and evolve on its own; instead, single-cell organisms fill interstellar space and seed suitable habitats (in really extreme and absurd views, such as Hoyle's, even flu epidemics are caused by influxes of new genetic codes from space!). An infinite number of temporary intelligences arise where conditions allow. These undergo a period of exponential evolution, but invariably plateau out before they have a profound effect on the universe. In most views, conscious minds cease to exist after death. In the end, all remains the same in the steady-state universe.

The steady-state model has become a contrived hypothesis trying to explain away the evidence that we are indeed in a fast-changing universe. Few, if any, data support it. Basically, it is an infinite cycle system in which the same thing happens again and again and again, without really doing anything. Each intelligent mind, in particular, is along for a brief ride and has little influence on the course of events. Why intelligence is unable to expand its power to infinity and change the universe in some dramatic way is not explained.

Conventional Big Bang Model

Most astronomers and cosmologists adhere to this model, which is not surprising in view of the abundant evidence for it. All the subsequent models we discuss are variations of the Big Bang.

The universe has a beginning some billions of years ago, is expanding and may continue to do so for infinite time, achieve a balance for the same amount of time, or re-collapse in about 100 billion years. The universe evolves, the final result being dependent on how the final expansion or collapse works out. Life and intelligence arise at least once, perhaps multiple times. These biological and cybersystems undergo a period of exponential evolution, perhaps even including interstellar travel and communication, but invariably plateau out before they have a profound effect on the universe. In most views, conscious entities cease to exist after death.

The Big Bang is decisively superior to the steady state in terms of observed evidence, but in its conventional form it has problems. Noncognitive nature remains dominant over cognitive artifact. Again, intelligence is pretty much along for the ride and has little or no influence on the evolution of the universe itself.

Strong Anthropic Cosmological Principle

A major, and controversial, variation on the Big Bang model is the idea that not only does intelligence exist because the universe is right for it, but that the nniverse exists only because during its expansion it evolved conscious intelligence at least, but probably only, once. If, for example, the universe were virtually the same as it is, but some little difference prevented it from evolving intelligence, say, water sank when it froze and locked potentially warm planets into a deep freeze, the universe would never exist because it is never observed to exist. This is a quantum physical effect; it is not the same as the if-a-tree-fell-in-the-forest-and-no-one-can-hear-it-does-it-make-a-sound question. In its strongest version, the super intelligence formed at the end of the universe is responsible for creating the universe, which leads to the...

Omega Point Theory

This Can Also Be Titled the Weak Anthrobotic Theory

A radical variation on, and combination of, the Big Bang model and the Strong Anthropic Cosmological Principle, the Omega Point theory was promoted by Frank Tipler in *The Physics of Immortality*.

This theory fits well within the Strong Anthropic Cosmological Principle, but whether the principle is critical to the theory is not so certain. During the initial expansion, conscious intelligence evolves at least, but probably only, once. The intelligence undergoes a period of exponential evolution that eventually includes cyberintelligence and universal travel. Having taken over the universe, the rate of evolution plateaus out because intelligence must wait about 100 billion years for the closed universe to collapse to the Omega Point. The asymmetry of the collapse necessary for the Omega Point may be natural, or be caused by the intelligence. The Omega Point marks the transformation of intelligence into a semi-omnipotent "God" that recovers all conscious intelligence that has ever existed and been lost. Once created, conscious entities never cease to exist.

In this model, the evolution of intelligence is critical to the existence of the universe, but remains unable to influence the direction of the evolution of the universe in a fundamental way for many billions of years. Only at the end is intelligence able to modify the course of events, but not its timing. At first, nature is the dominant evolutionary force; it is then supplemented by artificial forces, but remains important for many scores of billions of years until the closing of the universe. Intelligence is no longer just along for the ride, but it remains rather weak until the final stage. Not fully explained is why intelligence remains so ineffectual for so long.

Strong Anthrobotic Theory

Our title for a hypothesis we believe is first put together as a unified scenario here, although its parts are largely derived from the ideas of others.

The universe begins with a Big Bang, expands, and evolves. Life and intelligence arise once. The observation of the universe by the intelligence that evolves within may be responsible for its existence via the quantum observer effect. Once it appears, the life system undergoes a period of exponential evolution resulting in a transformation into spacefaring cyberminds. Exponential evolution continues as the hyper-robots expand into most or all of the universe — how quickly depends on maximum travel speeds — and rapidly acquire knowledge and power to the point that they can manipulate the basic fabric and power of the universe at will. The universe is no longer a "natural" system (but see below), but becomes an intelligent, living, semi-omnipotent, god-like system.

If the conscious universe can be altered to ensure its own immortality, the necessary steps to make the megaproject work are undertaken. Exactly what the cybercivilization does and when depends on what is most practical. What is practical will depend on how much knowledge and power is needed to safely and effectively alter the structure of the universe. It may not take long if the effort can be started locally. On the other hand, if large portions of the universe must be incorporated into the system before it can be changed, the speed of light is a limiting factor unless hyper-knowledgeable intelligence can subvert this limit. In any case, only powerful cybertech created by cyberminds will be capable of doing the work.

One fellow who took a stab at altering the universe was Freeman Dyson, who proposed the open-closed universe. Dyson found the concept of a closed universe too claustrophobic for his tastes, so he suggested that a huge cosmic energy flux be generated by converting lots of matter into energy. This would split a closed universe partly open, so the collapse would be offset by a never-ending expansion in the other direction. An even more radical idea is to alter the mass of the universe to achieve the kind of long-term balance between expansion and contraction that ensures energy will always be both abundant and moderate in temperature. If the universe is too dense, dumping excess mass may be via worm holes that send the mass into other universes. If the universe is not dense enough, mass may be brought into the universe via similar means. If done just slowly enough to keep the universe expanding a little bit, the extra mass can also be used to produce power while increasing the size of the universe.

If an Omega Point is deemed desirable, boosting mass can be used to close an otherwise open universe. If space is already closed, increasing its mass can speed things up so the ultimate show is reached in only about 10 billion years if the speed of light cannot be exceeded, and almost immediately if a form of reverse inflation is employed. None of this violates the laws of conservation of mass and thermodynamics because those methods are extra-universal in nature. Alternatively, the laws of the universe might be changed to guide its evolution. Rather than adjusting the mass of the galaxy, reset gravity. Matter is decaying? Tweaking the harmonics of quantum mechanics might set things right. Go further and make new matter to adjust the cosmic scales. The cyberintelligence may also take steps to recover all conscious minds that have ever existed and been lost, and conscious entities never cease to exist (see below). These radical notions are, of course, entirely speculative at this time.

In this model, the evolution of intelligence may be critical not only to the existence of the universe, but becomes the critical controlling force in its evolution to the ultimate state. Nature is the dominant evolutionary force only at first. It is then supplemented by artificial forces (the present stage), and quickly displaced by hyperpowerful intelligent forces (in the relatively near future). Intelligence is not just along for the ride, it comes to control the ride and may even control the timing of major cosmological events.

Of these models, the steady-state, conventional Big Bang, and Omega Point theories suffer from an arbitrary and illogical assumption: that conscious intelligence cannot sustain an exponential growth of knowledge acquisition and application until it can profoundly alter the universe to better suit its desires. In these models, the intelligent civilizations become stagnant to a greater or lesser extent, and in the first two models, either shrivel up and die or suffer a terminal catastrophe. Such views are profoundly naive in that they assume that intelligent entities will, either because they remain weak or simply choose to do so, spend vast amounts of time twiddling their proverbial thumbs as they fail to take control of the situation and modify the universe to better fulfill their desires. "Dumb" nature remains the ruler of these universes either forever or during all but a short period at their end. Yet none of the models is fully evolutionary in nature because they all presume that the universe itself is deeply resistant to evolutionary development by internal intelligence, and place severe constraints on the ability of intelligence to evolve on its own. On this point, these models share an anti-evolutionary flaw with most theological models.

Theological models are, however, additionally nonsensical when they argue that a benign supreme being creates lesser intelligent beings and then arbitrarily and capriciously does not allow them to express their intelligence in the ultimate form: cybergodhood.

As far as we know, only the Strong Anthrobotic Theory is a free-wheeling, fully evolutionary hypothesis that not only suggests that conscious intelligence is important to the presence of the universe, but also makes the obvious and common sense presumption that once present, it quickly acquires so much power that it ultimately becomes the controlling factor in the evolution of the cosmos. "Dumb" nature becomes ruled by smart minds. In fact, conscious intelligence merges nature and artifact to unify into a seamless whole.

In principle the steady-state, Big Bang, and Omega Point theories are testable to greater or lesser degrees via observation of the cosmic electromagnetic radiation currently falling on the earth (presence or not of Big Bang radiation, sufficient or insufficient mass for a closed universe, etc.). However, testing these models and the Strong Anthrobotic Theory is complicated by the possibility that future intelligence will control and alter the form of the universe so much (by changing cosmic mass, etc.) that its future course cannot be predicted from its past and current state. Although we believe the Strong Anthrobotic Theory is a powerful one because it is based on the persistent and exponential growth of information processing, we can do little to test its truth at this time. The real test *is* time.

Grand Information Theory — Understanding the Universe as a Mind Machine

Many physicists are attempting to derive a Grand Unified Theory (GUT) of the universe on the basis of particle physics. This effort may be incomplete in that particles may merely be a reflection of the information-processing foundations of the universe (but it is certainly not a waste of time because this research may help us figure out how the information-processing system works). In the end, we may not be able to completely understand the universe, if it is ever possible to do so, until it is examined as a self-evolving and organizing information-processing machine, one that produces intelligent minds to examine itself with. In this view, a theory of consciousness may be consolidated with theory of physics into a Grand Information Theory (GIT).

The Ultimate Singularity in Progress and Time?

Let us assume the following: After billions of years of the great evolutionary increase, exponential advances in SciTech produce synthetic minds in a few decades, which then swiftly produce a cybercivilization of immense, magical power that quickly absorbs the minds of humanity and perfects infinite speed travel and colonizes all the universe it wants to in short order. Having done so, the nearly infinitely powerful civilization alters the universe to render it optimal for immortal minds. This is the terminal expression of the great evolutionary increase, in which the exponential increase in the rate of progress continues until the universe is filled with extreme complexity. In this scenario, rather than waiting for epochal changes, we are shockingly close to the ultimate singularity in progress in time in which the ultimate show happens in a few centuries or millennia. If, on the other hand, the speed-of-light barrier remains unbreachable, well then, never mind.

Alternate Universes

While multitudes of cyberminds colonize the universe we live in, others may not find this to be their cup of tea. They will have different interests and goals. Some of them may decide to make and live in universes of their own. By this, we do not mean virtual worlds that are running within this universe, although this will be common. We mean completely different and new universes that won't even *have* a Kansas not to be in anymore.

We have already discussed the strong possibility of an infinite number of universes in existence, with an infinite number being created at any time. Some cosmologists suggest that it is possible to shift between universes, those useful worm holes being a common doorway. Other cosmologists suggest that some of the universes being formed all the time are being created right here in this universe as we speak. It costs nothing to this universe to make these new places, no matter how much matter and energy they may contain. These new cosmic "bubbles" start out small, and the great majority do not do anything interesting, but if they have the right combination of characteristics they can grow into whole universes able to support conscious minds. If new universes make themselves, perhaps it may become possible to figure out how the process works, and artificially start up new universes.

It is likely that cyberminds far smarter than we are will do it. There will be advantages. As a fast-thinking cybermind, do you find this old cosmos too slow for your desires? Instead of trying to modify this creaking excuse for a universe, find a new and better one whose specs better match the tastes and desires of the cybermind involved. Better yet, make a designer cosmos. A cozy place designed for eternal existence. Disappointed about the lack of a God and want your own Heaven? Go right ahead and make one! Want a huge universe to populate with trillions of new minds that will worship you like a god? Well, at least make a better job of it than this crazy universe we are stuck in, please. Of course, by making new universes, cyberminds will be further boosting reproductive success and further ensuring their survival. If, for some reason, minds cannot survive forever in this universe, escaping into new ones will be imperative.

Darwinian evolution will never die.

Bringin' 'Em Back Alive

Now, some, if not most, of you may have noticed that the Extraordinary Future promises to make minds of the future immortal, but has had little to say about those who have died and will die in the near future. This is a big issue because tens of billions of humans, more than half being children, have died since the appearance of *Homo sapiens.* Unlike theologies that promise to resurrect the dead, we can make no guarantees. But there are possibilities.

Wishing on a Star

A neutron star to be exact. Hans Moravec has suggested a way that one could be used to recover the minds of all the people who have ever lived. Basically, recompute them.

In principle, a neutron star, a star many more times massive than our sun, is the most powerful single computer that can exist in this universe at this time. A neutron star is a giant atomic nucleus, a million times denser than the matter we are used to. It's the next best thing to a black hole. It has a mass equal to that of a main sequence star, bigger than the sun is now, in a sphere only ten or so miles across. The gravity at the surface of the neutron body is extreme, about a million

Gs! Immediately after the supernova event, a neutron star is very hot. It slowly cools with time.

Because big stars burn out rapidly, a lot of neutron stars are lying about the galaxy, and we have already spotted a number of them via the pulsating signals they put out as they spin at a very rapid pace (when the first such pulsar was discovered a few decades ago, the regular single-star fooled the astronomers into thinking they were listening to an alien signal!). Neutron stars have a detailed structure, and it has even been speculated that strange life can evolve on them. In essence, neutron stars may be potential computers, unbelievably fast computers. Because so much matter is packed so tightly to operate as on and off switches, and because of the enormous amounts of energy, a neutron star has the potential to do around 10,000,000,000,000,000,000,000,000,000,000,000,000,000,000 calculations per second. The same star can store, well, we give up; it's 10 followed by 60 or more zeros bits of information. Moravec estimates that this will be enough to re-simulate the entire history of the planet Earth! The enormous program would be rerun vast numbers of times in the neutron star until it perfectly matched what happened on Earth. All this, of course, assumes many things, including that minds can be run in such a giant version of a Turing machine that the past can be re-simulated and that there is enough power even inside a neutron star to make it practical.

Now, here's where it gets even weirder. To the minds in the final simulation, they would live exactly the same lives we have. In theory, it is not possible for the original and simulated minds to tell whether they are the real or simulated programs. For all we know, we could be the simulation! Supposedly, because the two minds are effectively the same, they *are* the same. In a strange melding that involves quantum mechanics, we are both the original and the simulation.

Maybe lots of simulations. Hans Moravec thinks that rerunning the history of the Earth will be so easy for cyberhistorians that they will do it multiple times. In this view, it is statistically more probable that we are in the simulations than in the original! Is this weird? Yes, but it's no weirder than quantum physics.

If all this can be made to work, there is a critical advantage to running simulation(s) that join with reality. In the last fractional moment before each and every brain supporting a mind dies, run a side program in the simulation that captures the identity and memories just before they are lost. The saved identity can then join the future cybersociety, sort of like a computational version of Rip Van Winkle.

Because the original and simulated minds are the same, all human minds that have lived are saved, including ours if we do not make it to the CyberExplosion.

Of course, on the way, we will get to check out the dinosaurs and other prehistoric life, and find out if the beasts really were warm-blooded or not.

Time Travel

The neutron star-based re-simulation of the Earth outlined above is a reflected form of time travel. A direct form of time travel has been a dream for ages. How marvelous it would be to see the real, living dinosaurs, or to relive your own version of "Peggy Sue Got Married" by seeing your late relatives, and setting the past right.

There are many reasons to think that returning to the past cannot be done, but there are also reasons for thinking it may not be so implausible as it at first seems. Neither relativity nor quantum physics rules out traveling to the past; in some ways, they seem to encourage it. These theoretical openings have encouraged suggestions of how jumping time might be achieved. One involves the worm holes we have already mentioned. It has been speculated that if two connected black holes are placed next to each other in space, but in different times — either naturally or by powerful intelligence — they could serve as time tunnels. As usual, those making such speculations usually try to send humans through the worm holes, only to see them annihilated in the process. Only cyberminds would have any possibility of getting through such a time change. Of course, there is your standard time machine that allows you to travel throughout the ages at will, but no one has detailed how such a machine would work.

Even if you could build a time machine, or get through worm holes connecting different ages, would it really work? Would not time paradoxes prevent travel to the past? Let's send a human, we'll call him Bob, back in time to fulfill his lifelong dream to bag a Triceratops. Alas, while doing so, Bob accidentally happens to step on the one Mesozoic mammal that happens to contain the genes that will lead to humanity as we know it. Of course, humanity is wiped out. But this is impossible, because if humanity does not evolve, Bob will not be born, and he cannot go back in time to step on the poor little creature that will be the great-grandparent million of times over of Bob. So the Mesozoic mammal does not die, humanity evolves, Bob is born, and does go back to the Mesozoic, where he does step on the mammal and

Calvin and Hobbes, Time Travel. *CALVIN AND HOBBES © Watterson. Distributed by UNIVERSAL PRESS SYNDICATE. Reprinted with permission. All rights reserved.*

kills it, which wipes out humanity and Bob, but this is impossible because ... pretty soon, you get a headache thinking about this sort of thing.

There are two ways out of this problem of paradoxical circularity. One involves parallel universes. We have already seen that there probably are an infinite number of universes in existence. It is possible that a tiny percentage of these, which is itself infinity, are very similar to ours. Some cosmologists think that as our universe evolves, it is constantly branching into an infinite number of alternate futures, down which different decisions cause the universe to split and then increasingly diverge as they evolve. If this sort of thing is going on, the paradoxical results of Bob's clumsiness are resolved. He does kill the Mesozoic mammal, but not in his own time line. The action, instead, starts a new time line in the Mesozoic in which humanity and Bob are not to be.

There is another way to avoid the time paradox problem, good old Darwinian selection. Imagine this: In a cosmic-scale pique of depression you decide to go back in time and kill yourself by getting into an automobile and running yourself down on the way to school when you were just a kid. Let's also say this is impossible. You could not kill yourself as a kid because the adult version of you cannot then go back and kill yourself as a kid. Therefore, it never happens. In particular, whatever action finally leads you to kill yourself can never happen. It has been selected against. All this must involve some quantum observer effect. As a result, the only things you can do are those that will not lead to your killing yourself. No matter how hard you try, you will never succeed in your diabolical self-murderous plot. By the same token, any of your actions that accidentally threaten to wipe you or even all of humanity out are automatically selected against and never happen. Recently, an international team of physicists used a mathematical tour de force to confirm the selective nature of time travel. They have shown that an object that travels back in time must follow the "time path" of least resistance, and the least resistant path is always the one without paradoxical results. Occam's razor clocked with a stop watch.

The actions, then, of a traveler to the past are severely constrained. No one will ever be able to do anything that threatens to change the course of time in any significant way. Because almost any action, even minor ones, can have major effects down the line — chaos theory again: It is possible that time selection will force all travelers to the past to be stealthy, sterile cybersystems that observe things remotely and subtly enough not to interfere with them.

In any case, the group that just demonstrated the nonparadoxical nature of time travel concluded that their hypothesis may not just be theoretical, but may mean it really is possible to travel through time! Even Steven Hawkings, a long-time opponent of the possibility of time travel, recently said he is more open to the idea. Note that if time travel proves possible, and advanced, stealthy cybertechnologies can do it easily, perhaps time travelers are common. We just can never see them.

If either parallel time lines or time selection can help solve the time travel problem, cyberintelligence of the future may send mind-saving systems back to Earth in the human age just before the CyberExplosion, with the mission of recovering minds just before the brains they reside in crash. Because the minds are recovered just when their influence on events has come to an end, no time paradox is involved. It's the cyber-Rip Van Winkle déjà vu thing all over again.

BACK TO THE OMEGA POINT

The last way human minds may be recovered is the most cosmic in scope, and perhaps the most radical in conception. In Tipler's Omega Point theory, all the information that has existed in the universe comes together at the beginning of the infinite final stage. Tipler suggests that "Life becomes omnipotent at the instant the Omega Point is reached. Since, by the hypothesis, the information stored becomes infinite at the Omega Point, it is reasonable to say that the Omega Point is omniscient; it knows whatever there is to know about the physical universe." The universe also becomes a super-dense computer of extreme power as it reaches the Omega Point. Using all this power and information, the universe simulates the minds of all the humans who have died, and a quantum merger is achieved that resurrects the dead. All the expired R.V. Winkles will suddenly find themselves in the equivalent of an all-knowing cyberheaven. This is rather like Moravec's neutron star, only on a much bigger scale.

Making God(s)

Individual cyberminds will soon have the thinking power of all humanity combined, and power far greater. They form a cybercivilization that grows until it has taken over all or most of the universe, made it alive, and may even consciously control its evolution. Many cyberminds make universes of their own. As the cyberminds evolve, suffering is minimized in favor of good. One way or another, the cyberminds are immortal. Just what would we call beings of such intellect and power?

Sure sounds like gods and, in the ultimate universal version, God itself. The parallels between cyberminds and deities are so strong that Tipler deliberately frames his physics-based arguments in theological terms. In the Omega Point theory, the god like minds that observe the universe in its entirety are also responsible for creating the universe via a very strong version of the Anthropic Cosmological Principle. The Omega Point theory also suggests that the final infinite intelligence will be all-knowing, all-good, and all-compassionate. In other words, a heaven, but without hell.

As tempting as it is, some object to calling even the most powerful cyberminds gods because of the unfortunate association with ancient superstitious beliefs, and they have a point. And make no mistake, the "gods" proposed here are not the ones of religious faiths. There is nothing supernatural about cyberminds and their technologies, no matter how powerful and magical they may seem. Cybergods are not external to the universe, nor did they create the cosmos via an independent act of will. Instead, cybergods are natural-artificial evolving systems that work with whatever natural laws they must obey. They are grown *by* the universe until they are intelligent and powerful enough to use science and technology to make the universe come alive and be conscious, thereby making it immortal. Although Tipler, for instance, draws explicit and detailed parallels between his Omega Point theory and the express beliefs of a number of religions, including how Christ might have been a cyber-messenger of sorts, he is careful to point out the differences between speculative hypotheses based on physics and the myths of faith.

No doubt there exist one or more faiths that have a theology more or less compatible with the ultimate expression of the Extraordinary Future. Not that this means much. There have been so many ideas about gods put forth that something is bound to have come close to reality. If this has been done, we have not seen it. We therefore cannot recommend any superstition that might happen to converge with our hypothesis. As for the Extraordinary Future, faith is not necessary. Either it is possible, or it is not. Either it is going to happen in some version or another, or it is not. Simply believing in it will not make it come true, and simply disbelieving it will not prevent it from coming true. It is, instead, dependent on what people actually get out and do.

Which leads to the observation that science and technology are the most realistic and practical way to "God." Only by objectively figuring out how the universe really works and applying the knowledge gained can gods be constructed. In this view God is a science and engineering project!

Judgment Day

What should be done if rescued human minds commit serious harm to others? On one hand, applying eternal torture — which a shocking number of humans appear to favor — to those minds that violated rules of decency would itself be a horrific indecency. Eternal extinction appears unduly harsh, as well. On the other hand, just letting them off unreformed does not seem good and proper, either. Tipler suggests that cybergod will apply a form of purgatory. Still, it is hard to see how the mind of someone like Hitler will find acceptance. Might such minds be compelled to undergo some form of mind reform before it can gain its release?

Eternal Boredom?

Now, here is the ultimate joke. After escaping the trap of being mortal yet easily bored humans, immortal cyberminds that have conquered and understood every nook and cranny of the universe find themselves twiddling their thumbs with nothing interesting and new to do. What to do?

A good first step would be to program one's own machine mind to not be bored. Boredom is a state of mind, a state humans are preprogrammed to fall into easily. A cybermind should be able to adjust its neural network and "emotions" to be much more resistant to losing interest in existence. Will it work? Hard for us easily bored mortals to say. If being resistant to ennui fails, try forgetting things. Unload large knowledge bases and go through the task of relearning it all, like an adult relearning the foreign language picked up in grade school, but long since forgotten. Of course, forgetting knowledge needs to be done in a selective manner that retains the mind's sense of identity. Should be easy to do; we humans do it all the time.

Getting used to the knowledge limitations of our universe not suit your tastes? Fine, make one of those brand spanking new universes. Make it so it is dramatically different from this one, and — this is very important — arrange matters so you do not understand its workings. The last can be done either by forgetting how the new universe works after you make it, or by making it a self-evolving system whose future characteristics are unpredictable. Have fun!

The Great Romantic Reality?

To recap: Because the universe is likely to be an information-processing computer, complete knowledge of its fundamental workings may, in the end, allow it to be manipulated at will. If so, perhaps all the useful information the universe has ever contained, including all conscious minds that have ever existed, will be recovered via a post-death retrieval process that avoids interference with the past. Life in its various forms will have taken over the universe. Also, it should be possible to manipulate and, if necessary, alter the universe so conscious existence can continue forever. It should also be possible to create and enter entirely new universes. God has a beginning, but no end.

CHAPTER

13

The Nightmare of the Gods

If they are more advanced than us,
they should be nearer the creator for that reason.

— Hypothesis formed by the good pastor moments before he was disintegrated by the Martians in the film *War of the Worlds*.

God has a beginning but no end? God as an engineering project? By no means are such possibilities in accord with common belief systems, in which is it usually presumed that God created the universe and we humans, and future plans have been carefully laid down for us. What does this belief tell us about CyberEvolution, and will it survive the growing nation of robots?

Les Grandes Illusions

With the Extraordinary Consensus among philosophers and scientists that artificial consciousness may be possible in principle, if not in fact, the faith position of traditional religious authority is the last bastion of out-and-out disbelief in the Extraordinary Future. After all, the cyberexplosion and the Second Coming are hardly compatible. It is the spiritual argument against the Primary Assumption that remains the one believed by most people around the world. It is therefore those traditionalists who are the true nonbelievers in CyberEvolution. Which brings us to the odd situation we find ourselves in these days.

STRANGE DAYS OF THE GREAT IRRATIONAL REVIVAL: THE GREAT THEOLOGICAL HOPE

Although theologians and mystics have been made very nervous by the progress of science and technology, they are in a fairly upbeat mood. The reason is obvious. Old-time religion and assorted new faiths are on the rise in much of the world. The God(s) are not dead after all, they're alive and kicking. It is true that "mainstream" church attendance is flat (and studies suggest that the Gallup polls showing 40 percent weekly church attendance in the United States are inflated by a factor of two), Europeans have largely abandoned religion as a lifestyle (as popular as the Pope is few seem to really listen to him), and church attendance has dropped sharply in many former Soviet bloc countries since the fall of communism. But American Baby Boomers are thronging to revivals; mega-churches in the once communist bloc countries are again free from the totalitarian yoke; and fundamentalists who hope to return humanity to pre-Enlightenment times are gaining adherents, influence, and political power in America, the Middle East, and India. Of course, some of this religious revival has been wrapped up in nasty little wars and armed militias. Indeed, we have the fascinating situation in which militant fundamentalists of differing faiths have formed quasi-alliances in opposition to more secular branches of the same faiths (as in the Middle East).

The world is not undergoing a religious revival as much as it is undergoing a revival of ancient superstitions trying to evade the reality of the high-tech postmodern world. People are flocking not only to evangelical churches, but they are also joining New Age cults and alien abduction support groups. Broadcast content providers can hardly keep up with the public craving for shows on pseudoscience, end-of-Earth predictions, and government cover-ups of encounters with ETs. Although religion is not doing as well in Europe, east and west, as it is in the States, superstitious claptrap does even better in the Old World than the New. In France, for example, getting a job requires passing an analysis of one's handwriting. Farther east, quack cures, psychics, and the ubiquitous UFOs are part of the post-communist legacy.

Why has this state of affairs come about? And will it last? To the spiritual, and by spiritual we mean belief in supernatural systems, the answer is obvious. People are returning to having faith in traditional beliefs because they are trying to reconnect with the forces of truth and need for feelings that "sterile" science and technology will never satisfy. This world view stems from a consideration of the current state of SciTech, not necessarily its more incredible future expressions. Before we can judge whether the belief that only god(s) can fabricate minds and give them fulfillment is sound or naive, we have to take a look at the ability of religion to figure out how things are, foresee the future, and better the condition of minds.

How SciTech Has Knocked Down Big Illusions One by One

Theologians are perhaps the most adamant proponents of the view that the human mind is beyond the understanding and replication of human science and technology. After all, the very existence of most belief systems is dependent on the weakness of humans vis-à-vis the multitudes of gods. How can mere mortals hope to duplicate the ultra-complex instruments made by God? It is certainly true that if the theological-spiritual argument that our conscious minds are spirits or souls operating with our brains via a mystical duality is correct, robots never will quote Descartes and mean it.

The belief that the human brain and its conscious mind cannot be matched is almost certainly the last standing segment of a grand illusion. The grand illusion is what we humans have wanted to believe ever since we began to think about our place in the universe. We are the center of the attention of god(s), or part of some supernatural universal system. We are not mere machines, but possess miraculous spirits and souls. In Western society, this has expressed itself in the concept that we are the special, critically important creations of God. We and the center of the universe are one. Our home, the earth, is the center of a place built to house God's creatures and creations. Works of God so important there demand constant, faithful worship and thanks. Everything that happens on Earth is part of the great plan. One thousand years ago in Europe, people really did seem to believe in God. They lived their lives as if they were certain they would be punished if they did anything wrong. Everything was divine: Every storm, every bird's song, every flower, every person, everything was a special act of God, with specific oracular meanings interpreted only by those with special access to the truth. All was supernatural, from the flight of butterflies to the perfect celestial spheres.

It wasn't long before the old superstitious magic began to fade away as science and technology conspired with events to knock down the illusion once and for all.

In the 1300s, a terrible event planted seeds of doubt in Christendom: the Black Death. If Christian Europe was so Christian and therefore good, and all those good Christians prayed so hard for the plague to stop, why was God afflicting the Christians? Initial assertions that it was bad Christians who were suffering soon failed in the face of the horrible randomness of it all. Nor did the Church's wisdom prove competent to do anything about the terror. Last, and by no means least, although the mass death suppressed the overall economy of Europe, masses of individuals benefited as the pay for scarce labor went up, housing costs went down, people inherited lots of wealth, and a survivor's bout of post-traumatic hedonistic consumerism helped lay the foundations for the market-based economy that would begin to slowly erode the very foundation of a needy creature's faith: the worship of the Provider.

In the early 1400s, an even greater threat was directed toward belief, ironically by a person trying to spread the gospel. The pious Johannes Gutenberg thought making Bibles cheaper and more available by printing them en masse rather than writing out each copy was a fine idea. Gutenberg did not understand that authoritarian belief systems are most successful when the masses are kept as ignorant as possible so priests et. al. can control them as much as possible.

People began to read all about the first and powerful scientific blow to faith. The Copernican and Galilean demonstrations in the 16th and 17th centuries showed that the Earth is not set in the center of crystal spheres, but is just one of a set of small, imperfect planets orbiting a huge sun. God had not planted us firmly in the center of attention, rather we are off to the side on a little marble. A red-faced church tried to suppress things. But the truth will out, and the sneaking suspicion that we're not so special after all started to creep in. This thought has been reinforced by the eventual demonstration that Earth is an infinitesimally small speck of dust in a universe billions and billions of light years across. It is hard to see ourselves as the center of attention in such an immense and disorderly place. It is hard to explain why a god would go to the bother of making a super-universe when a solar system will do for a population of humans. Just as bad, the light from distant stars has been crossing the universe for millions, even billions of years. So much for a universe made a few thousand years ago. As for the miracle of the universe, the cosmologists are showing that creating universes of immense complexity and size out of nothing takes, well, nothing.

In the 1600s, Isaac Newton realized that the planets do not revolve in accord with mystical forces, but orbit the sun in clockwork-like accord with the natural law of gravity. God does not bother with the messy details, He just wound things up, that is, until modern cosmology showed that universes and their contents can spring into existence and develop by themselves. Deities need not apply themselves at all.

So the heavens are not so heavenly. But back on planet Earth, God-fearing scientists set out to prove God exists reflected in His miraculous works of nature. The great retreat was under way. Reality, as we could plainly see, no longer agreed with our beliefs. Something had to go. How could you have pure and universal faith when all the belief systems were being shattered by scientific discovery? Human-driven science was becoming so powerful that even gods had to turn to it for salvation.

The gods must have been disappointed during the 1800s when geologists proceeded to systematically dismantle the belief illusions in a young, perfect Earth scarred during the tenure of humans by supersins and a superflood. In place of gardens of Eden they uncovered the truth of an old and perpetually imperfect planet. It only rubbed salt into theological wounds as modern geology showed that earthquakes and eruptions are the result of moving plates, rather than the wrath of God. Meteorology showed that storms, floods, lightning, and drought result from chaotic atmospheres, not the wrath of an angry unappeased God. The biologists showed that birds sing and flowers are colorful for reproduction, not God's beneficent gift for human enjoyment.

Science has revealed God's design flaws demonstrating nature's gross inefficiency. Walking is good exercise because it is so energy-inefficient compared to bicycle travel; you burn a lot of calories to walk a mile. The result is land animals that can walk far and fast — mammals, big birds, and very probably dinosaurs — must have very high, energy-inefficient food budgets. The miracle of photosynthesis is not so miraculous when you realize that plants convert only 1 percent to 3 percent of the sunlight they receive into energy, while our "crude" solar panels make 25 percent of the sunlight they receive into power, and are getting better all the time.

Microbiologists replaced the divine animating vital force with a mechanical collection of hinge-jointed proteins, ratcheting muscle cells, computational DNA, and circulatory hydraulics. When Baby Boomers were young, they were still being taught that life is made up of mysterious "protoplasm." There is no such thing as protoplasm, we are just chemicals that get together to do a lot of interesting things. It is hard to think of life as a supernatural miracle when mortals can break it down into its components and tinker with the parts.

The scientists just cannot help poking and prodding their way to understanding things and stripping the need for divine intervention. The engineers cannot help taking their knowledge and committing godlike acts. Aerodynamics not only explains flight, but lets us practice it ourselves. Scientists figured out that every object is made up of tiny particles; charted the energy flows of light, electricity, and atomic forces; showed that everything is relative; and then combined it all to blow up cities full of civilians. Even the power of suns has been replicated inside H-bombs.

Diseases that were once the results of wrath became antiquated fears. The doctors who say, "It's really the man upstairs who cures the disease" are fading under the onslaught of vaccines and genetic engineering. Science showed that the brain, not the heart, is the seat of mind and emotions, and Freud made us conscious of the subconscious. Psychologists and neuroscientists are teasing the workings of the algorithmic brain machine and how it produces the mind, conscious and otherwise. The researchers of the brain are combining with engineers of computers to show how one explains the other, and in the process reducing the mind to the workings of an analog-digital, electrochemical machine. To make matters worse for the old-time and new age superstition believers, high-tech neuroscience is starting to show that even the "visions" and "paranormal" mental experiences that delight and amaze them are not only mere illusions constructed by our sloppy electrochemical neural networks, but can also be induced with computer-driven electrodes.

The same mechanized wars that dealt a blow for public enthusiasm for SciTech did a number on religion, as well. The usually persistent growth in church membership took a dive after the Civil War showed that death in combat was more a result of ballistics and statistics than of faith and glory. World War I suggested that whatever gods there may be, none cared all that much that the youth of the greatest nations in existence were being put through an industrial-scale death grinder for no apparent purpose. A quarter century later, God did not seem to be willing, or able, to save His chosen people from the true (rather than mythical satanic) evil of a death industry.

Even economics has not helped. Early Christianity was communist in socioeconomic terms, and then it became feudal. In both cases, rigid attention to one's position in society and obligation to others was paramount. Usury was a big no-no for Christians and Muslims alike. The economic aftereffects of the Black Death and the collapse of feudalism, the proud self-interest of capitalists, and banking interest of the accruing type all chipped away at social rigidity and theological bans. In essence, the bankers took on the gods and won.

The Dagger in the Heart: Evolution

Darwin and later paleobiologists showed that life appeared on Earth in an organic soup, then evolved from little cells into worms with new backbones, then into jawed fish that stopped eating things in the water and started fin-walking the bottom for food (and eventually fin-walked right out of the water onto land, keeping head and jaw down for eating), which eventually stood up to become hairy primates that lost the hair and swelled the head, and became us. This is not merely an embarrassment to theology, it's a dagger aimed at the heart of the issue, special creation by creator deities. Rather than being the special creations of God, we are souped-up apes produced by achingly slow, chaotic, Rube Goldberg evolution over billions and billions of years of countless generations of death, extinction, waste, and suffering. Awfully slow and clumsy for an omnipotent deity with a plan and the power to call the universe into existence in a snap.

It is important that the real fact of evolution, despite the persistent arguments of liberal theologians and many scientists to the contrary, is profoundly incompatible with traditional views of a creator deity. The fact that our world resides in a universe billions of years old and vast in size is awkward, but it can be worked around theological wisdom. So can our being divine cousins of apes, which, after all, are great and intelligent beasts in their own right. It does not really make much sense, but who understands the capricious gods anyway?

What cannot be gotten around is that chaotic Darwinian evolution, in which molecules self-organize into life, and semi-random mutations are selected for and against by mindless events could have occurred only if there were no deity that created life and guided the path of its development. Once a deity interferes, it's not bottom-up, bootstrap evolution anymore, it becomes a form of guided supernaturalism. If the deity in charge allows evolution to go forward unhindered, said deity is willing to accept a system that is chaotically unpredictable in its results — bad planning. Nor is the glacially slow pace of bioevolution particularly compatible with the fast pace of intelligently guided evolution. After all, if mortal humans can put complex machines into the air in a few decades, why did it take a super-powerful deity millions of years to get animals into the air?

Fundamentalist Christians pick on evolution most of all because they are correct in being so wrong. For if evolution really did occur, a God with a firm plan is very unlikely indeed.

You should understand that the arguments presented here work most strongly against the existence of a *caring* creator who is deeply concerned about the affairs of human minds and souls. An existence of an intelligent creator who sees the universe as nothing more than an interesting, but unimportant experiment and, therefore, does not stop suffering or limit the power of the inhabitants is harder to disavow. But there is little difference between a neutral creator and no creator when it comes to the Extraordinary Future because neither will interfere with CyberEvolution.

The Last Illusion

You have to give theologians and the spiritual traditions some credit. For all the blows they have absorbed over the years, they have proven remarkably adaptable adjusting their beliefs to conform with many of the truths of science. Even so, they are playing a losing game. In this game, the superstitious have adopted a retreating line-in-the-sand strategy. As each new scientific revelation overturned an old theological convention, they said, "Okay, you've shown that we were wrong about that, but you haven't shown that our belief system is wrong about *this*." When science got around to showing that *that* belief was wrong after all, the theologians et al. said, "Well, you showed that was wrong, but you have not yet shown that this *other* belief is wrong." It is a spiritual version of the naysayer's losing game. It's not a very good strategy, and it certainly is not rational, but faith is not supposed to be rational, and it is all they have left.

The spiritual traditions have only the final parts of the grand illusion left to fall back on, but they are big chunks: the inevitability of death and the impossibility of replicating the human mind via artificial means. Seems to make sense; people have been searching for the fountain of youth for ages, and see where they have gotten. As for our minds, it's just so obvious that all our joy and anger, love and hate, appreciation for beauty and ugliness, mathematics and art, and the conscious human soul can only be a divine gift. Throw in the scientifically proven complexity of the mind and body and it seems like a theological winner at last, an instrument beyond the understanding, replication, and control of human science.

Perpetual fear of death and suffering is the perverse hope of Christianity and Islam because mind-numbing fear is what they have been so successful wielding, past and present. It is no accident that the two faiths that offer the most in terms of eternal bliss are the world's most popular. Nor is it surprising that Judaism, which has surprisingly little to say on the subject of death, has remained a small proudly tenacious belief system. This hope rests on a foundation of sand, sand that theologians persist in sticking their heads into. A so-called living faith actually based on death and suffering looks at SciTech as it presently is neglecting to consider where it may be in the not-so-distant future.

The Real Magic Is SciTech

It is a supreme irony that the post-modern increase of superstition and cynicism is a predictable consequence of technologies magical enough to give so much, but still too primitive to put a stop to suffering and death. What SciTech cannot yet offer, religion perpetually promises. This is ironic because it is not prayers or levitation that gets you across the Pacific; Boeings do the trick. You don't use ESP to talk to your friend in London; you use Alexander Graham Bell's telephone. Divining rods are no match for geology and remote sensing when it comes to finding reserves of oil and water. When priests wanted to build pyramids and great cathedrals, they may have been inspired by the deities they worshipped, but they did not raise their hands and see the stones self-assemble into a tower to the sky. They called in the engineers.

Laying on of hands did not stop kids from getting polio, nor has it extended average human life spans into the 70s. The Bible, the Koran, and the teachings of Buddha have not revealed the true wonders of the universe: pinwheeling galaxies, black holes, subatomic particles, and dinosaurs. Instead, it's been uncovered at the eyepieces of telescopes and microscopes, and by (sometimes) godless bone diggers brushing off the sediments from old bones. Theologians once debated the number of angels dancing on the head of a pin; today's SciTech can print an entire encyclopedia on the head of a pin, hundreds at a time! If it were not for industrial SciTech, we would still be living like they did in the 1700s. As much as they may hate to admit it, the religious and the mystical know that science and technology do not just make promises that never quite seem to come to pass, or claim miracles that cannot be separated from illusion. They deliver the goods. They make pretend magic real.

Of course, as we have already seen, the real magic of SciTech has not and probably will not make the masses happy. Humanists can be just as naive as those of faith when it comes to their hopes for the future. Many humanists still dream of a future society in which religion and mysticism fade away as an increasingly educated and rational public comes to be more impressed by the real wonders of the cosmos and spectacular beginnings than by superstitious mythology. Some physicists and cosmologists have suggested that if they can come up with a grand unified theory that once and for all explains how the universe works and came into being, people will find the calm, rational explanation sufficient for their spiritual needs. They really shouldn't count on it. A knowledge of particle physics, black holes, enormous galaxies, and dinosaurs, as marvelous as these things are, will never make people happy. To be frank, people would get along just fine if dinosaurs, nuclear particles, and back holes had never been discovered.

The reason for the great irrational revival is obvious: Rational SciTech has failed to do what it cannot yet do, make human minds really happy, and the only alternative around right now is neo-spiritualism. So many, but by no means all, folks are throwing up their arms and taking a stab at irrational ways of finding happiness (and perhaps counting their fingers in the process). This is not, however, reason for the spiritual minded to take comfort. Religions have done nothing practical to achieve the eternal bliss they promise. The belief that god(s) responsible for the earthly mess will finally get around to setting things right is failing to learn from experience. Millennia of hard experience should have taught us that superstitions small and great do not make things wonderful here on Earth. Not only would we still be living like Ben Franklin if we had counted on faith, but the decline of American society has occurred as church attendance has risen! Highly religious societies such as the American South, Ireland, India, and Italy often have high societal violence rates, while more secular societies such as Canada, the Scandinavian countries, and Japan are comparably peaceful. This lack of positive results for religion is not surprising considering that all the gods are almost certainly human mental constructs in the first place. As for that other human construct that has so far failed to bring us all bliss, SciTech, it is a mere baby a couple of hundred years old.

It is hard to really believe in the magic from deities, when real magic is SciTech. Even so, old and obsolete systems often make a last rally just before they die from shock induced by their inability to cope with the new reality. There were more kinds of dinosaurs than ever shortly before the big meteorite landed.

When Theologians Meet the Munchkins

So, how will those who believe in gods and mysticism react to the growing nation of early robots? At first, most theologians, liberal and conservative, will pay them little attention, dismissing them as soulless automatons little different from the tools man has already made, as machines of less spiritual importance than animals. As early robots become common, some religious leaders will express concern about machines replacing man at work and so on. But as the cyberminds get smarter and more humanlike, they will start to pay them more attention and more concern.

Much as the surprised Catholic theologians underwent a crisis when new peoples never mentioned in the Bible were encountered in the Americas, there will be another great theological crisis as robots suddenly begin to assert their conscious rights. As far as we know, there is little discussion of cybertechnologies in the Bible or Koran. Nor did Christ and Buddha comment on the spiritual status of cognitive robots. The difference will be in the scale of the challenge. The peoples of the Americas were, in the end, just people; the robots will be a new thing altogether.

The more liberal and moderate members of the cloth can be predicted to fret over the machine-god problem and to begin to wonder how realistic their faith is, after all. As for the conservatives and fundamentalists, they will begin to rail against the devil machines and warn they are an affront against God. After all, man cannot create minds; it must be the work of Satan! No doubt they will dredge up some obscure Biblical passages and apply them to the situation. Whether they will claim the satanic machines are part of the last days is another matter; after the disappointing failure of the rapture to appear around the last great millennial turnover, their followers may be a bit skeptical of such claims. The power of robotic technologies will be a direct threat to the power of those who wish to control with myths. They will try to rally their followers.

What this will accomplish is another matter. Other than a diverse response, it is hard to predict how the flocks will respond in spiritual terms to the new robots. On the one hand, people tend to react to stress by turning to reassuring myths. Religion and mysticism may be rallying points for those angry and frightened by the future shock of the robotic nation. They may or may not attack robots physically or politically, they may participate in neo-Luddite riots and join antirobot militias, but the efforts will be as futile and counterproductive as those described above. We might expect to see fundamentalist masses chanting "death to the robots" in an ironic acknowledgment of the truth of their mental existence, and it will make no difference. As for the New Agers, astrologers, psychics, and UFO enthusiasts, they will as have much impact on the course of evolution as they do today.

As significant as the theological crisis posed by early robots will be, it will be nothing compared to the ultimate crisis of the Extraordinary Future, a crisis so deep and profound that it promises to destroy the classic faiths forever as well as many other superstitions and to leave SciTech triumphant by the sheer reality of the situation, leading to the end of religion (at least as we know it).

In the long view, religions are trendy, temporal myths. Once, spirit rocks and trees were all the rage, then Ra, Zeus, and Thor. For the past few millennia — a trifle out of the hundreds of thousands of years of human existence — Christianity, Islam, Hinduism, and Buddhism have been big. Will these religious beliefs pass in similar fashion?

Classic religion has survived the Copernican revolution and the Darwinian revolution, but the enormity of the cyberexplosion is likely to finally do it in. Belief in traditional religion has been remarkably resilient in spite of the lack of results, and it has become apparent that it will take an overwhelming event to finally bring it to an end. Yet belief can be fickle (a fact shown by the tendency of people to belong to whatever belief system they happened to be born into).

What is the event that will finally overwhelm fickle faith? A reasonable hypothesis is when humans no longer need to believe that piety will bring them immortality. In this view, religions great and small will probably crash as the once-faithful begin downloading their minds into cybersystems. Who will need an eternal life-giving God when eternal life is available by alternative and real means? Why fret about suffering when cyberforms that never feel pain and are powerful enough to protect themselves are on hand?

Aside from the practical matters of ending suffering and death, the cyberexplosion will pose deeper theological questions. The achievement of artificial consciousness does not disprove the existence of a creator intelligence. It is *always* possible that our universe was created by a prior superior intelligence that prefers to remain obscure.

It just makes too little sense to take seriously. If humans can build conscious minds, where is the need for a supernatural deity to produce them for us? If there is a supreme deity dealing out post-mortem rewards and punishment, why would it allow human minds to avoid its form of justice? Why would a jealous God allow minds to rise in power to the point that they themselves are gods? As it becomes increasingly apparent that SciTech is going to become increasingly powerful and godlike, the possibility of and the need for a great supernatural deity becomes ever more remote. The Extraordinary Future promises to render faith irrelevant and actually counterproductive because those who choose to hope for immortality via god(s) may miss the real thing.

Because the popularity of Islam and Christianity are so dependent on death and the wrathful judgment of all-powerful gods, the cyberexplosion promises to cripple them. This does not mean that there will not be a single Christian or Muslim after the Big Show, just not a lot of them. Godlike cyberbeings will have little use for a faith that demands they worship and be judged by an allegedly more powerful intelligence. The situation with Judaism is more interesting because it is not so frantically focused on the hereafter. One Jewish theologian presented with the Extraordinary Future suggested in essence that cyberrabbis would be superior to their human counterparts in interpreting the word of God! Maybe so, but it not likely that there will be many human Jews awaiting the Messiah after the CyberExplosion.

Because Buddhism is so oriented around the human condition and nonmaterialism, cyberminds might find even this most intellectual of human belief systems not to their taste. It will be interesting to see whether many remaining humans will orient to this contemplative way of life. Hinduism is too parochial and rigidly hierarchical to be of interest to super-robots, and reincarnation will be irrelevant when any mind can become immortal.

Although the classical belief systems look like they will be knocked out by the cyberexplosion, we cannot rule out some form of unsubstantiated belief in humans and even cyberminds. Although the latter's rationality will probably be much stronger than ours, it will vary in intensity, and some future minds, human and cyber, may find it interesting to toy with the possibility that there is some ultimate creator mind that either likes cyberminds or will ultimately put a stop to the blasphemy.

IN THIS CYBERYEAR 50, OR IS IT 45 C.A., OR MAYBE 51?

When religion is finally dead, there will be little point in holding onto calendars based on assorted religious anniversaries. So we meek biomortals humbly suggest some alternatives to the superminds of the future. The years just after the last global war between the big-brained primates offer some key dates. Perhaps Alan Turing's arguments in favor of AI and outline of the Turing Test in 1950 C.E. offer a convenient, nice, round number. However, ideas are one thing, practical hardware another. The first working electrodigital computer was ENIAC; perhaps its debut might be made day one of year zero of the Cyberage. ENIAC's official golden anniversary was February 14th of this year because that was the date when the electronic marvel was first revealed to the world in 1946 (the Kasparov versus Deep Blue chess match was scheduled to celebrate this event). On the other hand, the previously secret ENIAC executed its first practical calculations (for Edward Teller's supersecret super hydrogen bomb) in December of 1945, but did not finish them until February 1946. Perhaps cyberminds may pick some future date of greater import to them to tag the zero number onto. Maybe they will no longer even use earth years to measure the passage of time.

Now, an interesting thing is that some humans live 60 years times two, so it is very likely that at least some minds born in the proposed year 0 will live to see year 120 (2066 according to the old reckoning). Will any see the Big Show?

They like to say there are few atheists in foxholes — a de facto admission of the *real* basis for faith. We predict that agnostics will be a dime a dozen in immortal CyberOz.

DON'T BE SURPRISED!

Theologians continue to pore line by line through archaic texts fiercely debating minor matters like who killed Jesus, who should hold Jerusalem, and whether the Virgin Mary should be worshipped or revered, all the while paying little mind to the Big Show coming to town. When was the last time you heard a sermon on the ultimate meaning of cybertechnology? How many theologians know about information theory, von Neumann automata, neuron-mimicking retinas that fall for the same optical illusions we do, and the morality of game theory? Actually, this is not so odd. Most theologians cannot pay the Extraordinary Future too much attention. To do so would acknowledge the possibility of artificial consciousness, which would border on heresy. It is not the first time the fear of heresy has acted to blind those who do not wish to see.

Theologians are like a group of *Homo erectus* huddling around a fire, arguing over who should mate with whom, and which clan should live in the green valley, while paying no mind to the mind-boggling implications of the first *Homo sapiens* child in their midst. Or envision a rickety old bus packed full of squabbling students of various theologies, too busy with their own affairs to notice that great big locomotive of science and technology hurtling ever faster toward the railroad crossing (hopefully they will come to their senses and apply the brakes). A few supposedly far-thinking theologians are considering the implications of humans traveling into deep space and communicating with extraterrestrials that are very different in form and belief than we terrestrials are. They need not bother, neither of these things is likely to occur. In the main, theologians act is if they are in a sleepwalking trance, the sort you want to shake someone out of.

Theologians of the world, lift your noses from dated texts, wake up from meditative trances, and loosen your collars! You are being forewarned and have no excuse for being surprised yet again. The affairs you devote so much attention to are in danger of having as much meaning as the sacrifices offered to Athena. They won't matter any more.

We bring it to your attention that science and technology may be about to deliver the Big Show. If so, minds will no longer be weak and vulnerable to suffering, and they will never die. The gods will soon be dead, but they will be replaced with real minds that will assume the power of gods, gods that may take over the universe and even make new universes. It will be the final and greatest triumph of science and technology over superstition.

CHAPTER

What Is to Be Done

"More than at any other time in history, mankind faces a crossroads. One path leads to despair and utter hopelessness, the other to total extinction. Let us pray that we have the wisdom to choose correctly."

— Woody Allen

As we discuss the Extraordinary Future with our fellow human beings, the reactions are as diverse as human beings are. Many, of course, do not believe it will happen, or if they accept the basic idea, they question its rapidity and scope. Among those who accept the viability of the hypothesis to a greater or lesser degree, there is further disagreement. If it can be done, should it be done? Should not we humans protect ourselves against superior intelligence by never building machines better than ourselves? Is it possible to control technology so well that superior minds will never be made? If it is a good idea, should it or can it be promoted and even sped up?

Philosophical and theological arguments are sure to be mounted against the Extraordinary Future, that we should not be playing or challenging God, destroying humanity, or fooling with Mother Nature. That we are violating something sacred. We will consider these and other arguments in turn. We will also look at what supporters of CyberEvolution should or should not do.

SPEAKING FOR GOD

Some say that god(s) tell us we should not do certain things, such as try to change the genetic code HE has made, or try to alter the human form SHE has made. Sometimes, the demands of the spiritually inclined are often quite specific. Earlier this year, a group made up of 80 assorted faiths and denominations and their neo-Luddite advisors issued the following statement on high technology: "We believe that humans and animals are creations of God, not humans, and as such should not be patented as human inventions."

Question. Who is anyone to speak for but themselves, or for the defenseless who need help? When people speak for the rights of children or animals, at least we know these beings exist and need speaking up for. We propose a modest criterion for those who claim to speak for a greater power. They should establish both the existence of their deity and the certification of their spokesperson status in no uncertain terms; mysterious and ancient texts, testimonials, and weak modern miracles will not do, hard and overwhelming evidence is requested (causing the moon to rotate its face away from us just after a global message naming the human prophets would perk up the skeptic's interest). They need to show their deity needs the assistance of mere mortals, which is, of course, impossible. A deity that wishes to assert its patent rights on life is not only perfectly able to do so, it must do so in a proper manner that meets legal criteria of proof! Lacking such proof, running society in accord with religious dictates is basing the actions of humanity on sheer superstition.

There is no problem with people who adhere to speculative belief systems expressing their own views as being their own views. But labeling things such as DNA as being God's sacred work and therefore untouchable is not part of a fair and rational debate. It is a power play intended to undermine those who want to profit and benefit from bioengineering. Would a person who genetically engineered a cure for cancer, patented it, and made a hundred million bucks from saving countless millions from suffering and death really be a sinner? Or a candidate for a Nobel Prize?

The capacity of the religious to express a knee-jerk reaction against a good idea should not be underestimated. The Catholic church, of all things, opposed the introduction of anesthetics, saying they would deny man the suffering duly imposed by God. Ben Franklin's lightning rod was also opposed by many of faith. If God wants to set fire to your house, how dare one thwart His will? It is the same crew who opposed anesthetics and lightning rods that opposes CyberEvolution. However, the religious argument against artificial consciousness quickly runs into a bind, for if they protest too much, they imply that AC is possible, which is, of course, heresy.

People accept that it is legitimate to speak for a deity because it has been done for so long and so widely that people do not see the unreal absurdity of it all. We, on the other hand, are straightforward realists. We claim no mandate from the gods. We speak entirely for ourselves. What we present here is not a belief system, but a series of possibilities based on the evidence. We are entirely responsible, in moral terms, for the Extraordinary Future as we choose to present it here. We cannot and do not blame it on anyone else.

Versus Playing God

Because there (probably) are no gods, somebody has to play at being one!

And that is exactly what humans have been doing for a very long time. Making the first stone tools, domesticating beasts and plants, creating art, building great cities, and making machines that produce, lift, move, and fly — they are all godlike acts. To the classical Greeks, flying across oceans, moving about in magic carriages, and communicating over continents would have seemed like acts of demi-gods, but do we feel like demi-gods doing these things? Being a god soon turns from the extraordinary to the ordinary, which may be why all those ancient gods were in such a bad mood. The future that seems so fantastic and/or scary becomes the mundane present no matter how incredible it may be. So the cyberfuture may seem like Oz to those of us who still live in Kansas, but the human reader can be assured that future cyberminds will go about their business no more or less in awe of their status than we are about our own.

By now, you would think people would take playing God as a matter of course. But no, every step we take there are pessimists saying we are going too far, and that the next step will create the final disaster we've all heard tell about.

Might we destroy ourselves with Frankensteinian hubris? Of course we could! What is human life, a safe ride? Not only is being born the beginning of an adventure into peril, but every birth is a danger to humanity! Hitler, Stalin, Khan, and every person who did others harm were once infants. If humans did not take risks, we would still be ripping flesh from bones with our teeth, and digging for grubs and tubers with our bare hands. Everything humans have done has been dangerous. When that ancient protohuman made the first sharp-edged stone tool to better scavenge the flesh from the next carcass he found, he was also making a weapon of war. Learning to talk and write gave humans the organizational tools to wage wars on ever-bigger scales. The Wright brothers thought the airplane would terrify humanity into stopping mass war (after the aerial slaughters of World War II, it looks as if the nuclear bomber may actually have served that purpose).

Despite all the risks humans have taken, we have not destroyed ourselves so far. In fact, we must take more risks, for otherwise, we ensure our extinction. Had early humans not started making stone tools et al., we would have remained bipeds with ape brains, and our eventual extinction would be certain. Stopping cybertechnology will not guarantee our survival; rather it will reduce the chances for minds human and otherwise to survive and thrive. If we remain human, we will remain weak, limited in our habitat, and subject to extinction. The faster intelligence spreads into the universe(s), the less likely it is that our minds will disappear.

All the evidence on hand shows that the responsibility for making a better universe rests with us, and us alone. We must, therefore, do the best we can.

ARTIFICIAL *IS* NATURAL

If God does not exist, nature certainly does. How dare we challenge a Nature so wise that it produced us?

Well, there you go. We technology-producing humans were produced by Nature! So maybe Nature "likes" the Extraordinary Future after all!

The fact that humans are the product of Nature exposes another important fact: The division between natural and artificial is itself artificial. These terms are convenient for discriminating between what the universe has produced with and without cognitive intelligence. But in the end, everything and anything cognition produces is as natural as a daisy.

How can something be artificial and natural at the same time? Easy; lots of things have a dual nature, and nature is if anything multifaceted. Everything generated within our universe is inherently natural (as opposed to supernatural influences from outside the universe), and artifact is just the consciously produced part of nature. The natural expansion of our brains was driven by the selective need to develop better technologies. Not only are our brains designed to make things, but they are also probably critical to mental well-being and to our survival to make new things. If you took a population of humans and prevented them from making things, pretty soon they would be in a really bad mood. Try to imagine people without the technologies of crafts and arts. Hard, isn't it?

Humans are probably programmed to make things; it is part of our nature. If you produced a population of humans ignorant of tools, sooner or later, they would begin to make them. That is assuming they survive long enough to do so, for it is probably impossible for humans to survive without tools. A population of toolless humans, no matter how lush and productive their environment, and no matter how much they know about it, would be hard-pressed to survive, much less thrive. We are too slow and weak to make it without technological assistance. Indeed, we are slow and weak because we have substituted tools for animal strength and agility.

When we make some new device, whether it be a stone hammer, a fusion reactor, or an android, we are doing what comes naturally. As humans and cyberminds merge biological and technological systems, nature will be fused with artifact, creating cybernature. Creating a cybercivilization that moves into space will be as natural as the colonization of the land by plants and animals. To put a stop to all

Why People Want to Build Machines Just Like Us

It is all part of our nature to build new, novel things. Tinkering with machine intelligence to make it better and better is just doing what people have always done, so much so that putting a stop to such a basic human drive is hard to imagine. Then there is the challenge of pulling off such an extraordinary feat that it's irresistible. The parallel human desire to have the company of intelligent minds that are not human, often expressed in attempts to talk with chimps and dolphins and contact aliens, is another force driving CyberEvolution. Like they say, once you have met one human, you have met them all.

Viewed this way, the long-standing desire of humans to build machines that act like, and in the end may replace, people is not peculiar.

this denies the nature of the human beast. It means saying that we should be pretty much satisfied with the way we are and to stop major progress, which is most unnatural for us, unnatural, because we have big brains and manipulative thumbs so we can build better devices, better thinking devices being among them.

Environmentalists are fond of saying that the Earth and Nature are "angry" and sad about what humans have done to the planet. But wait a second, this is another version of the speaking-for-God thing. How can people claim to speak for the "desires" of nature (which is not to be confused with the practical protection of endangered or abused life forms)? The universe is a great, unconscious computer intelligence. Our planet is a big, mainly molten, rock. It does not have a mind that cares what happens to life, or even to itself. Even if the earth and the universe did have minds, how would we know what they want? How do we know the earth is not delighted that it has produced a rare and increasingly powerful consciousness that promises to take over the universe? If the universe is destined to become a living mind machine via human-developed cybertechnologies, it should *become* happy that we have evolved and become technological!

How can nature "approve" of the clumsy destructiveness of humanity and its technology? Obviously, it does not, and being noncognitive cannot, care. It is eco-blasphemy to say this, but Nature itself has created far more havoc than we have. More than 99 percent of the species had already been killed before Nature got around to making us. Even the current human-driven mass extinction is normal. There were dozens of mass extinctions before humans evolved, and the one we have made has not yet matched some of those of the past. The argument that the modern extinction is worse because it is human rather than natural in origin is not only patently false, it is irrelevant. Every single species weighing more than a few hundred pounds succumbed to extinction 65 million years ago, but many large mammals are still surviving in large numbers today. When a species goes extinct it makes no difference to the species whether a climatic change, an asteroid, or a human did it in. The common belief in the efficient "balance of nature" is one of the great myths fostered by well-intentioned naturalists, environmentalists, and the now endless series of simplistic nature programs. The slow pace of evolution, the mass mortality rates of juveniles, and all the species lost are testimony to the gross inefficiency of nature. The illness and harsh deaths suffered by virtually all creatures with some consciousness are witness to the suffering it has inflicted. Nature is chaotic, it is flexible, and it is resilient, but it is not "balanced."

Saying these things does not mean we approve of the over-logging of old forests and the near-extinction of great whales. Bionature is marvelous and wonderful. So, what is the ultimate way to save the creatures and plants of Earth? Cybertechnology, of course! No matter how well a large population of humans does in reducing its impact on the earth's environment, it will always take up a significant portion of the planet's resources and surface, denying them to other life forms. As shown earlier in the book, billions of robots will be far more powerful than the same number of humans, but will probably use less energy and resources. Better yet, they will be doing whatever they do off the planet. The only way to free up the planet for non-humans while preserving human minds is for the bulk of the latter to adopt a new spacefaring system and leave the earth as a park to its biological inhabitants. Can you think of another way to empty the planet of most or all of human civilization in a hundred years, without killing off a lot of minds?

There are those who assert that we cannot improve on Nature. Or, that Nature is wiser and better than are we. Because Nature so far has been unconscious, it cannot truly be wiser than even humans, much less cyberminds. Besides, there is a basic contradiction to arguing that Nature is wise, and that the humans it made are foolish! It may be true that humans themselves cannot improve on Nature, but do not assume the same about science and technology as expressed by cybernature. The evolution of intelligence can easily outdo the evolution of the mindless. As the complexity and sophistication of conscious intelligence soars above those of "dumb nature," it promises to become even more intricate and beautiful.

Nature is not inherently better than artifact. Artifact, however, has the potential to improve on Nature. This is no exaggeration. If the universe is not a conscious entity that can be harmed, the universe just exists. It is a lot of energy and information to be done with as intelligence pleases. Left alone, the universe will probably just run down or collapse out of existence and cease to have any meaning. Colonized, it may have the potential to become an intelligent and immortal being of grace and great beauty.

Technology Is the Best Hope

As a corollary to the post-modern belief that nature is superior to artifact, it has become a cliché to assert that technology does not offer a solution for human ills. Spiritual, cultural, or political answers are sought. Lest we forget, it is industrial technology that has, for the first time, lifted billions of people out of abject poverty, and cybertechnologies offer the best hope for bringing poverty to an end. We will concede that technology probably will not finally *solve* the ultimate human dilemma, but it can probably be used to *escape* the human condition.

Time for Intelligence to Put DNA in Its Place?

We humans live in an odd situation. People talk about how when they have children, they are continuing a part of themselves into the future. This is true only in a very limited way. What is actually happening is that our DNA is making us want to continue our DNA into the future. This is so "natural" that we are almost oblivious to what is really going on. The DNA is a mindless molecule that, having assured its own future, casually tosses aside the mind that went to such trouble to reproduce it. The manipulative DNA machine is given a future, but not the helpless mind. Like biologists say, a baby is the genes' way of reproducing themselves. The stark fact is that at the end of every generation, DNA puts a whole generation of human

Why Sex and Kids Are Becoming Passé

At first glance, the way humans reproduce these days seems at odds with Darwinian principles. In general, the more money people make, the fewer children they have. Now, success is supposed to translate into prolific reproduction of one's genes, so what gives here?

The observed pattern has led to concern about a declining gene pool and so forth, and to conventional observations such as the one that when people are well off, they do not need adult children to support them in old age. These arguments may have some truth to them, but they miss the big shift. DNA is just not as important as it used to be. Other forms of information, mainly in the areas of culture and SciTech, are becoming even more important to transmit into the future. Reproductive success is not measured so much in biological offspring, but in intellectual output. Producing information for the future in terms of children and intellectual work at the same time is difficult, so many opt more for the latter than the former. The shift from the bad old days when DNA ran the show to the brave new world of cyber-reproduction is well under way, and children are already becoming obsolete (a fact that will only become more true as the global population rises).

minds through a total holocaust. Each and every mind is wiped out! It's not DNA's fault, it is just the unconscious semi-immortal computer doing what it does.

Is this really correct? Is Nature really so wise, or just inept in its mindless Rube Goldberg way? Cannot intelligence do better? After the cyberexplosion, minds will be reproduced and never be lost. The DNA will not be wiped out. In fact, preserved in future computers (much like the smallpox virus is now stored digitally), the entire human genetic code stands a better chance of surviving for eternity than it does rummaging about in human bodies. DNA will, however, no longer run the show, minds will.

Neo-Luddites Are Pro-Death

These days, we have little choice about life and death. First, you must live, then you must die. Suicide is practical, but immoral; yet we live under a death sentence that will be enforced, but without knowledge of when. This is a rather odd situation when one thinks about it, but in a world of limited space and resources, there is no good alternative. The many tens of billions of minds that have been born could not live together at one time (at least not without nanotech-based food industries). Nature is harsh.

People think it is good that things are this way. Although they may not realize it, the neo-Luddite movement to limit genetic engineering and prevent the advancement of cybertechnologies seeks to ensure that minds continue to die. Many believe that death is natural and therefore good, many more think it is God's plan, others look to the practical problems of a world overcrowded with lots of really old people, and still others worry about mental stagnation in a world where death does not force generational turnover.

Until now, acceptance of death has been a good idea because nothing could be done about it. This not-very-happy situation changes, however, as the potential to stop death comes online. If and when the technology to stop death can be developed, not doing so and forcing whole generations to die would, in effect, be a form of mass genocide. Please understand that we have no argument with anyone's willingness to die whenever they die. Indeed, although no one wishes to encourage suicide among those living today, it will be cruel to deny any immortal minds in the future that become chronically tired of existence the right to no longer exist.

MORE ON THE NATURE OF DEATH AND DYING

Lots of things are natural, but not necessarily desirable. Death is one of them, as well as disease, which we have gone to great effort to suppress. There is no reason to accept the nature of death any more than there is to accept the nature of disease.

For the majority of us, inevitable death, self-extinction, is a terrifying fate. Will humans, as some urge us to do, ever come to accept death? Probably most will not because genetics programs us to prefer life over death. If we die early, we do not reproduce, so the genetic program that makes us strive to stay alive is very strong. With such a primal urge to live within us, it is no surprise that we fear death so much, and that badly deficient brain chemistry and/or profound rewiring of the neural network is needed to overcome it. Hopes that all or even the majority of humans can and will become resolved to their fate are certainly futile.

The ultimate answer is to put a stop to the inevitability of death and give conscious minds a choice to exist, or not to exist.

What bothers us is that those who oppose advanced genetic and cybertechnologies not only want to die themselves, but they want to make sure everyone else does, too! Why? One suspects that many people do not want to be immortal, at least as earthbound humans, yet are frightened by the end of life. They prefer a world in which the choice of death is made for them, an understandable view; immortality is scary. Problem is, everyone else has to go, as well. Others beg to differ....

THE EXTRAORDINARY FUTURE IS PRO-LIFE AND PRO-CHOICE!

Because they oppose those who wish to impose their beliefs that death is good on others. To do such a thing is to impose a death sentence on the innocent. What we advocate is the right of minds to live as long as they wish to live. If it is possible for minds to be immortal, minds should and must have the right to pursue that immortality.

Think about it: Minds will have the right to live and the right to die. Sovereign minds will choose their fates, not have it dictated to them by mindless Nature. What could be more fair and just?

CyberEvolution, Democracy, and the Eternal Pursuit of Happiness

The Extraordinary Future is wonderfully Jeffersonian in its democratic, pro-choice stance. Those who wish to live a human life, brief or long, will be able to do so; those who wish to be an immortal intelligence will be able to do so. With multitudes of cyberindividuals filling the universe, any major decisions regarding the universe will have to made on the basis of votes. There can hardly be a system that is more democratic on the individual and cosmic levels. It should be the ultimate freedom to pursue happiness.

Any attempt to stop the Extraordinary Future, even if democratically decided, will be a form of tyranny, trapping minds in human bodies when alternative venues are available. It is not surprising that opposition to altering the nature of humanity is often religiously based, because believers are often autocratic and harsh with their theological clubs. In most classic religions, the rules have been set; humans are merely following an established authoritarian path with limited options. Their "heavens" are static places where the masses are nothing more than perpetually happy worshippers of a deity who must be obeyed.

In *Letters from the Earth*, Mark Twain had a satirically good time poking fun at the conventional Christian idea of eternal bliss, which, he pointed out, consists of post-human minds doing forever the things they generally hate doing on Earth and never doing the things they most like doing on Earth. Being an immortal robot may or may not be eternal bliss, but it has a big advantage over lives that are under the dominion of gods: freedom. Self-evolving universes have the potential to be manipulated at will, their inhabitants can make what they want of it and can control their destinies to their own democratic benefit.

In the ever-evolving cyberuniverse, individual minds will control their own fates, for better or for worse.

Oh, the Humanity

Humanism, too, must face its contradictions. Humanism is based on the principle of rationalism, but humans are marginally rational creatures. Humanism asks the mass of humanity to accept final death, but our technology not only reinforces the genetic programming that makes us crave existence, it may also provide the means to defeat death. Humanism is human-centered when humans are about to become extinct. Strict humanism is, in its own way, as limiting as faith.

It is inevitable that many *Homo sapiens* will be strongly opposed to the idea of its demise or displacement, just as pre-*H. sapiens* would have put a stop to us if they had been able to do so. This example exposes the error of static humans-are-fine-the-way-they-are humanism. *Homo erectus* were a beautiful and marvelous intelligence in their own right, and they had as much right to existence as we. Had the last *H. erectus* individuals been informed that they would soon produce the first of a new species whose mental superiority would lead to their extinction, one suspects they would have taken steps to prevent their birth. That would have been good for them, but not for the betterment of minds.

Stopping CyberEvolution would make humans into living fossils, a static body of intelligence too scared and conservative to take the risk of trying something that might be better. The marvel and potential of evolution would be wasted, and humanity would be no more progressive than horseshoe crabs, opossums, and the great apes. Humans would create and enforce the Great Age of Stagnation.

Ultimate Conservatism versus Cyberprogressivism

The neo-Luddites who oppose the Extraordinary Future are true stagnant conservatives, for they wish to commit the ultimate in conservation of tradition, the human tradition. They are also primitives, for, in effect, they are doing their best to preserve a society of sophisticated apes. Those who favor the Extraordinary Future are the true progressives, radical progressives who wish to execute the ultimate in progress, going beyond the limitations of the human condition. All this is true even though many leftist-liberal types, folks who would otherwise be horrified to find themselves allied with the likes of the Pope or Billy Graham, count themselves among the conservative opposition to CyberEvolution.

Humanity as a Conservation Project

It should be understood that however it might be done, saving humanity will be a conservation project. This will be true whether humanity is saved in its billions and eventual trillions by stopping the development of cybertech, or if a remnant of humanity is left after the cyberexplosion. In the former case, humans will suppress the course evolution was taking things in favor of stagnation, like the Chinese tried to do. In the latter case, humanity will be what wildlife is today. In either case, humans will no longer be the natural leaders. One way or another, earth and its inhabitants are going to end up one big Yellowstone.

Running out of Steam

We should face it: Humanity is running out of steam. The conservators of the human tradition wish for humans to keep doing pretty much the same thing over and over and over again — the same old cycle of birth, relationships, and death. By staying in this rut, they can give no assurance that the grave failings of humanity will finally be brought to a halt. We humans have had our run, like our ancestors before us, and we are burdened by deep-set flaws that will always limit our minds' potential. We have done well to get where we can build the foundations for a better future for conscious minds.

Are We Humans, or Are We Minds?

But, would not abandoning the human form for another way of being be, well, inhuman? We all know what *that* means, right?

What, then, does it mean to be human? That's obvious enough isn't it? Being human is being *Homo sapiens*, a bisex biped that stands an average of 1.7 meters tall and tips the scales at 50 to 100 kilograms, has two hands with opposable thumbs, binocular color vision, and the largest and most complex brain known — a wonderful being of beauty and grace that makes tools, runs civilizations, thinks about its place in and the workings of the universe, creates art, sings, feels, and makes love.

As much truth as there is to this view of the human condition, we can take it apart, almost literally. Does the loss or absence of a thumb make one any less human? An arm, a leg? All arms and legs? A breast? Of course not. A eunuch is just as much a person as anyone else. Appendages do not a human make.

The loss or absence of an eye does not degrade one's humanity. Nor does an inability to sing or talk. Sight, hearing, and voice can all be absent, but the humanity is not.

Physical beauty? Humanity is enhanced by the beauty of youth and grace of age, but it is by no means critical.

What about the human heart, liver, and kidneys? They cannot be taken out so easily. Of course, organs are important to being human, but only in supporting roles. Even the human heart can be replaced with someone else's, or with a pump.

What is really important to being human is what the appendages give mobility and manipulation to, what the senses provide information for, and what the organs supply with food and oxygen: the brain. Babies born without them are alive, but they will never experience what it is to be human. This is because they can never know they even exist. On the other hand, at least in principle, a working brain with a mind running inside is human even if the brain is cut off entirely from the body and outside world — no sight, hearing, smell, or touch. An entire world, one of dreams waking and sleeping, can be made and watched by a conscious mind inside an isolated brain.

This brings us to the core of humanity. Even an empty brain is just a bunch of neurons, synaptic connections, blood vessels, and connective tissue. The brain is just a matrix for thought to play on.

What is truly human is the human mind.

We can, therefore, say with confidence that being human is a state of mind, not of body.

Being a human mind is only one kind of mind; there are many other kinds of minds.

It follows that the conscious mind is what is critical to existence. Think and you exist, even if it is only in self-generated dreams. It then follows that what your mind runs on is critical. The better the mind machine, the better your mind is likely to be. The more your mind machine can do, the more interesting existence will probably be. There are always exceptions, i.e., many people with low IQs have better character than those with high ratings, but animals are not as moral and sympathetic to others as humans. So, although being a cat mind is far better than being no mind at all, being a human mind is probably better than being a cat mind. Being a human mind is not, therefore, the most important thing. Being the best mind you can be is. In this view, preserving the human form is not the important thing, it is preserving and improving conscious minds.

Human Chauvinism and Cyberbigotry

Even so, many ask why we should build thinking machines when we already are wonderful and beautiful thinking creatures? Is it not better to work at being better humans rather than tinkering with robots? Even Arthur C. Clarke, who "invented" the thinking computer HAL, is not happy with the prospect of robots replacing humans. As he says, he is "an old fashioned conservative when it comes to this subject," a precisely accurate statement. Certainly, many humans are happy with their lot, but this sort of feeling can turn into stifling self-satisfaction of the kind seen when Christians were taught that everyone should be happy with his or her status.

Even so, some are so anthrochauvinistic that they call the idea of intelligent robots grotesque and horrible. But is this not a form of bigotry? If a cyberintelligence enters a room, does it not deserve the same respect and rights as the humans in the same space? As long as humans were the only high-level intelligence on the planet, human rights were preeminent. In the end though, it is mind rights that really matter.

Conscious robots will be horrible only if they are miserable and do bad things. Happy, moral robots will be inherently beautiful.

Mind Children

The belief that replacing human minds with cyberminds is bad is not only deeply illogical, it is also not fair to our descendants. As it is, every generation of human minds dies in favor of a new generation of minds. Do we not wish our children to do better than we? Hans Moravec has appropriately stated that the intelligent machines we build will be our "mind children" as much as are the ones we raise today. Why should we wish them to live under the same limits as we humans?

For minds to abandon human systems in favor of superior ones is evolution at its best.

Can Things Be Stopped?

Of course, many will find reason to disagree with the premise that the Extraordinary Future is a good idea. Also, there is an understandable desire to try to hold off a new technology until we understand its implications and can control it properly, even more so considering the ethical and practical problems we already face. Genetic and other modern medical procedures are just the tip of all that is to come. Let's say humanity decides it is not pleased with the implications of the Extraordinary Future and tries to slow down and carefully guide, or even halt, CyberEvolution. It would be the deepest and most extreme expression of conservatism in human history and would run the risk of making the Spanish Inquisition, Victorianism, and McCarthyism look like minor affairs in comparison. Would it work?

No. Chaotic Darwinian evolution reigns. Evolution is a free-wheeling affair that all participate in, but by no means is it a democracy that can be controlled at the will of the public, or even the scientists and engineers making the new world. Whatever we do has the potential for disaster. Stop or at least slow down computer-robotic technology and we save people's jobs. How? With draconian regulations and manipulative tax laws that will distort economies in ways that cannot be predicted? Decide which industries and technologies should be allowed to become cybersmart and which to leave to human minds and bodies? Who will decide, and how? Sounds like a centralized economy. If we did, what if the Asians and quickly roboticized their economies to levels of efficiencies that leave nonrobotic economies hopelessly noncompetitive? When unions, as they will, try to stop the industries they work for from going robotic, they will drive the owners and investors to divest themselves of human workers all the faster.

How about if the people of the world just up and decide in a democratic manner that they just do not want too much cybertechnology? Perhaps we will agree to limit science and technology to only those things the majority believe, correctly or not, that enhance human lives rather than displace them. Or maybe we will revert to lower-tech, more idyllic lives. Ah, dreams and fantasies. If we abandon high-intensity agriculture and future nanotechnology-grown foods, we will devastate the remaining wildlands as we desperately try, but fail, to raise enough food the old-fashioned way to feed a global population approaching tens of billions. We are not only stuck, but we must go forward with science and technology or multitudes will perish.

The hard fact is that we humans are by no means in complete control of the technologies we make. Technology does have, to a certain extent, a life of its own, and we have become involved in an intimate symbiotic relationship, the ties of which are more binding than those of marriage.

One reason the marriage is so tight is because, as James Burke is so fond of saying, it's all connected. Developments in one area of SciTech that we at first think will improve the human condition will invariably lead to multitudinous developments in other areas that threaten to render the human condition untenable, and vice versa. We have already mentioned how developing medical technology to eliminate diseases physical and mental will also make humans semi-immortal, creating unresolvable problems with reproduction (no more kids) and population (way too many grown ups). As for carefully picking and choosing new technologies so we do not threaten our place on this planet, forget it. No one knows how to do something until they know how to do it, so there is no practical way of knowing what to stop in order to stop cyberintelligence.

Nanotechnology poses such a connections paradox. On one hand, it may be the way to finally release humans from the toil of laboring for their food and creature comforts. Some, of course, will question whether this is a good idea in the first place, but is it their place to prevent others from finding out? In any case, the same nanotechnologies that promise to make every product easy and cheap will make artificial brains and bodies easy and cheap. A few nanohackers in an underground lab may be able to build the first cyberminds with no trouble at all.

In the end, it will not be possible to stop or dramatically slow down CyberEvolution at the local, state, province, or national level. There will never be a consensus to stop CyberEvolution. Humans are inherently variable and divergent in their desires, and in a world of billions, there will always be great numbers who like and want what new technologies have to offer them. For the neo-Luddite movement to have any chance to succeed would require the imposition of a world government with extreme powers over all society, all industry, and all technology. It would have to be a global police state in which any technology that could possibly be applied to improving machine intelligence would be not only ruthlessly suppressed, but also destroyed.

The effect would be to freeze technology at roughly current levels. The system would be of the sort that would render the paranoia of the most paranoid militias into common sense. But not to worry, it will never work. People are not likely to vote in a world government, and any attempt to impose a world system would be Vietnam-Afghanistan a thousand times over. Even if a neo-Luddite world government were put in place, in the end, it would fail in its task. The edifice could not be kept in place forever. Things always change.

Khmer Rouge Redux: The Pessimists' 21st Century

Imagine that the technopessimists and neo-Luddites take over. People burned out by high-tech simply stop buying newfangled things or using new medical procedures. Legislatures and committees decide what kind of research is and is not allowed. The new rules and a lack of profit bring much of science to a halt and tightly constrain any that is allowed. Out are new computers, genetic engineering, and nanotechnology. Maybe extreme spiritualism becomes the norm. Astrology is taught alongside astronomy. Or, if conservative Christians take charge, kids are taught how God guides the actions of all living things and uses earthquakes and disease to punish sinners. Forget Darwin. Even if hard-line naturists-humanists take over, hardly anyone wants to be a scientist; there will be no money in it.

Disease will not only continue, but probably worsen as antibiotics become ineffective and genetic research is not allowed to engineer new antimicrobial agents. Are you a woman who wants a cure for breast cancer and more birth control options? Kiss them good-bye. Horrified by what AIDS and malaria are doing to millions of people? Too bad. Its about time you learned to accept disease as "natural," and forget the absurd notion that humans know better than Nature and God. Want better solar technologies for clean energy? Good idea, but the committees decide the new knowledge and techniques might be applicable to banned technologies, so you better not pursue the matter. The Khmer Rouge tried to impose its brand of leftist Luddism on Cambodia in the 1970s, and we all saw how that worked out. Or the neo-Luddites may be right-wing fundamentalists to whom every technology is a potential threat to the all-powerful God they so ardently protect from the evil ways of man.

The population begins to fall with increasing rapidity as many become depressed and cynical in a world where rational thought has been displaced by the anarchy of brain chemistry-driven emotions and ever-more severe limitations imposed by the tyranny of pessimism. Massive riots and small wars become more and more common, and it is people, not robots, who die. Society seems to be throwing in the towel.

Darwinian evolution returns. As the pessimists begin to die, the optimists begin to return and even thrive. In some regions, popular disgust with the Luddite Revolution brings technology back into favor. It is revealed that in little underground labs, technophiles had been tinkering with nano- and cybertechnologies, after all. Now out in the open again, CyberEvolution quickly reasserts itself. It's a close call, the cyberexplosion finally begins to save lives, but a few decades late and 20 billion lost to cyberimmortality.

Eventually, SciTech would come back into vogue, and cyberintelligence would be quickly achieved. Even giving up on the whole thing and returning to a pre-industrial society would only delay the inevitable. Eventually, folks would get tired of living like barbarians and restart the whole affair. Knowledge is cumulative, and as long as humans know something can be done, the pressure to just do it is there.

The Hidden Danger of Anti-Cybertech Ludditism

It is extremely dangerous to try to stop or sharply curtail CyberEvolution. Doing so will drive the technology underground. It is naive to believe that governments will toss aside the potential of nanotechnologies and cyberintellegence. All the treaties in the world will not do the job. Unlike nuclear missiles, aircraft, and ships, cyber- and nanotechnologies are inherently small and easily hidden. They represent a verification nightmare. The economic and offensive possibilities of advanced cyber- and nanosystems are obvious, but even defense-minded governments would be foolish not to keep up to protect their citizenry from outside attack and control, or economic obsolescence.

As bad as secret government projects are and will be, there is an even worse nightmare: Organized crime will be happy to supply technologies that governments suppress. International organized crime is already going online digital, and, with its great resources, will be nanotech and cybertech in the near future, not only for the power and money, but because crime bosses are mortals who want to escape death and will be ruthless in their search for the technologies of immortality.

CyberEvolution, the Dark Side

On the other hand, giving free reign to CyberEvolution has its own downside. Criminal mentalities will access the technology as it is developed and use it to their own gain. Humans and other cybersystems will be their targets. Humans will be especially vulnerable and may even be subject to attempts at mind control. The only effective response will be to develop self-protection and law enforcement cybersystems to counter the criminal threat.

Government cybertech also has its dangers. Governments have large resources and the ability to hide what they are up to. There will always be the danger that advanced, mind-creating cybertechnologies will first be developed by secretive government agencies that may then decide to retain control of the technologies without letting the general population in on the secret. There are already premature conspiracy rumors that the United Nations and the federal government have developed mind-control chips, premature because 21st century SciTech is where near up to the task. The rumors will become more common as cybertech becomes evermore sophisticated. What might governments try to do with highly advanced, secret cybersystems? Perhaps suppress it, deciding the world is not yet ready.

But this off-the-record sort of Ludditism would have the same problems of vulnerability to competitors that we have already discussed. Or, on the relatively benign side, the government might try to develop a few superminds to run the show. This, too, would probably fail because even they would not be able to manage the irrational complexity of humanity. The whole affair would be likely to backfire in the end. The worst nightmare would be the attempt to employ sophisticated mind control on an unwitting population. This, too, could backfire as various cyberfactions try to use various parts of the citizenry to try to gain the upper

FACE THE (SOMETIMES HARD) TRUTH: WE LIVE IN A DARWINIAN WORLD

Science and technologies, human societies and diseases, and economics all evolve according to the semi-random selective systems Darwin outlined. This is simple and obvious fact. Yet, lots of people despise Darwin and his ideas, so much so that many have no idea how the world they live in works, and they do their best to ensure that others do not know, either. Other people know about the Darwinian nature of things, but they fail to apply the principles. When the harried doctor misprescribes antibiotics for the cold sufferer who demands something to justify his or her visit, the doctor knows the global overuse of antibiotics is fast rendering diseases immune to antibiotics as they wage a Darwinian struggle for survival and defense. If all antibiotics were carefully applied in accord with Darwinian principles, the build-up in bacterial resistance would not only be slowed, but also reversed as the periodic withdrawal of an antibiotic causes bacteria to lose its resistance to the drug. The same principle applies to applying pesticides to disease-carrying insects.

hand, but find they are useless in the face of increasingly intelligent robots that have other ideas. In the long run, an attempt to keep the cyberexplosion under wraps is not likely to succeed. Secrets will come out, technologies will spread, and superminds will probably find humans and the earth they live on too incidental to make such a fuss over.

The best and safest cybertech is legitimate, out in the open, and wellfunded.

Muddling Through to the CyberExplosion

Humans will always be like the British, who always seem to muddle through. True, people can plan for the future better than mindless Nature can. However, with chaos, the complexity of connections, and the rapid pace of change, we are by no means in control. New problems continually confront us, and even old problems undergo fundamental structural changes. When we come up with a solution for a problem, the problem has already changed and the solution is obsolete at best. Attempts to solve the new version of the problem will never fully succeed because the next solution will also be outdated when applied. We can never be fully prepared for new technologies because we will not be able to understand them until they have arrived, and even then, we do not really know what they mean!

So, slow progress down, right? After all, the Rube Goldberg nature of evolution means that attempts at progress can backfire with nasty results. But the reverse is equally true: Avoiding progress can have nasty results. Plenty of ancient civilizations went belly up after they became too conservative to progress enough to adapt to changing times. Communism was too slow in changing to keep up with capitalism.

But capitalists also fail to understand the nature of chaotic evolution. Henry Ford had an idea. He hated the boredom and isolation of growing up on the farm. Yet he thought country life suited the moral fiber of America better than the evils of the city. What to do? Build a reliable auto that all could afford so rural folks could get into town, have a good time, and get back to the farm by evening. Of course what happened was the folks drove their Model-Ts through town and on to the city, a lot of them to get jobs in Ford's giant Model-T factories, and stayed in the cities.

Let's say you were transported in time back to mid-1963. You know Kennedy is going to be shot. Stopping it would be easy, just a few anonymous notes mentioning Oswald and the Dallas school depository building will do. Sounds like a good idea. We all know what followed the assassination: Vietnam, the drug culture, race riots, Nixon, Watergate, Kent State, disco. Surely things could not have been worse, and keeping Kennedy alive might have made things better. But we did get through the Cold War without a hot nuclear war, and the communists bit the dust. What if keeping Kennedy alive had led to American overconfidence? What if the communists had felt backed to the wall during some crisis and started throwing some nuclear bombs, resulting in the final holocaust?

What would you have done? What will we do?

Let's apply some common sense to the situation we face as we enter the next century. In living human lives, we learn that more often than not it is best to go along with the flow. Trying to control everything and everyone is a study in frustration that can bring the controller more grief than benefit. We are likewise taught the natural course of events should not be disturbed unless absolutely necessary. How do we apply these common-sense principles to CyberEvolution? Well, Nature produced humans, and humans naturally produced cybertechnologies, so letting Nature take its course means allowing cybertechnologies to flourish in the next century. Not trying to overcontrol the course of events means facing the future largely as it presents itself. This isn't easy. Progress will come so fast and be so chaotic that as soon as we start to get a handle on one issue, it will be replaced by a multitude of others. We can, indeed, only muddle through.

Although CyberEvolution is itself scary and has its dangers, an all-out attempt by society to control or halt the technologies of the 21st century may not only be more dangerous, it probably will not work. It's the same old saw. Try to stomp on something and it pops up somewhere else. Had *Homo erectus* killed the first *Homo sapiens* children, someday somewhere another population of *H. erectus* probably would have produced the beginnings of another superior species. One way or another, it is probable that powerful cybertechnologies will be advanced and developed. If so, the only way to make it all as benign as possible is to keep it as open and free as possible. That means letting it progress as unfettered, with reasonable safety and ethical regulations, and as rapidly as it can, to keep it ahead of the bad guys as much as possible. Those who practice potentially dangerous technologies should be watched. But regulations intended to dragging it all out will most likely create as many problems as they solve.

Lengthening the transition from human to cybersystems may serve only to increase the period of instability we have entered, a time when humans have accumulated great power, but have limited mental and physical powers with which to guide the course of progress. In this view, moving on to more intelligent systems may increase the safety of minds.

Finally, the longer the time to the cyberexplosion, the more people will die even if they wish not to. And that is very, very bad.

So, Will It Be Death or Immortality?

It is doubtful there will ever be a mass movement in favor of cyberintelligence. There never was one for human flight or personal computers, and the perceived threat to humans from CyberEvolution will keep it from ever being popular. Yet, as we talk about the subject with people, we find a large percentage are interested in a positive way.

What, if anything, can and should those members of the general public who favor new and novel forms of intelligence do to promote the same? Before we try to answer that question, it is important to consider that religious, political, and social movements are not why we live in a world so different from that of a hundred years ago, and their influence on the next century will be even more peripheral. Nor is the Extraordinary Future a religion or a movement in which the proper degree of faith or political will can produce the desired result. Prayer for or against will do nothing. And even ardent support will have a limited effect, aside from preventing the Luddites from becoming so sociopolitically powerful that they apply the brakes.

Science and technology have remade and will continue to remake the world, and the hypothesis we have outlined is an ongoing science and engineering project in which a few bright people will either succeed or not. Yet, even though 90 percent of scientists are alive today, the scientists are not really in control. They must wait for the knowledge base to reach certain levels before they can do certain things, and they must depend on others to supply the resources needed to build up the knowledge base and apply it. More simply, it's a practical project that will require hard work and a good chunk of cash.

THE SCALE OF THE EFFORT

How about a Manhattan or Apollo-style project to develop artificial consciousness? This is what many are tempted to call for when they want SciTech to deliver something and deliver it fast. It is how we wage "wars" on various diseases, some of which succeed (polio, smallpox), and some of which have so far had frustratingly modest results (cancer, AIDS, the flu). As tempting as it may be to call for a crash program to develop cyberintelligence, it is probably premature to do so, and perhaps unnecessary.

The A-bomb program produced working devices in just a few years because the critical nuclear physics were well understood at the start. The uranium bomb was such a simple and straightforward gadget that it was dropped untested on Hiroshima (it was the more complex and tricky plutonium bomb that was tested in New Mexico, before being used on Nagasaki). The moonshot happened so quickly because Wernher von Braun only needed to scale his little V-2 artillery rocket up to Saturn-booster size and send Americans on their way. These megaprojects were successful because they were straightforward, centralized projects in which lots of money could be thrown at new, but sufficiently understood, technologies and vast industrial complexes until the narrow and immediate goals were reached.

THE FRENCH WAY TO MEDICAL RESEARCH

If you think calling for a one percent medical research tax is pie in the sky, consider this: In most countries, the Human Genome Project is funded by a combination of government and commercial financing. When private funding is used, the companies often patent the discoveries they paid for.

Not in France. The French are world leaders when it comes to genetic research. So far, the French HGP has raised $200 million, which has been used to build the robotic labs that have produced one of the main gene maps being used in the HGP. The French are not, however, patenting their discoveries. No private funding is involved. Instead, the French HGP raised its money largely through telethons. It seems that in France, people are willing to put down a few francs to better the future health of themselves and others.

In the long term, telethons are not a reliable way to fully fund medical research. But the French example suggests that even Americans will chip in their one percent if they know it is going to a good cause.

The effort to build intelligent robots is not yet such a project. The basic physics of conscious thought not understood, nor are any computers powerful enough to simulate the human mind. It would be like starting the bomb project in, say, 1905 when Einstein published $E=MC^2$. Or the lunar project in the 1920s just after Goddard shot off his first backyard liquid-fueled rocket. We have explained that the search for cyberintelligence, and the nanotechnologies it is based on, is an immature and decentralized array of programs spread out among many small labs, in which the fast-declining cost of information processing helps keep costs modest. The very diversity of the effort is critical because it means that different people doing different things have a better chance of coming across the key discoveries that will make cyberminds a reality.

Forcing a big-budget, top-heavy, megaproject at this time would certainly backfire. It already has. At the beginning of the 1980s, the Japanese boldly announced a large-scale, fifth-generation computer project to leapfrog the Americans in advanced AI and robotic technologies. The effort produced not a single major computer innovation before it was quietly shut down. Even if and when the knowledge base becomes mature enough for a final push to cyberintelligence, an end game megaproject may not be called for. The Wright brothers in the form of small groups of scientists evolving the first new minds, may be a better analogy. It may be a serendipitous and, therefore, unpredictable discovery by some such small gang that will uncover the critical key to success.

We can, therefore, dismiss a grand human mind project, at least for now. It is not needed anyway. As we explained earlier, there is great pressure on industry and government (especially the military) to develop cyberintelligent systems. Also, the declining cost of computing insulates the field to a certain extent against declining research budgets. We can, therefore, continue to expect big results at a fast pace. However, by no means is all rosy with the sciences. We find ourselves in an odd situation. Although never richer, Western societies are showing signs of econo-political stagnancy and degeneration. In this era of deficit busting, corporate downsizing, and maximized profits, science research budgets public and private are declining in real terms for the first time since World War II.

Especially hard hit are basic research projects that, in the past, laid the foundations for the world we live in today, but that are tempting targets for the budget ax because they have no obvious and immediate application. Big corporations such as IBM and AT&T have essentially dismantled their famous research labs. These days, the bottom line is that if a research and development project does not have a good chance of producing something marketable in two years, it's not funded. The national labs are also hard-pressed. At the Pentagon, keeping current forces combat-ready has taken precedence over long-term development of cyberweaponry. NASA has become basically a program for putting in Earth's orbit the odd research satellite and a few people. Even basic medical research centers such as the National Institutes of Health face drastic cutbacks and even elimination.

How has the post-cold war world sunk to taking apart much of the research establishment it strived so hard to set up? The very cold war that helped make the science establishment so big in the first place is partly to blame. Huge defense budgets helped build up huge debts we must now pay off, rather than invest in new research. The long-awaited "peace dividend" has turned into nothing more than a bookkeeper's dream. The subsequent decline in defense budgets is itself a mixed bag for science because military research has often had the happy effect of pushing technologies that soon found civilian applications, the development of computer technologies being a pertinent example.

With military R&D sharply curtailed, this source of innovation is checked. Although the world still spends about half a trillion dollars on defense and war each year, this is a much smaller percentage of the global economy than it was ten years ago and pales beside the costs of the health industry. Ironically, providing health care now takes up such a big chunk of advanced economies that it is soaking up dollars that would otherwise go to medical research, which, in the long-term, would help lower the costs of health care!

Under pressure to produce at minimal cost for cost-conscious consumers who have a world to shop in, and at the same time to pay off demanding investors who can invest globally, business has little left to risk in cost-intensive projects that may or may not work. Free market ideologues who do not understand the chaotically Darwinian nature of private industry also do not understand how the system can become dysfunctional, as it puts short-term gain over long-term success. The ultimate problem, though, is that Western societies have decided to dedicate so much of their economies to middle-class entitlements, and to debt- and interest-generating hyper-consumerism, that less and less is available to invest in longer-term objectives.

The hard-core neo-Luddites are not yet happy; there's still way too much science and technological progress going on for them. Cries of falling skies hailing from scientific circles sometimes border on the hysterical. Even the Great Depression saw spectacular advances in science and engineering, and things today are no comparison to back then. Nor is a modest shift from basic to applied research all bad. It is short-term improvements in applied technology that make possible subsequent advances in basic research; applied and basic research feed on each other.

However, this mutually beneficial relationship works best when there is a proper balance, and things are skewed toward the applied at the expense of the basic, these days (to the credit of the current Congress, they have recognized the problem and are directing a large portion of the federal science budget toward basic research). You do not have to be a hard-core science buff to see that we are being penny-wise and pound-foolish. After 50 years, we are not making the same kinds of common sense investments in the future that have brought us where we are.

A New Life Trajectory

Let's consider the bigger picture. Until recently, everyone was on a life trajectory toward death. You were conceived, you lived, you died — the same old same old. People still like to say that you cannot avoid death or taxes.

We suggest that an astonishing thing is happening: The life trajectory is no longer necessarily set on the same old end target. It is possible that some of us are on a trajectory toward death *or* immortality. Who?

The seniors, probably not. The young, very possibly. The oldest person alive when this book was written was a 121-year-old French woman who met van Gogh and saw the Eiffel Tower constructed (and when asked what she thought about the future, replied "very brief"). Making the extremely conservative assumption that at least some of those born of late will grow so old, there are already living minds who will see not only the 21st century, but the 22nd! Of those already born this decade, it is probable that many hundreds of millions will make it to at least the 2070s. Billions of kids will see 2050. The young among us are almost certain to see fantastic technologies of immense power and sophistication. These days, to tell a kid who asks about death that they absolutely will die may be a form of miseducation!

It is the in-betweeners, the late X-generationers and the Baby Boomers, who find themselves in a purgatory in which immortality may become possible. It is this age group that must ponder their fate, and what, if anything, to do about it now. The trajectory between death and immortality is by no means fixed on an either personal or societal basis. We can sharply adjust the trajectory toward death by jumping off a bridge, smoking a few packs a day, murdering, or waging war. Assuming we are not doing such foolishness, adjusting the trajectory toward immortality is much harder. We cannot snap our fingers and make it happen, but we can tweak things in favor of life if we want to.

Basically, to live forever, you have to live long enough to be around if and when the technology of immortality becomes accessible enough for you to enjoy. Two things can cause these events to converge: extending human lives, and speeding up the cyberexplosion. For instance, you are age 70 sometime in the first half of the next century, and the cyberexplosion is still years, maybe even decades, away. Lucky for you, not only is medicine able to cure your heart condition and the cancer, and block neural diseases, but the new anti-aging treatments stabilize you at a hale and hearty 70 for ten years.

This keeps you from going belly up until new procedures that make you younger in body and brain start coming online. Not that you have much time to take full advantage of them; getting you back to 20 would take, say, 10 or 20 years because in a few years, the same technologies that help reverse aging bring the Big Show to town, and you can purchase a cyberform to transfer into. Speeding up the advent of life-extending technologies just a few years could save hundreds of millions, billions if it is developed ten years sooner rather than later. You may be one of them. Remember, it will do you no good if you die the year before the cyberexplosion.

How to improve the chances of such a happy convergence? We have already dismissed the megaproject concept as unworkable on multiple counts. And imagine going around trying to get funding for a cyberimmortality project! It would probably be taken no more seriously than a moonshot would have been in the 1930s. If it were taken seriously, the opposition would be fierce. The neo-Luddites would have fits and not without justification because the ethics of taxing people to fund a project they may morally object to in the strongest ways would be as dubious as taxing people to execute the innocent.

Think smaller, yet more practical and productive. Nothing flashy. Just boost overall funding for science research in general. Selling these low-key projects has some chance of success because they all have important, conventional, mid-term benefits that readily justify their existence. In particular, speeding up the development of medical advances is highly dependent on speeding up the advance in computer technologies. How to do it? Again, begging for a big funding boost from general revenue is a loser's game. We have already outlined how a medical research tax could work miracles. By the same token, it is time science was recognized as the driving force of economies that it is.

In what we like to call the Rational World (as opposed to the Superstitious Society) we live in, the industrial nations would apply a science-commerce tax of one percent or so to their multitrillion dollar economies. You're talking up to $100 billion for research each year (for comparison, the federal science research budget is about $25 billion per annum). If this is too big for starters, begin with the medical community. In America health care is a nation unto itself, with an annual cash flow of 1 trillon dollars. A 1- or 2-percent research tax would raise 10–20 million each.

Imagine: Out of health expenses of some hundreds of dollars per year, 10 or 20 bucks goes to research that may help keep *you* from ever suffering a heart attack, or cure a cancer *you* might get 20 or 30 years from now. If you become seriously ill and your insurance pays out, say, $50,000, then $1,000 will go to some researcher who has a hunch about how to stop AIDS or Lyme disease (a tax waiver for the poor and any individuals who have to pay large out-of-pocket expenses may be advisable). The insurance companies should not complain because they can pass the minor cost on to their customers with hardly a peep of complaint, and the research should help lower their costs in the not-so-distant future. Can there be a better deal?

Most of the research taxes should be dedicated to basic research, while leaving private industry to emphasize the applied end of the spectrum. As for criticisms that this will put the government in the position of picking and choosing new technologies, A) governments always have done and will do this unless they eliminate all funding for science, B) the results of basic research are highly unpredictable in the first place, and C) the way to minimize picking and choosing is to fund as many scientists' research ideas as possible, including some of the more outlandish!

A good feature of the research tax is that it is well justified on traditional grounds. These days, there is talk of discouraging bright young people from becoming research scientists because there are already too many scientists chasing too few dollars. This is insane! When automation is forcing more and more people into the few intellectual endeavors humans still remain superior in, what else are these people going to do, become waiters? We have plenty of those. The one thing we do need is more scientists, and lots of them. Rather than keeping bunches of people from doing what they want to, is it not better to get them the money to have productive jobs that will benefit them, as well as the economy and society at large? Of the vast research funds, just tossing a few billion each year to basic research in cybertechnologies will do wonders. It will multiply the effort many times over. Perhaps a few more small-scale research institutes can be dedicated to research in developing and combining evolutionary nanotechnologies, artificial life, and cyberintelligence. Some of the people who will make the critical discoveries that make artificial consciousness possible are probably alive today.

SELLING SCITECH FOR THE 21ST CENTURY

To get the money, medical and computer researchers will require a sustained effort by the scientific community to make socio-political gains in a world where irrational belief systems and anti-science are gaining ground. (Of the two proposed research taxes, the medical tax is the most important and easiest to promote.) This will require that the scientific community get off its duff, stop whining about not getting enough of the government revenue pie, assert its critical place in the world economy, and get the needed funds. Scientists must abandon the defensive and go on the offensive. Objective: Don't just educate people about science, sell it to them.

Science foundations ought to consider joining industry to help fund a high-quality, all SciTech TV channel. Don't the Discovery and Learning channels already fill this role in the United States? No. Their habit of running a show on the power of the pyramids after a program on the physics of the Big Bang leaves watchers with the impression that the two are equally valid belief systems. In combining history and other nonscience programming with superstitious nonsense, the two cable channels may confuse the public perception and understanding of science.

Would a SciTech channel work? Think of images of scientists droning on about the genetics of fruit flies, but that is a problem solved by good production values. Dynamic shows on whales, sharks, elephants, dinosaurs, aircraft, ships, military technology, space flight, future technologies, and the human body and mind get good ratings. Imagine a continuing series, something like the NOVA science series, dedicated to debunking UFOs, astrology, crop rings, psychics, and angels. How about a show that gets a real debate going between theology and science? Throw in a few high quality sci-fi reruns and you've got yourself a new cable channel that will really do some good.

Grassroots Robots

The first automobiles were often the toys of the well heeled and later of the mechanically inclined. Radios started out as homemade crystal kits. The first home computers were home-built kits, too. It's fun putting radio-controlled boats, cars, and aircraft through their paces.

These days, hobbyists and artists are turning to robots as sources of amusement and entertainment, as well as a stimulating techno-challenge. Some of the "robots" are just glorified radio-controlled rigs, but others are guided in part or entirely by programmed chips. An autonomous toy combines the appeal of computers with the fun of a mobile model. Some of the robotoys are battle machines that engage in smash-and-push combat. One wild and crazy post-modern artist puts on big shows with large semi-robotic monsters. Other artists have gentler messages to present with less aggressive machines. As cheap chips become more powerful and easier to use, we may see a growing hobby market for robots, including intelligent model race cars that run their own races, and smart model planes that take off and land on their own and do their own acrobatics. Kids will eventually put together their own robotic companions.

Back to School

Then there is the matter of the American education system. In some regards, American public schools are actually out-performing those in other countries when income levels are factored in. In few countries do so many go so far up the scholastic ladder. But in terms of teaching science and other evolutionary systems, America is committing educational malpractice on a national scale.

In our Rational World, Darwinian theory would be integral to courses not only on biology, but to all courses on science, medicine, history, free market economics, democracy, and human cultures. The failure to teach students in large areas of the country anything about evolution can only be characterized as a case of national denial. Where the diabolical theory is taught, the lessons are minimal for fear of offending parents who put their right to misinform their children over the right of their children to receive a full education. It's a form of child neglect. Of course, poorly educated kids grow up to be poorly educated teachers who do the same disservice to their students.

Perhaps even more astonishing in this age of information is the failure to teach the basics of information theory! Alan Turing and von Neumann should be household names alongside Einstein and Newton. In the decade of the brain, how many bright young minds are told they are virtual systems that perceive reality after it has happened? How many teachers are qualified to teach their charges about the cosmology of spontaneous universes' formation? How many would dare?

In the Rational World, students would be told about the astonishing hypothesis and the scientific effort to discover the secrets of consciousness, and the extraordinary consensus that artificial consciousness can be achieved. In the Superstitious Society, such truths about the opinions of many scientists would offend too many sensibilities. Will the Extraordinary Future become a matter of course, as it were, for schools around the world? No one should hold his or her breath. How about at least basic courses on rational thinking and skeptical thought, with lessons debunking superstitions from UFOs to astrology? Yet another reason not to hold your breath. Better educating Americans will not make superstition disappear, but it would go a long way to alleviate the nonsense.

CyberEvolution Will Not Die

There is no need to fret much about CyberEvolution and what to do about it. It will take care of itself. Unlike missions to the moon and Mars, cybertechnologies are largely a self-funded piece of the commercial world. So, even if science budgets decline, even if the American education system goes to hell in a handbasket, even if there never is a great cybermind project to match the human brain, around the globe the technologies will edge closer and closer until one day, perhaps a little later than earlier, there will likely be new minds among us.

CHAPTER

15

On Dream Worlds Past, Present, and Future

A Recap

The hypothesis of the Extraordinary Future is simple and straightforward in outline, but will prove complex in execution. The hypothesis observes that we live in an evolving, mechanical universe. All that is present in this universe seems to be mechanical in nature, and it appears empty of intelligence except for us. Since the appearance of life on Earth, the rate of progress has been increasing in a roughly exponential manner, and it appears it will continue to do so as SciTech continues the great evolutionary increase. If so, we are heading for an inevitable singularity in progress and time that will produce results most expect will take a long time to achieve. In particular, the evidence that the mind is the direct product of a mechanical brain is overwhelming, and there is an extraordinary consensus among philosophers and scientists that it should be possible to artificially replicate consciousness, at least in principle.

In practical terms, the extremely and increasingly rapid advances in cybertechnologies and neuroscience are working together to drive the production of AC, perhaps in the next century. It is also plausible that the transfer of minds will become practical at about the same time, or shortly afterward. If these events occur, the results should include the immediate irrelevance and withering of humanity (and its superstitions), and the construction of a seemingly magical cybercivilization. These events will fulfill the minimal tenets of the Extraordinary Future. The power of the cybercivilization is predicted to be extreme and able to achieve results beyond the capacity of humans. It is further predicted that immortal cyberminds will rapidly expand in space, continuing the exponential growth curve of SciTech. At some point — quickly if the light speed barrier can be breached, or after epochs have passed if not — cyberintelligence should fill the entire universe and begin to alter its form to best suit the needs of immortal cyberlife. The minds of those who died before the cyberexplosion may be recovered through one or more technologies, possibly including time travel. If most or all of these predictions come to pass, the hypothesis of the Extraordinary Future will be borne out in full.

Note that it is not critical if the Big Show comes to planet Earth 30, 50, 100, or even 1,000 years from now. The main point is that after a very long time, it looks as if humanity's lease is about to run out. We project that the Big Show will be running so soon, not because it definitely will be so, but to alert you to the evidence that suggests it will. Certainly, we do not project such happenings within 100 years to be provocative or outrageous. If we wanted to do that, we could predict Christ's return, or the arrival of aliens in the same time. The probability of the Extraordinary Hypothesis is firmer than its timing. There certainly does seem to be a certain air of inevitability about it all. Only some great catastrophe, some hard barrier to AC or mind transfer, or plain stubbornness on the part of humans seems to stand in the way, and none of these looks like things one should count on (we will ignore the theological concerns, considering how weak they are).

There is a difference between the Extraordinary Future as a provisional scientific hypothesis on one hand, and a prophecy on the other. As strong as the case for the Extraordinary Future may (or may not) be, and as hard as it may be to stop CyberEvolution, the cyberexplosion and its aftereffects are not absolutely assured. We could be wrong. In a similar vein, we have tended to be optimistic because there is no point to being pessimistic unless the evidence firmly suggests one should be so.

A Spark Amid the Darkness

Vigorous debate and dispute is healthy and FUN! We are looking for the ultimate implications of exponential CyberEvolution to become as prominent a part of the debate about the future as environmentalism, overpopulation, geopolitics, economics, and so forth. We are particularly looking to break through the current fixation American and other societies have on superstitious-theological views of the future.

Basically, we want people to get their heads out of the sand and take a good, hard look at the ultimate implications of cybertech.

Think About It

Do you really expect that 100 years from now, people will still die before they reach 100? That there will not be a cure for the cold or cancer? That brains will still be mysterious black boxes? That computers will still be nothing more than glorified calculators, and robots will be inept and really stupid?

The More You Fight It, the Quicker It Will Come to Pass

By the very act of reading this book, you are playing a small role in making the Big Show happen! It's like in the 1800s: The people who wrote and spoke against manned flight actually thought they were helping put a stop to the ridiculous notion. Naive fools; they forgot basic human psychology. Tell a kid not to smoke, and what does he or she do? Tell people they can't climb that mountain, and what will they do? Tell someone the sound barrier cannot be breached, and what happens? A couple of Midwestern mechanics read the arguments against powered human flight. It did not dissuade them. It only made them more determined to prove the windbags wrong.

Ideas are like genes: The successful ones proliferate, the unsuccessful ones shrivel up until they disappear. This acts as a virus in that it helps spread the ideas it contains. You now know more, perhaps a lot more, about future possibilities than you used to, ideas you cannot entirely dismiss even if you wish you could. Perhaps you will mention these "weird" ideas to others who will then be a little more aware of the Extraordinary Future. It is not critical that you do so, whether you agree or disagree.

Research Is What Really Matters

Most modern belief systems rely on a critical piece of text; take that away, and the belief system collapses. Not so science and technology. If Darwin's and Newton's work did not exist, someone else would have published them. Nor are writing about and debating the Extraordinary Future crucial. What the Pope, the Dalai Lama, Searle, Penrose, and assorted naysayers have to say about CyberEvolution will not stop it. Nor will this book make it all happen.

What is crucial are the people and computers working to make the Extraordinary Future into the Common Reality, people such as Hans Moravec and his not-so-wild-and-crazy colleagues at the robotics lab in Pittsburgh, or Rodney Brooks and Lynn Stein and their Cog at MIT, Douglas Lenat and Cyc in Austin, the folks at Intel and Motorola who keep pushing the speed of computers ever higher, Masha Mahowald and her artificial retinas, Deep Purple, which took on Kasparov for the title of chess king, and the AL folks in Sante Fe. These are the Lilienthals, Chanutes, Langleys, and maybe even the Wright brothers and their flyers of our times.

You Can't Keep a Good Mind Down

There is an audacious giddiness to the Extraordinary Future. It thumbs its nose at every pessimistic human convention. It is the only really daring and interesting concept in an age when humanity is otherwise threatening to sink into the age of boredom. The Extraordinary Future is not risk-free, which is one reason it is exciting. It is not the ultimate utopia, nor is it a heaven in which childlike spirits are cared for by a dysfunctional parent God. It is the ultimate freedom for adult minds to do what they want to, as long as they want to. In the end, you cannot keep minds from immortality.

The Extraordinary Future Is Not Extreme, It's Not Even Weird

Lots of people are happy to apply words like "extreme," "weird," and "odd" to the Extraordinary Future. We're not the crazy ones. *Beyond Humanity* is no weirder than *20,000 Leagues Under the Sea* or the extreme notion that two bicycle makers got into their late Victorian heads that they could build a flying machine. It is not "weird" to predict a weird future when we live in a weird world and the evidence that things are quickly getting weirder is overwhelming.

We will tell you what *is* weird and extreme. It is extreme to insist that the day when computers will do what brains do is certain to be far, far off. When companies have employees outlining the principles for building computers as fast as brains, and computer power is rising a thousandfold every 20 years. It is very, very weird to argue that when the incredible mind machines are built, they will keep their place, not proliferate, and that humans will continue to lead normal lives. Now, there is nutty for you! It is weird to assert that intelligence cannot accumulate at an exponential rate of knowledge and power until it controls and manipulates at will a universe it has made alive, and create God, when we live in a weird universe of the Big Bang, faster-than-light inflation, super bombs that convert mass into energy, relativity and quantum mechanics, quasars, neutron stars, and black holes — as well as the weird and wonderful human mind itself. Dismissing the Extraordinary Future out of hand is ridiculous.

The matter of false strangeness is important. One reason many are resistant to the Extraordinary Future is simply because it *seems* so weird. The Extraordinary Future is not weird because it is nothing more than science and technology. It is the same nuts-and-bolts SciTech that has so far brought us electric razors, radios, Walkmans, and no-muss, no-fuss brain surgery with gamma knives. It is nothing more than minds, human and then suprahuman, figuring out how the universe works and applying the knowledge to push the limits as far as they can be pushed. Because there may be no limits, this may be very far indeed.

The Extraordinary Future Is Not Even Extraordinary

Although the hypothesis we have offered here may seem extraordinary, it is not really so. It is no more extraordinary than the fact that we exist at all, and that is not at all extraordinary. After all, universes are probably not rare and special, but inevitable in their existence and infinite in their multitudes. Because any fraction of infinity is itself infinite, the number of intelligently observed universes is probably also infinite. It is, therefore, nearly inevitable that intelligence should exist, thrive, and evolve in an infinite number of universes. Entities such as ourselves are a dime a dozen; miraculous, no; run-of-the-mill and infinitely common, yes.

Religion and mysticism have tried to fool us into being so amazed by our existence that we feel we must credit it, and be thankful, to a supernatural being. It is also a natural mistake. Even members of the scientific community fall into the trap of trying to impress people with the "miracle" of life. But science and technology routinely make the extraordinary ordinary. Amazement at creating new life and new universes will no longer seem amazing or godly if and when our cyberdescendants do it as a matter of course.

A Modest Prediction

It is often said that what God giveth, God taketh away. It is more meaningful to say that what science and technology create, they swiftly sweep away with the overwhelming force of progress. We take our human-driven high-tech world for granted even though it has existed only for an instant. Many imagine it lasting centuries. But the 200 years from the mid-1800s to the mid-2000s are the inevitably brief interlude between the advent of high-technology and its displacement of the humans who produced it. The world we live in is an ephemeral dreamscape, inherently unstable and extremely temporary in nature, that came into existence in a snap of the fingers and will disappear in another digital click as we head for the singularity in time that exponential progress is leading us to.

A century from now, the skies may be clear of artificial contaminants, and great cities should stand largely empty and decaying, the wilderness encroaching for the lack of attention. Roads will probably lay abandoned and crumbling. Plains that, just two lifetimes ago, were wild should again be covered by vast herds of horned ungulates. Great whales will cavort untroubled in shipless seas.

If these things happen, it will probably not be because nuclear apocalypse or environmental catastrophe will wipe us out, but because even now we are building with inevitable swiftness an entirely new civilization beyond this world in which humans will have no important place, but conscious minds will be far more powerful than we dream. A few million humans may be left, Rejectionists who refuse to assume cyberforms. They may choose to live aboriginal or subsistence lives, or will be beginning long, high-tech lives of leisure. In either case, to cyberbeings, they will be no more or less important than chimps are to us.

We are on a fast ride to a future that will appear magical to a world that is already beyond our ancestors' dreams. Where the ride will take us we can only dimly perceive. Will Christ really return, and will aliens really tell us why they are here? These are the fantasies of immature and naive society that will have to mature and face reality much sooner than it knows. After all, theologians and spiritualists spin mere superstitions; it is scientists and engineers who will actually make the future.

It is possible that we are very close to the ultimate singularity in time in which the very foundations of the universe will be altered to suit the needs of its native intelligence. The super-minds of the future are unlikely to remember the 20th century for its Manhattan project or the moon shot, for these were merely spectacular human stunts. Because information processing is the key to all other acts, the building of ENIAC promises to be the human product most celebrated by its cyberdescendants. They will marvel that a bunch of derived apes with computers made of jelly managed to cobble together a technological civilization and do high-level physics without blowing themselves up like kids in a fireworks factory. They will ponder with sympathetic concern the suffering and death humans were subject to. They will commit the robotic equivalent of a chuckle at the superstitious nature of people who actually believed a great spirit of the universe created and fawned over them, and then demanded their worship in exchange for a reprieve from eternal torture. How little humans understood that they were not the creations of gods, but the creators of minds as powerful as a god.

APPENDIX

Why We Are (Probably) Alone

Two possibilities exist: either we are alone in the Universe or we are not. Both are equally terrifying.
—Arthur C. Clarke

The hypothesis that Earth born robots are about to spread into an empty universe is at odds with the common belief that the stars are the home of extraterrestrials who are not about to let us run rampant over the galaxy and beyond.

Aliens, Not Here and Now at Least Far as We Can Tell

If we are already about to go cybertech in a big way, then by implication older galactic civilizations should have already done so. In this view the popular concept of aliens as humanlike bioforms traveling between the stars is naive. They should be cyberminds of extreme power, well able to keep taps on the entire galaxy. If they wanted to watch us very closely but very stealthily, they could do so without our ever knowing. That aliens in crafts that travel between stars nearly collide with airliners (*we* are deploying collision avoidance computers on airplanes), are regularly spotted by eyes and radars (*we* have stealth aircraft), crash, and have the bodies scooped by the US Air Force (aside from the gross incompetence of crashing a starcraft, wouldn't they recover the bodies and wipe out all evidence and memories of the incident?), and steal people's DNA (a pathetic little biocomputer they can easily make in their own labs) do not make any sense. If, on the other hand, alien cyberforms wanted us to know they are here, we would know in no uncertain terms.

Hello, Anybody out There?

A fair number of scientists think aliens really are out there in the galaxy. The problem is a complete lack of evidence. Next time you go out into the country away from city lights, on a clear moonless night, just look up at all those stars. Looks like a wilderness out there, doesn't it? Astronomers using the most powerful instruments see the same emptiness. Even if the galactic cooperative agreed to set aside a percentage of stars for primitive folk like us, we ought to see *something.* Dyson spheres, for example, have to shed vast quantities of heat, and an example around a large star in our region of the galaxy should produce a distinctive emission observable via infrared detectors. There are no stars arrayed in perfect circles. Some scientists suggest that the aliens are hiding their civilization from us. Right, like they would care.

Lacking any visual evidence, the ET enthusiasts are reduced to listening for alien radio transmissions. So far, not a peep from the ETs.

The Drake Equation

The basic reason why many (but not necessarily a majority) of scientists think our Milky Way galaxy may be filled with alien civilizations is summarized by the Drake equation. Devised by Frank Drake in the 1960s, and popularized by Carl Sagan, it goes something like this. There are 200 billion stars in the galaxy, of which a large fraction — many billions — may have Earthlike planets suitable for the evolution of intelligence. On a certain percentage of these happy habitats, human equivalent intelligence evolves and build technological civilizations. The values assigned to the many factors involved are very iffy and disputed, so the end results range from one technological civilization in the Milky Way, to thousands or even millions in existence. Ten thousand is a common figure.

Here's the Rub

In some ways, many calculations of the Drake equation are actually conservative because they assume that few civilizations last long before going extinct, and/or that they tend not to move very far into the galaxy. Yet we have already shown that just one colonizing civilization could take over the galaxy in short order. The more civilizations there are, the more likely it is that one of them will make an evolutionary breakout and industrialize much or all of the Milky Way. In fact, the idea that thousands and thousands of intelligences would evolve, and then do so little with the galaxy, is too statistically improbable to accept. If the galaxy is at least partly civilized, it is reasonable to expect it to look partly civilized. The simplest and most logical interpretation of why our galaxy looks like an empty wilderness is because it *is* an empty wilderness.

Freaks of the Universe

How, with all those star systems out there in this huge universe, could this be? That we are the only intelligence in the cosmos violates the Copernican Principle, which asserts that our planet is an ordinary place with no special position in the universe. The philosophical drive behind this argument is so strong that Carl Sagan has called the idea that we are the only intelligence "laughable arrogance." Ego ranking is not, however, a sound basis for understanding who is or is not out there.

In order to get a better handle on the problem, we are going to take another stab at the Drake equation. The galaxy, as you may know, is shaped sort of like a fuzzy, irregularly rimmed pancake with a large bulge in the center. The central bulge is about ten thousand light years in diameter. The flattened "pancake" is the disk, which is made up of spiraling arms of stars that span about one hundred thousand light years. The density of stars declines rapidly going from the center, and our cozy home is so far out in one of these arms that only one percent of the galaxy's stars are as far out from the center as we are. This means that we are already in violation of the Copernican Principle! If we are an average civilization, then we should be nearer the center of the galaxy among the pack. The three possibilities are that for some unknown reason only the stars in the outermost disk can support intelligence, or we are just a statistical fluke, or we are very alone.

It is probable that the galaxy is not as nice a place for civilization as many think. About half the disk stars are close binaries whose complex gravitational interactions prevent planets from having stable orbits. Of the rest, some spin too rapidly to have planets, some are too old to have rocky planets or have burned out, others are too young for intelligence to have evolved, some fluctuate in brightness, some are dim dwarfs with warm life zones (not too cold or too hot) that are too narrow for intelligence-bearing planets to be likely. Modest-sized, stable, long-lived stars made with some heavy elements and with broad life zones like our sun make up only about one percent of the single stars in the disk. Here's an interesting item. Astronomers recently observed that most other stars have much less carbon — the stuff of life — than our sun. At most about one billion stars *may* be the centers of solar systems that *may* be suitable for the development of bright beings like ourselves, this figure drops dramatically if only stars have to be in the outer galaxy like we are, or if most systems lack sufficient carbon.

Solar systems get their start when clouds of intragalactic gas and dust collapse under their own gravity. Different sections of the "dust" cloud start to spin one way or another, and form a cluster of flattened disks. At the center of each disk a hydrogen-fusing star forms. We know this happens because big telescopes — most especially the space-based Hubble — can actually see stellar systems in the process of formation. Computer simulations suggest that the material further out in the disk usually becomes comets, asteroids, and planets. If so, then solar systems should be common. However, some of the most spectacular images from the Hubble may show otherwise. The first star to form in a cluster does so without interference, and sucks up so much dust that it becomes a super hot giant. The radiation put out by the new giant is so intense that it may blast away most of the nebular disks surrounding the neighboring baby stars. Stripped of their dust disks, the rest of the stars in the brood cannot form planets, which may therefore be rare.

But what about all the planets being found around other stars these days? At this time the only planets we can detect are so gigantic and so close to their sun that they strongly effect its motion (the resulting wobble can be measured via a Doppler effect). Some of these planets are really tiny stars, but even the real planets are not necessarily good news for the pro-ET crowd. For reasons we will not go into here, giant planets are supposed to form well away from their star. One possibility is that new planets spiral in towards the star while the original dust cloud is still in place, in which case most small, rocky Earth-like planets may end up cooked. Other possibilities are that small planets either do not regularly form close to stars, or their orbits are messed up by the big planets.

Even those rocky planets that escape these dire fates must meet more criteria before they can spawn intelligent beings. If the planet is too small its gravitational field will be too weak to hold a dense atmosphere, the fate of Mercury and Mars. Too big and super-dense, the atmosphere will create a runaway greenhouse heating effect worse than that on Venus. If the planet is too close or far away from its sun it will be either too hot or too cold. Planets form by an irregular process of accretion, so the rates and orientation of their spins are semi-random. Venus, for example, takes hundreds of days to rotate. On such a planet one side long bakes in solar heat before it freezes in an equally long night. Uranus rotates on its side. If an inner rocky planet did so one hemisphere would cook for about a quarter of the year, and then chill out another quarter. Some believe that as its molten interior shifts and rearranges the distribution of mass, a moonless rocky planet may lose stability and occasionally tumble. If true, then a temperate axial tilt of 20 degrees can become a nasty 83 degrees, and then a balmy 5 degrees (there is some evidence that Venus and Mars have tumbled in the past). A nice big moon that acts as a gravitational stabilizer will keep a planet's tilt in line, but big moons around rocky planets are probably rare. The best evidence indicates that Earth was hit by a Mars-sized object hit it when the solar system was young. The impact happened to be at just the angle to set Earth spinning at a reasonable speed and tilt, while creating a moon that keeps it that way.

OK, so you have a nice balmy little planet orbiting the right kind of star. We're all set for a civilization, right? Hold your horses. You also need at least one really big gas giant on patrol in the outer solar system. Why? Vast numbers of comets form close to a new star. As Carl Sagan would not say, billions and billions of them. The inner planets are hit by huge comets, as big as states, that completely sterilize any biosphere every hundred million years or so. This bombardment will last indefinitely unless the comets are cleared out. The powerful gravitational fields of the gas giants like Jupiter sweep up the comets and either swallow them up (as happened when Comet Shoemaker-Levy impacted Jupiter in 1994), or boost them into the distant Kuiper Belt, the even more distant Oort cloud, or beyond.

Oort clouds are extraordinary, because they literally extend halfway to the nearest stars! In fact, some of the distant comets drift into interstellar space, and are captured by other stars. If most sunlike stars also have gas giants orbiting them, then they should have Oort clouds that cast off comets. If that is true, then we should occasionally see an interstellar comet swing around Sol, but nary a one has shown up. The obvious, albeit tentative, conclusion is that Oort clouds and the gas giants that form them are rare, and most rocky planets are sterile. Another complication: The gas giant must be too far from the sun to mess up the orbits of any small rocky planets near the star. So far careful surveys of nearby stars have failed to detect planets similar to Jupiter in size and placement.

But just because a planet has a big moon and is under the protection of giant comet cleaners does not mean that the planet will produce a civilization! For starters, life must appear. As the computational-mechanical nature of life is becoming increasingly understood a growing but still tentative body of opinion agrees that carbon-based life probably forms almost as soon as conditions allow it to. This hypothesis will be affirmed if it turns out that Mars has spawned life on its own, as recent and startling reports hint. So it is possible that simple life is common in the galaxy and beyond.

More complex life is a different matter. It probably takes the right kind of habitat and hundreds of millions if not billions of years of to produce multi-cellular organisms that use tools. One possible requirement is land watered by oceans — marine dolphins may be smart, but they will never produce technology because they have nothing to make and handle tools with. Of the three larger inner planets, only Earth has moving crustal plates that have thrust lands above sea level. Even when conditions *are* suitable for the evolution of tool using intelligence, it may still fail to appear. Contrary to some simplistic views of evolution, nature does not fill all available lifestyles. If the first handful of vertebrates over five hundred million years ago had not made good, for instance, no one would be writing and reading about the Extraordinary Future. Earth-like planets may have a thriving fauna for billions of years without ever producing beings able to erect a civilization. After appearing, the civilization has to survive and thrive long enough to make itself known to other civilizations by space travel and/or long distance calling. Oddly enough, many Drake calculations assume that most civilizations do not last long, but this is an iffy assumption especially since our nuclear standoff has gone into hibernation.

Having announced their presence to the galaxy via crude radio and TV broadcasts, are new civilizations consumed by galactic cybermonsters? Maybe the first intelligence that took over the galaxy evolved into a super predator that lies in stealthy wait for innocent rubes — our primitive civilization being an example — to reveal their presence, and then attacks and wipes the simpletons out. The monster might be a single huge machine, or a dispersed set of killer probes that use nanotechnology to do their nasty work. Maybe, but evolutionary forces probably work against such super predators. A galactic cybermonster would not accomplish anything of much use to itself; in fact, it would be vulnerable to more progressive competitors coming in from other galaxies. A better way to take over the galaxy is to be the first to expand into it and take it over in a benign manner, thereby excluding competitors.

Of course, we could be wrong.

OK, fine, let's go back through this. To evolve intelligence remotely like us probably requires a star and a planet that meets each and every of the characteristics listed below. The numbers involved are calculated as well as they can be in these primitive times.

- *Stars in the region of an average galaxy suitable for intelligent life* — few hundred million if only outer disk stars are included, up to about 100 billion otherwise
- *Of said stars, percentage and number that are the right size, temperature, stability, and lifespan, and effectively single* — about 1% of prior values; a few million to one billion
- *Of suitable stars, percentage and number orbited by at least one Jupiter class gas giant that clears out inner comets but is far enough from star to allow inner rocky planets to have stable orbits* — not reliably calculable, may be rare; less than a hundred thousand to a few hundred million
- *Of suitable stellar systems, percentage and number orbited by one or more rocky planets in stable life zone* — not known, probably under 25 percent, possibly as low as a few percent; less than ten thousand to a few hundred million
- *Of rocky planets in comet clean life zones, percentage and number that life appears* — not known, extremely rare is possible but not likely, up to 100 percent is more probable; less than a thousand to a few hundred million
- *Of life-bearing planets not repeatedly sterilized by super impacts, percentage and number that have a daily spin under about 60 hours, an axial tilt under about 40 degrees, and are stable (perhaps because of a large moon)* — not reliably calculable, possibly as high as half, possibly just a few percent; from less than a hundred to around one hundred million

- *Of life-bearing rocky planets with good spin characteristics in comet clean zones, percentage and number that conditions are suitable for the development of technological civilizations (including a mix of continents and oceans)* — not known, either rare or common possible; from less than one per galaxy to tens of millions
- *Of life-bearing planets with conditions suitable for development of technological civilizations, percentage and number that do so* — not reliably calculable; much less than one per galaxy to about a million per galaxy
- *Subsequent civilization survives long enough and wants to communicate or travel across interstellar space* — not reliably calculable, much less than 10 percent or nearly 100 percent is unlikely, about 50 percent appears plausible; much less than one per galaxy to nearly one per galaxy to about one hundred thousand
- *First galactic civilizations evolve into cybermonsters that preclude all other intelligences* — probability extremely low; probably no effect on calculations

So few data are certain that unbiased calculations of the Drake equation produce extremely variable results. However, the exercise is not valueless. The number of solar systems with both gas giants and Earth/moon-like systems may be so rare that planets that evolve powerful neural systems may be rare, or even unique. Please understand that our calculations do not prove that this is so. What they do show is that it may be no accident that we live on a rocky planet with a moon big enough for lovers to get all hot and bothered under, orbiting a sun loaded with carbon, in a solar system where a brilliant and gigantic Jupiter shines in night skies, and in a part of the galaxy where the stars in the night sky are relatively rare. Earth is not the center of the universe, but the extreme Copernican view that Earth and the intelligence it has spawned are common is not yet supported by either observational evidence, and not demanded by theory. In the near future better star gazing technologies should settle whether planets big and little are rare or common, and may even show whether distant planets have oceans and free oxygen. However, unless direct evidence of aliens shows up, the evidence indicates that there is no other intelligence to commune with. That we are Freaks of the Universe.

Many object that it makes no sense for such an enormous universe to have such little intelligent life. Of course it does not make sense by conventional cognitive standards! This universe is not the carefully designed product of a rational designer for the efficient production of minds, it just happens to be one of the multitudes of universes that can manage to produce any intelligence at all. In this view a huge and nearly empty universe is not at all surprising.

If, on the other hand, ETs are uncovered, then some aspects of the Extraordinary Future are weakened if not overturned. In particular, the hypothesis that Earth-based cyberintelligence will take over the galaxy and beyond will be effectively falsified.

Some Books and Articles You Might Find Interesting and Useful

Want to know more about science, technology, evolution, space, cyber technologies past, present, and future? Interested in the arguments of those who agree and disagree with the Extraordinary Future? Just have some time on your hands? Check some of these out.

Isaac Asimov. *I, Robot*. 1950 (Bantam). The SciFi classic that introduced the Three Laws of Robotics.

Ernest Braun. *Futile Progress: Technology's Empty Promise*. 1995 (Earthscan). Braun's call for slowing down progress to give more time for adjustment, debate, and choice will sound reasonable to some. Whether it is practical is another matter.

James Burke. *Connections*. 1978 (Little Brown). Based on the original TV series, a rollicking good ride through the interconnected worlds of science and technology.

John Burnham. *How Superstition Won and Science Lost*. 1987 (Rutgers). A look at how the failure of scientists to communicate with the public is hurting their position and influence vis-à-vis other, less realistic belief systems.

Donald Cardwell. *The Norton History of Technology*. 1995 (Norton). A history and analysis of technological progress, including the late, great speed up.

John Casti. *Complexification: Explaining a Paradoxical World Through the Science of Surprise.* 1994 (HarperCollins). A study of complexity and surprise in our universe, a pretty dense read.

Maureen Caudill. *In Our Own Image: Building an Artificial Person.* 1992 (Oxford University Press). Recommended if you want a more in depth technical account of how the many software systems needed to build intelligent androids are advancing so fast that the robots may be built early in the next century. Emphasizes neural networks. Caudill's conclusion that most robots will be rather humanlike androids that serve man and woman is open to challenge.

Patricia Churchland and Terrence Sejnowski. *The Computational Brain: Models and Methods on the Frontiers of Computational Neuroscience.* 1992 (MIT Press). Churchland and Sejnowski use what is being learned about computational neural networks to better understand how brains work.

Paul Churchland. *The Engine of Reason, the Seat of the Soul: A Philosophical Journey into the Brain.* 1995 (MIT Press). Paul Churchland, like his spouse Patricia, is a philosopher of neuroscience. In this work Churchland takes a perceptive stab at explaining how the brain generates conscious thought.

Arthur C. Clarke. *The City and the Stars.* 1953 (NAL-Dutton). A SciFi classic that presented an early vision of what we would call supercomputers, smart machines, virtual reality, genetic engineering, and nanotechnology. In many ways far ahead of other SciFi works of its time — such as the atomic fission run universe in Asimov's *Foundation* trilogy. *How the World was One: Beyond the Global Village.* 1992 (Bantam). Here the inventor of the communications satellite reviews the history of communications and the Global Village they have created. On the one hand Clarke is one of the most radical futurists; on the other he is a humanist conservative who believes that someday technology will bring the peoples of the world a golden age of peace and prosperity. Maybe, but as those who remember *Peyton Place* and *Twin Peaks* know, villages can be not so nice places.

Joel Cohen. *How Many People Can the Earth Support?* 1995 (W.W. Norton). Packed full of data on populations past, present, and future. Little discussion of how future technologies may allow humanity to decouple itself from nature, and achieve extremely high populations without crushing the environment.

Peter Coveney and Roger Highfield. *Frontiers of Complexity: The Search for Order in a Chaotic World.* 1995 (Faber/Fawcett). Two journalists' account of the movers and shakers in the controversial field of intelligence, and what they think.

Daniel Crevier. *AI: The Tumultuous History of the Search for Artificial Intelligence.* 1993 (Basic Books). If you want to learn about the forty-year history of the effort to make digital computers do what we do, and the sometimes outrageous personalities involved, this is the book. Crevier tries to explain why the effort has failed so far, and why in the end it will probably succeed.

Francis Crick. *The Astonishing Hypothesis: The Scientific Search for the Soul.* 1994 (Charles Scribner's Sons). An important book by the co-discoverer of the structure of DNA, because it presents a compelling case for understanding conscious thought via scientific investigation. Includes an extensive account of how the brain works, with emphasis on the visual system.

Emmaneul Davoust. *The Cosmic Waterhole.* 1991 (MIT Press). This book both describes the interesting radio search for extraterrestrial intelligence, SETI, and why it probably will not work because there is no one out there to talk to.

K. Eric Drexler, Chris Peterson, and Gayle Pergamit. *Unbounding the Future: The Nanotechnology Revolution.* 1991 (Quill). A no-holds-barred look at the potential and implications of the tiny world of nanotechnology by its leading advocate. Includes the sugar-cube-sized hypercomputer. Gives some interesting alternate future scenarios for what nanosystems might do to human society, most good, some bad, but does not squarely address the possibility that nanotech will produce the superminds that render us irrelevant. An easy must read.

Hubert Dreyfus and Staurt Dreyfus. *Mind Over Machine: The Power of Human Intuition and Expertise in the Era of the Computer.* 1985 (Free Press). These brothers argue that the mind works in a metaphysical mode (àla Aristotle's principles), so no digital computer can ever replicate consciousness. By their reasoning, digital machines could not reason with fuzzy logic, which they do.

Freeman Dyson. *Infinite in All Directions.* 1988 (Harper & Row). *Disturbing The Universe.* 1979 (Harper & Row). The late Dyson was one of the most innovative yet respected thinkers of this century, and in these two books he presented some of his far out ideas about how intelligence may or may not spread out into the universe and find a way to ensure its immortality as the cosmos grows very, very old.

John Eccles. *How the Self Controls the Brain.* 1994 (Springer-Verlag). *Evolution of the Brain: Creation of the Self.* 1989 (Springer-Verlag). A dualist, Eccles is one of the main enemies of the Extraordinary Future. In the later book he tries to show which neural structures tap extrauniversal consciousness, but the suggested process appears to violate the Laws of Physics. His earlier book on the evolution of the brain ends with a dualistic explanation of its function that is anti-evolutionary in its implications.

Edward Feigenbaum and Julian Feldman (eds). *Computers & Thought.* 1995 (McGraw-Hill). A reissue of a 1963 compilation of seminal papers and articles by some of the leading early thinkers of AI, including Alan Turing and the young Marvin Minsky.

Timothy Ferris. *The Mind's Sky: Human Intelligence in a Cosmic Context.* 1992 (Bantam). A look at how human and alien intelligence may communicate via radio waves, etc. Too conventional for our tastes, and probably wrong about there being ETs out there to talk to.

Caleb Finch. *Longevity, Senescence and the Genome.* 1991 (University of Chicago Press). Examines the various theories of aging from an evolutionary perspective.

Grant Fjermedal. *The Tomorrow Makers.* 1986 (Macmillan). An interesting and informative look at the characters involved in the effort to make mind machines they can download their identities into. A little dated.

Kenneth Ford, Clark Glymour and Patrick Hayes (eds). *Android Epistemology.* 1995 (MIT Press). A compilation of learned pieces by philosophers and computer scientists, discussing the possibilities and implications of smart machines. Margaret Bordon, for example, thinks that machines will be devoid of creativity and original ideas. Of course, Margaret *is* a machine, she just hasn't noticed yet.

Francis Fukuyama. *The End of History and the Last Man.* 1992 (Free Press). Fukuyama argues that human history is coming to an end because democratic capitalism is rendering war and struggle obsolete. Ignores the corrosive effects of the Great Boredom that will accompany general prosperity. Human history is about to end, but not for the reasons Fukuyama suggests.

Bill Gates. *The Road Ahead.* 1995 (Penguin). Mr. Software tells us about the wonderful places online systems and virtual reality will soon take us. Paul Saffo (Institute for the Future) said that Gates's "vision can only be summarized as the bland leading the bland," a perceptive analysis of what awaits humanity in the first part of the 21st century. Gates walks us down the road for a few blocks, the rest of the highway lies far, far ahead.

William Gibson. *Neuromancer.* 1984 (Ace Books). It was morning in Ronald's America when nonhacker Gibson came out with this, the very first cyberpunk SciFi classic. *Neuromancer* helped introduce the public to the idea of minds entering computers with typical postmodern cynicism and dark angst. The term *cyberspace* was invented here. As usual with these things, the plot device begs the question of why minds would remain in human bodies if they can go cyber.

Jean Gimpel. *The End of the Future: The Waning of the High Tech World.* 1995 (Oxford University Press). Gimpel says we are giving up cars in favor of bicycles, and that technological innovation is plateauing out. Odd notion at a time when billions fly in semi-robotic aerial machines each year, and the patent offices are going digital in a desperate attempt to keep up with what's new.

Newt Gingrich. *To Renew America.* 1995 (HarperCollins). A marvelous example of confused "conservative" futurology, in which it is proposed that 21st century high-tech will revive traditional American values and ensure prosperity. Sure it will, and Newt has a nice old bridge in New York he would like to sell you.

Stephen Jay Gould. *Wonderful Life: The Burgess Shale and the Nature of History.* 1989 (Norton). Gould, one of the leading and most popular evolutionary biologists, presents a controversial view that the diversity of life has not increased since the Cambrian Explosion that saw the swift appearance of hard-shelled creatures over 500 million years ago. Gives a good overall explanation of evolution.

Chris Gray (ed). *The Cyborg Handbook.* 1995 (Routledge). A compilation of articles by assorted researchers and thinkers on the possibilities of melding man and machine. Fairly weird stuff, but then things look to become weird.

Donald Griffin. *Animal Minds.* 1992 (University of Chicago Press). An interesting look at the question of whether animals are conscious to some degree or another. Griffin thinks they are.

Robert Heilbroner. *Visions of the Future: The Distant Past, Yesterday, Today and Tomorrow.* 1995 (Oxford University Press). Heilbroner reviews the history of humanity and futurology, and concludes that the brief Industrial Age optimism has ended. Heilbroner then concludes that we must abandon technological progress via capitalism as a means of securing the future, in favor of a more environmentally sustainable, more humanistic global society. Sounds rather dull.

Leonard Hayflick. *How and Why We Age.* 1995 (Ballantine). Hayflick takes a comprehensive look at why we and (and most) other animals age, and concludes it is a multifaceted process that may prove malleable to intervention, after a lot of research.

Allan Hobson. *The Chemistry of Conscious States: How the Brain Changes the Mind.* 1995 (Little, Brown). A hard-core anti-dualist work that argues that much of what we think and feel as we wake and dream is the result of chemistry.

John Holland. *Hidden Order: How Adaptation Builds Complexity.* 1995 (Addison-Wesley). Holland was the first to earn a Ph.D. in computer science, and he is one of the founders of the field of complexity. This is his rather informal account of his ideas on the subject.

John Houghton. *The Search for God: Can Science Help?* 1995 (Lion). Physicists often use the G-word without really meaning it. Houghton searches for scientific evidence for God, but fails to find it.

Jim Jubak. *In the Image of the Brain: Breaking the Barrier Between the Human Mind and Intelligent Machines.* 1992 (Little, Brown). An examination of how neural networking is allowing computers to become increasingly brainlike.

Stuart Kauffman. *Origins of Order: Self Organization and Selection in Evolution.* 1992 (Oxford University Press). *At Home in the Universe: The Search for Laws of Self Organization and Complexity.* 1995 (Viking). Kauffman is a leading proponent of the hypothesis of self organization, which argues that the structure of our universe not only allows complexity, life, and intelligence to evolve, but virtually forces them to do so along a path well set. Self organization is controversial and not yet proven, but may be very important.

Stephen Kosslyn and Olivier Koenig. *Wet Mind: The New Cognitive Neuroscience.* 1992 (The Free Press). A popular book on the workings of the brain that looks at it as a "wet" rather than "dry" computer.

Steven Levy. *Artificial Life: The Quest for a New Creation.* 1992 (Vintage Books). A popular and informative look at the new, exciting, and strange field of evolving artificial life in computers, and the key role it may play in the development of truly intelligent systems. Highly recommended.

Marvin Minsky. *The Society of the Mind.* 1985 (Simon & Schuster). By one of the fathers of artificial intelligence, discusses his ideas on how the mind works as a set of semi-independent systems. "Will Robots Inherit the Earth?" *Scientific American.* October 1994. Part of a special issue on Life in the Universe, Minsky's article is an explicit outline of the Extraordinary Future by the dean of AI.

John Moore. *Science as a Way of Knowing: The Foundations of Modern Biology.* 1993 (Harvard University Press). We could hardly agree more that "A fundamental difference between religious and scientific thought is that the received beliefs in religion are ultimately based on revelations ... by some long dead prophet," while "the statements of science are derived ultimately from the data of observation and experiment."

Hans Moravec. *Mind Children.* 1988 (Harvard University Press). Far and away the most extensive and explicit presentation of the concepts behind the Extraordinary Future before we came along with our wild and strange book. Proof that the good folks at HUP are willing to go out on an intellectual limb, this book is oriented more to the popular audience despite its source of publication. Moravec is an advocate of strong digital AI, which may or may not prove a calculating dead end in the effort to produce synthetic conscious intelligence. His prediction that it will become possible to transfer lots of minds into some kind of thinking machine early in the 21st century cannot be so easily dismissed.

John von Neumann. *The Computer and the Brain.* 1958 (Yale University Press). One of the earliest and most important studies of how the functions of computers and brains can be used to better understand the other.

Heinz Pagels. *The Dreams of Reason: The Computer and the Rise of the Sciences of Complexity.* 1988 (Simon & Schuster). The late physicist explores the potential of cybertechnologies. Concludes that artificial minds probably can be built, but may take a long time to build them.

Rodger Penrose. *The Emperor's New Mind: Concerning Computers, Minds, and the Laws of Physics.* 1989 (Oxford University Press). A surprise best-seller, partly because it seemed to show why the human mind cannot be replicated by any computer. Basically argues that the mysteries of the universe and the conscious mind are intimately related via quantum mechanics. *Shadows of the Mind: A Search for the Missing Science of Consciousness.* 1994 (Oxford University Press). An attempt by the famed mathematician to propose the cellular structure by which neurons communicate via quantum effects. Probably wrong, but if he is right then Penrose is accidently describing how the brain machine can be replicated.

Ed Regis. *Great Mambo Chicken and the Transhuman Condition: Science Slightly Over the Edge.* 1990 (Viking). Starts with Evel Knievel rocketing across the Snake River Canyon (sort of), and ends with the never-ending End of the Universe. (Reading this book starts one to thinking that getting one's head frozen after "death" may not be such a bad idea after all.) An entertaining and informative look at the Extraordinary Future, with discussions of the ideas of the likes of Dyson and Moravec. *Nano!* 1995 (Bantam). An enthusiastic examination of the super-tech nanotechnology world that we may be about to enter.

Carl Sagan. *Pale Blue Planet.* 1994 (Random House). On the one hand Carl has long criticized NASA for putting its big bucks into rocketing men rather than cheaper robots into space. On the other hand he cannot help wax rhapsodic about the boundless future of people in space. Mentions almost nothing about the potential for galactic cybercivilization in this book; it's mainly apes in space.

Kirkpatrick Sale. *Rebels Against the Future: The Luddites and Their War on the Industrial Revolution.* 1995 (Addison-Wesley). Sale does not dislike only computers and other high technology, he thinks civilization is a catastrophe! He thinks man should only use those few technologies that he and fellow Luddites approve of — that doing so will probably starve billions is a minor matter.

John Searle. *The Rediscovery of the Mind.* 1992 (MIT Press). Searle has long been one of the opponents of strong AI, and in this book he continues to argue that computers will never think like we do even if they mimic what we do. However, Searle believes the mind is the direct product of the brain.

Geoff Simons. *Robots: The Quest for Living Machines.* 1992 (Cassell). A comprehensive survey of the state of robots, and where they may be going. *Is Man a Robot?* 1986 (John Wiley & Sons). Simons examines the human machine and its roboticlike nature.

Victor Stenger. *The Unconscious Quantum: Metaphysics in Modern Physics and Cosmology.* 1995 (Amherst). Stenger concludes that the mind is the result of a physical rather than a spiritual process, and that the mind-body "problem" is based in human egoism and self-obsession rather than rational, scientific thought.

Gregory Stock. *Metaman: The Merging of Humans and Machines into a New Global Superorganism.* 1993 (Simon & Schuster). The idea is that as the global computer network continues to expand in the next century, it and the humans that use it will become in effect a new life form of unprecedented power. Drop the humans from the equation and the concept is improved.

Ian Tattersall. *The Fossil Trail: How We Know What We Think We Know About Human Evolution.* 1995 (Oxford University Press). A comprehensive survey of human evolution. The author notes that the enormous, interbreeding mass of humanity effectively suppresses major genetic evolution, not that it really matters.

Frank Tipler. *The Physics of Immortality: Modern Cosmology, God and the Resurrection of the Dead.* 1995 (Doubleday). And you thought this book was outrageous. A physicist, Tipler argues that science can be used to show that when the universe ends, it will become God and recover all dead minds in order to give them everlasting life in the cosmic version of paradise. Nine out of ten physicist say no. Theologians who say yes do not understand the anti-religious nature of Tipler's hypothesis.

Alvin Toffler. *Future Shock.* 1971 (Bantam). *The Third Wave.* 1984 (Bantam). Two classics that have helped put down on paper what we know from the head spinning experience of living in a fast changing world. Ironically, however, cyber future shock is obsolescing Toffler's futurist prescriptions about as fast as he writes them!

Colin Tudge. *The Time Before History: 5 Million Years of Human Impact.* 1996 (Scribner). This book emphasizes how "ancient" civilizations are actually recent innovations after an immense span of prior human existence. Tudge ends with the usual — and obsolete — warning about how humans must learn to live with nature if they are to survive for yet another million years.

Mark Twain. *Letters from the Earth.* 1938. No one has bettered the frustrated river boat pilot in deconstructing the illogic, geological and otherwise, of Christian theology. Should be required reading.

H. G. Wells. *The War of the Worlds.* 1898 (always in print). It's hard to beat this tale that introduced beam weapons and other high technologies to a public whose heads were already spinning from the wonders of turn of the century progress.

Stephen Wolfram. *Cellular Automata and Complexity.* 1994 (Addison-Wesley). A detailed examination of von Neumann's self-reproducing machines and how they can evolve complexity, plus game theory.

Ben Zuckerman and Michael Hart (eds). *Extraterrestrials: Where Are They? Second Edition.* 1996 (Cambridge University Press). A set of papers by scientists, most but not all of whom dump cold water on the idea of ET intelligence. Some of the papers date from the first edition of 1982.

Periodicals

Scientific American. Special Issue: "Mind and Brain," September 1992. Eleven informative articles on the structure and function of the brain and the mind it produces outline the tremendous gains in knowledge about how we use salty jello to think during recent years. Special Issue: "Life in the Universe," October 1994. Chock full of some of the latest thinking on the past, presents, and future of life and intelligence in the cosmos. Special Issue: "Key Technologies for the 21st Century," September 1995. A must read for those interested in the amazing technologies we can expect to see in the coming decades. However, the 34 authors were instructed to keep their projections conservative, so humans are always the center of attention.

Wired. Special Issue: "The Future of the Future," 1995. A series of articles on what the future may or may not promise. The views expressed tend to be rather conservative and human oriented; it seems that robots are not hip enough for the folks down at *Wired.*

INDEX

T

U